Cambridge Lower Secondary

Complete
Physics

Helen Reynolds

Anna Harris

Elliot Sarkodie-Addo

Second Edition

Contents

Stage 9

How to use your Student Book

Welcome to your Cambridge Lower Secondary Complete Physics Student Book. This book has been written to help you study Physics at all three stages of the Cambridge Lower Secondary Science curriculum framework.

Most of the units in this book work like this:

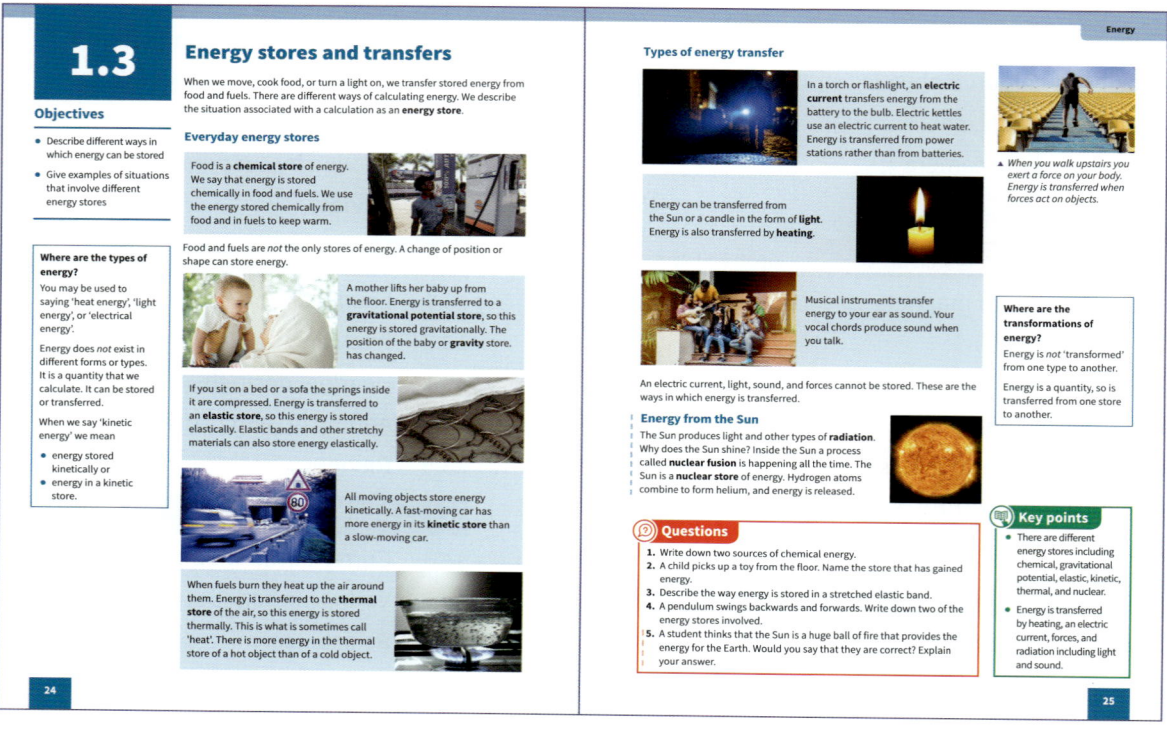

- Every page starts with the learning objectives for the unit. The learning objectives are linked to the Cambridge Lower Secondary Science curriculum framework.

- Key words are marked in **bold**. You can check the meaning of these words in the glossary at the back of the book.

- At the end of each unit there are questions to test that you understand what you have learned. The first question is straightforward and later questions are more challenging. The questions are written in the style of the Cambridge Checkpoint test, to help you prepare. Answers are available in the Teacher Handbook which is available in print and digitally via Kerboodle.

- The key points to remember from the unit are also summarised here.

These units cover the Physics topics in the Cambridge Lower Secondary Science curriculum framework.

In addition, many of the units help you think and work scientifically, put science in context, prepare for the next level, and test your knowledge.

Thinking and working scientifically

Thinking and working scientifically is an important component of the curriculum framework.

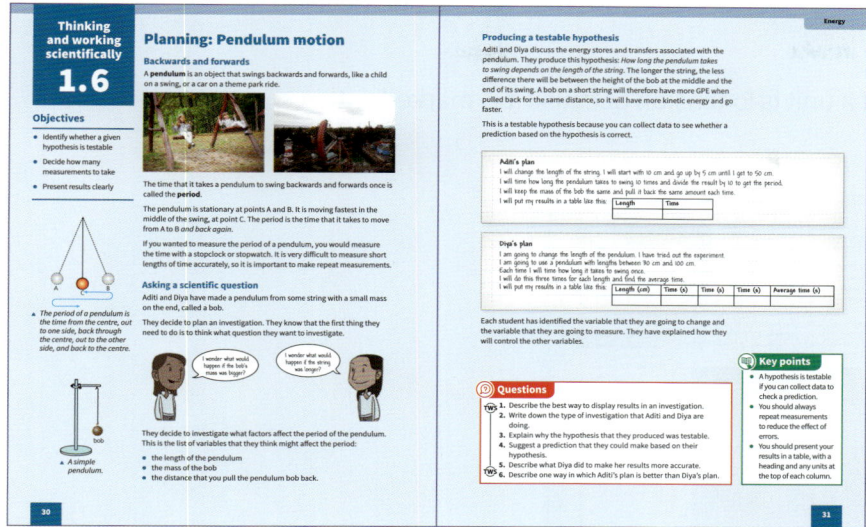

The Thinking and working scientifically units and features will help you learn:

- how to understand and apply models and representations
- the importance of asking scientific questions and planning how to answer them
- how to carry out enquiries such as fair test investigations and field work
- how to analyse data, draw conclusions, and evaluate your enquiry.

 Questions which test your Thinking and working scientifically skills and knowledge are marked with this icon.

On pages 8-19 you will find a dedicated Thinking and working scientifically chapter which introduces essential skills which will be useful throughout every stage of the curriculum framework.

Science in context

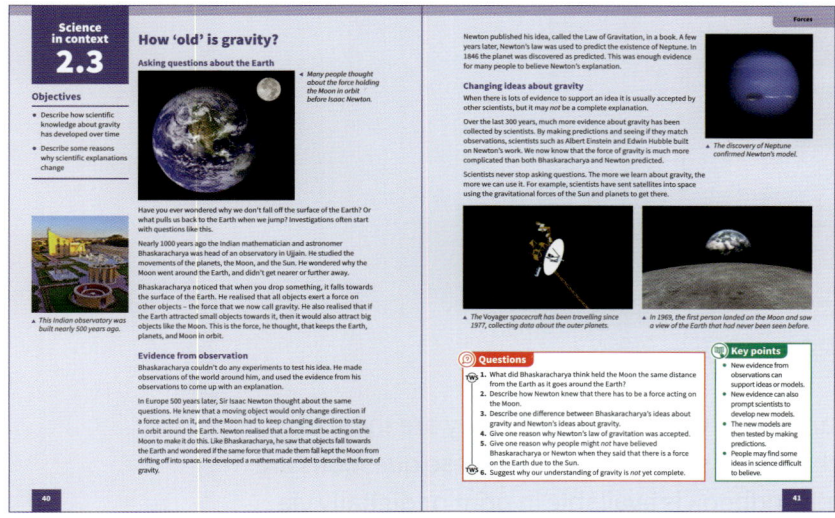

Science in context units will also help you learn:

- how scientists throughout history and from around the globe developed theories, carried out research, and drew conclusions about the world around them
- how science is applied in everyday life
- how issues involving physics are evaluated
- the global impact of the use of physics.

Extension

Throughout this book there are lots of opportunities to learn even more about physics, beyond the curriculum framework. These units are called Extension because they extend and develop your science skills further.

You can tell when a question or part of a unit is Extension because it is marked with a dashed line, like the one on the left.

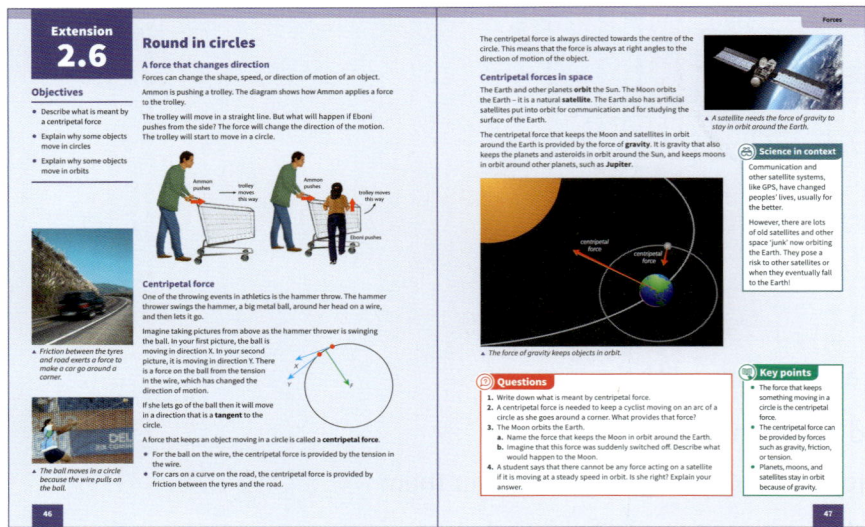

Extension units will not be part of your assessment, but they will help you prepare for moving onto the next stage of the curriculum and eventually for Cambridge IGCSE Physics.

Review

At the end of every chapter and every stage there are review questions.

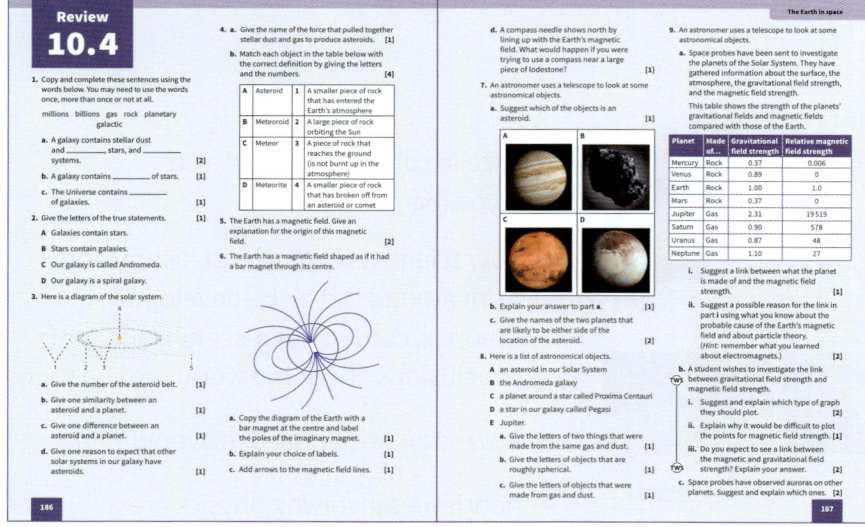

These questions are written in the style of the Cambridge Checkpoint test. They are there to help you review what you have learned in that chapter or stage. Answers to these questions are available in the Teacher Handbook. The Teacher Handbook is available in print or digitally via Kerboodle.

Reference

At the back of this book, on pages 244-249, there are reference pages providing further information that will help you while you study.

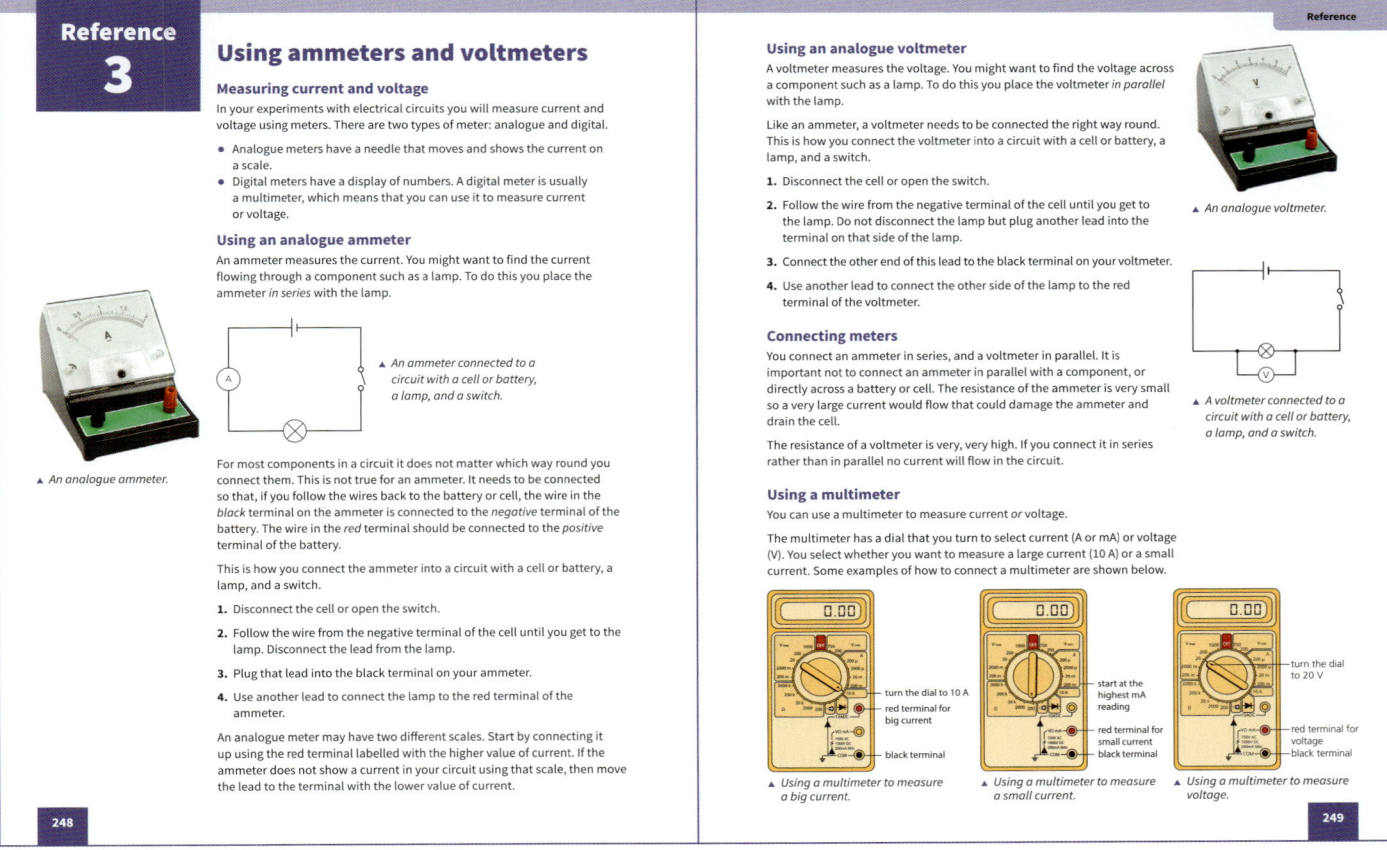

They include information on:

- how to choose suitable apparatus
- how to work accurately and safely
- how to use ammeters and voltmeters.

Objectives

- Recognise that there are many ways to find answers to questions in science

- Understand how to decide on a question to investigate

- Understand that there are some questions that science cannot answer

Part of making a prediction is to think about what might happen if your hypothesis is wrong. Your investigation should be able to show the difference between a correct and an incorrect hypothesis. Your conclusion will say whether the evidence supports, or does not support, your hypothesis.

How does the type of games I play affect the battery life of my mobile phone?

▲ *Subira's question can be answered by doing a fair test.*

Asking questions

How do scientists answer questions?

We can ask lots of different questions about the world. Why does the battery last longer in some mobile phones than others? What might mobile phones be like in the future? Which mobile phone is best?

- There are questions that science can answer
- There are questions that science cannot answer.

What makes a question 'scientific'?

Scientists make **observations** and ask questions such as, 'How do fossil fuels form?' or 'Why are there are so many different animals on Earth?' These are scientific questions.

A scientific question is a question that you can answer by collecting and thinking about **data**. Data can be numbers from measurements, or words from observations.

Hypotheses and predictions

When they have a question, scientists may produce a **hypothesis.** A hypothesis is a scientific theory or proposed explanation made on the basis of evidence that can be further tested. A **prediction** is what you think will happen in the future. Scientists base their predictions on a hypothesis. Then they do an investigation or make further observations to collect data to see if their prediction is correct.

- A hypothesis is **testable** if you can:
- write a prediction based on the hypothesis
- collect data to see whether your prediction is correct.

Types of investigation

Scientists do **investigations** to collect data. There are lots of different types of investigation, for example:

- a fair test
- making a model
- a field study

- a survey or set of observations over time.

Fair testing

In science, anything that might change during an experiment is called a **variable**. The thing that you deliberately change to see whether it affects the outcome of the experiment is a variable. Anything that is affected as a result of your change is also a variable.

In some situations, scientists design an experiment to try to answer their question. To be sure of the answer, they must make it a **fair test**. In a fair test, the scientists change one **variable** to find out what effect it has, and they are careful to keep all the other variables the same.

The quantity that you change is the **independent variable**. A quantity that changes as a result is called a **dependent variable**.

Making a model

Sometimes it is not possible to do a practical investigation to answer a question – maybe what you are looking at is too big or small or dangerous to experiment on. Scientists can also make **models**. As well as helping to answer the question, a model can also be used to predict or to explain. Two types of model are a **physical model** and a **computer model.**

- A physical model is useful for very large-scale or small-scale systems. You may have used a physical model of the Earth and the Sun to explain why we have day and night.
- A computer model uses a computer program to find answers.

Field study

A **field study** is an investigation into plants or animals in their natural habitat. When doing fieldwork, it is important that you make observations without affecting what you are looking at.

A survey or regular observations or measurements

To answer some questions, a scientist might make lots of observations or measurements, or do a **survey**. They might do this over a long time, or all at the same time but in different locations. Sometimes a scientist uses data that other scientists have collected before.

Questions that science can't answer

Scientists cannot answer every question. They cannot answer questions about opinions, or questions for which the answer does not depend on data.

Science cannot answer Sanaa's question.
It could tell you:

- which phone battery lasts longest
- which phone can access web pages fastest.

But an investigation, a field study, observations, or a model will not tell you which phone is best. This is because different people will have different opinions about what is important: some people want a big screen, some people want a good camera, some people want a tough case.

In a fair test, you change the independent variable, measure the dependent variable, and keep all the other variables the same. The other variables are called **control variables**.

What types of food do chimpanzees eat?

▲ Thulani's question can be answered by making lots of observations of chimpanzees in their natural habitat. She would collect data and then choose the best way to display it.

Which country uses the most fuel?

▲ Mosi's question could be answered by collecting data from lots of different countries and comparing them.

Which phone should I buy?

▲ Science cannot answer Sanaa's question.

Objectives

- Describe how to plan a fair test
- Describe how to plan other types of investigation

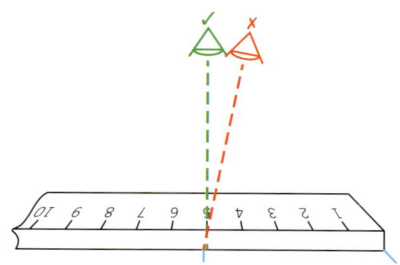

▲ *You should look straight at a scale to make an accurate measurement.*

not accurate
not precise

accurate
not precise

not accurate
precise

accurate
precise

▲ *Readings can be precise but still not accurate.*

Planning and carrying out investigations

How do scientists get the data they need to find the answer to a scientific question? They need to make a plan to collect accurate and precise data.

Planning fair tests

Before carrying out a fair test, you should make a plan. This helps to make sure that you have all the equipment you need to get useful results, and that you do not forget to do something, or do anything dangerous.

Selecting equipment

You need to select equipment that enables you to make measurements of your independent and dependent variables. You may also need equipment that will help you to control the other variables.

You should think about which equipment is most appropriate and how to use it appropriately. For example, you may need to decide whether a measuring cylinder or a beaker is better for measuring volume, or how to measure length accurately.

Accurate and precise data

The measurements you make in an investigation are called data.

It is important to collect data that is **accurate** and **precise**.

Accurate data are close to the true value of what you are trying to measure.

Precise data give similar results if you repeat the measurements. The repeat measurements in each set are grouped closely together. Precision is also determined by the smallest division of the measuring instrument you are using.

Reliability

You need to be confident that your data is **reliable** when you make a conclusion. Data is reliable if you have taken enough measurements. How many are enough?

- You need to have a big enough **range** of values of the independent variable. The range is the difference between the biggest and smallest values. If your range is not large enough you may realise after the investigation that you cannot be confident in your conclusion.
- You need to repeat your measurements. We usually make three **repeat measurements** for every value of the independent variable. You can do fewer, or more.
- You need to deal with **anomalous results**. An anomalous result is a measurement that is very different from the others in a set of repeat measurements and might be a mistake.

Risk assessments

A plan should also include a **risk assessment**. This explains how you will reduce the chances of:

- damage to equipment
- injury to people.

Risk depends on the probability of the damage or injury happening, and the consequence if it did. You can reduce risk by:

- reducing the probability of something going wrong (e.g. keeping glass objects away from the edge of the desk)
- reducing the consequence if something goes wrong (e.g. wearing safety goggles).

Sometimes risks appear very small. You should still note them and say that they are not significant when you write your plan.

You should know the meaning of any hazard symbols, and consider them when you are planning your investigation.

▲ *Hazard symbols for flammable (left) and corrosive substances (right).*

What should a plan include?

Your plan should include:

- the scientific question that you are trying to answer
- why you have chosen to do a fair test
- the independent and dependent variables
- a list of variables to control, how you will do that, and the values each control variable will have
- your hypothesis: the scientific reason on which to base your conclusion
- your prediction: what you think will happen
- the list of the equipment you will need
- how you will use the equipment, step by step, to collect accurate and precise data
- how you will record your results
- a risk assessment, even if the risk is very low.

Planning other investigations

Not all investigations are fair tests. Some questions are better answered using a field study, making observations, or using **secondary data.**

Secondary data have been collected by other people. If you are planning to use secondary data you need to be confident that:

- the information is reliable
- the observations are accurate.

What should a plan for other investigations include?

There are some similarities in the plans for fair tests and other investigations. You should include:

- the scientific question that you are trying to answer
- why you are using this type of investigation
- how you are collecting precise and reliable data or observations
- the sources of secondary information, if you are using it
- your hypothesis: the scientific reason on which to base your conclusion
- your prediction: what you think your data will show
- the list of the equipment you will need, if you are doing it yourself
- a risk assessment, even if the risk is very low.

Objectives

- Describe how to record data from a range of investigations

- Describe how to deal with anomalous results

- Describe how to calculate the mean (average)

Collecting and recording data

How do scientists collect and record the data that they need to answer scientific questions?

Using tables

Measurements are easier to understand if they are in a clear table.

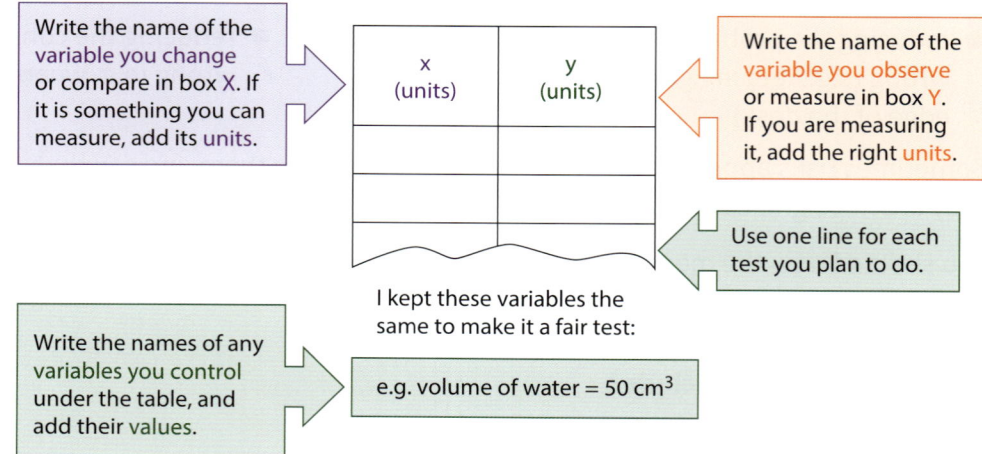

Write the name of the variable you change or compare in box X. If it is something you can measure, add its units.

Write the name of the variable you observe or measure in box Y. If you are measuring it, add the right units.

Use one line for each test you plan to do.

I kept these variables the same to make it a fair test:

e.g. volume of water = 50 cm³

Write the names of any variables you control under the table, and add their values.

Using the correct units

If you use the wrong units for your measurements, your calculations will be wrong.

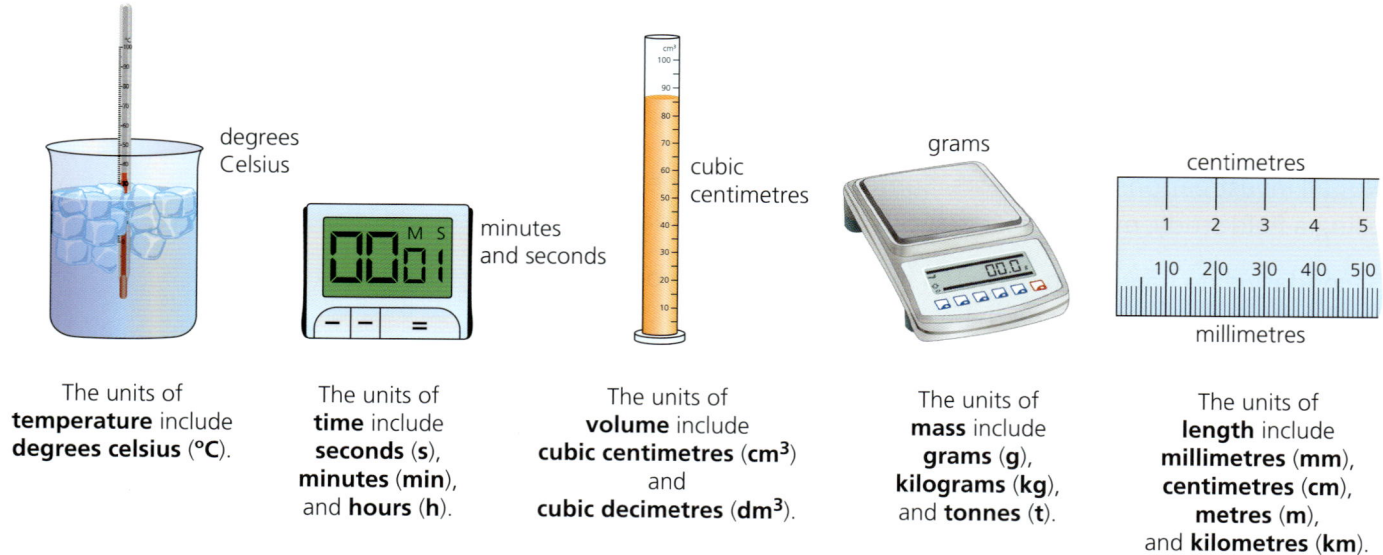

degrees Celsius

minutes and seconds

cubic centimetres

grams

centimetres

millimetres

The units of **temperature** include **degrees celsius (°C)**.

The units of **time** include **seconds (s)**, **minutes (min)**, and **hours (h)**.

The units of **volume** include **cubic centimetres (cm³)** and **cubic decimetres (dm³)**.

The units of **mass** include **grams (g)**, **kilograms (kg)**, and **tonnes (t)**.

The units of **length** include **millimetres (mm)**, **centimetres (cm)**, **metres (m)**, and **kilometres (km)**.

Recording repeat measurements and calculating the mean

In most fair tests you should find the value of the dependent variable more than once, and usually three times. This is a **repeat measurement**. You should record all your repeat measurements in your results table, always to the same number of decimal places.

When you have finished the repeated measurements, you should:

- Check for any **anomalous results**. Do not erase them.
- You can repeat the measurement, and if one is very different from the other two put a line through it and ignore it. Use your new measurements.
- When you are confident that you do not have any anomalous results, calculate the average (**mean**) of the measurements.
- To calculate the mean you add up the all the repeats and divide by the number of repeats.

For example, three students find the time it takes to draw a table. Jamil takes 75 seconds, Abiola takes 35 seconds, and Karis takes 73 seconds.

Abiola's result is anomalous because it is very different from the others. Jamil and Karis find out why. Abiola's table is very messy. She did not use a ruler. They decide to leave it out of the mean. The mean is (75 s + 73 s)/2 = 74 s.

Your average should be rounded up to the same number of decimal places as in the data.

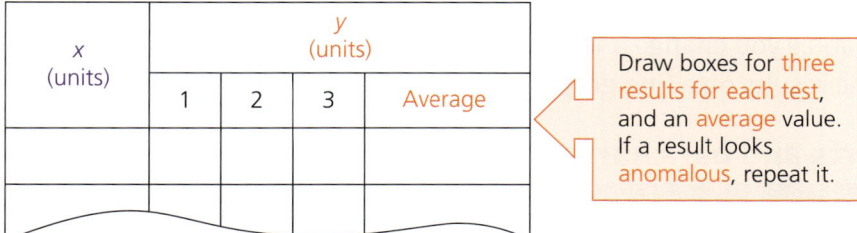

Draw boxes for three results for each test, and an average value. If a result looks anomalous, repeat it.

Recording measurements and calculated quantities

There are some experiments where you will need to calculate quantities using the measurements you have made.

In this situation, you should make another column in your table.

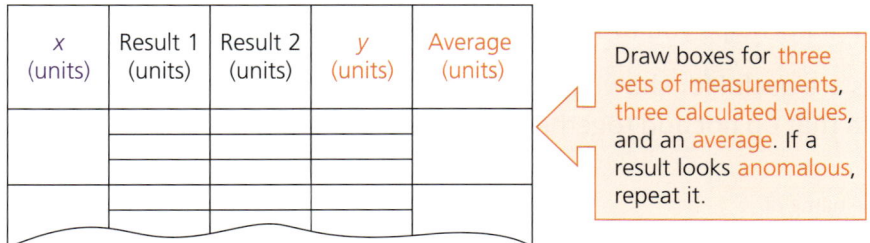

Draw boxes for three sets of measurements, three calculated values, and an average. If a result looks anomalous, repeat it.

Recording observations

In investigations where you are recording your observations, you need to adapt your table. In some cases, you may want to include images or diagrams, so the boxes in your table should be large enough to include them.

Objectives

- Describe how to decide which graph to plot
- Describe how to draw a bar chart, a line graph and a scatter graph
- Describe how to draw a line of best fit

Drawing graphs

How do you know which type of graph to plot? What is the best way to plot graphs? The way that you display the results of your investigation depends on the type of data that you have collected.

Types of data

If the values of the variable you change (*x*) are *words*, then *x* is a **categoric** variable. There is no logical order, like size, for the categories. Names are one example.

Variables like shoe size are **discrete** variables. They are numbers, but there are no in-between sizes. The number of paper clips in a pot or people in a room are discrete variables.

You can only draw a **bar chart** or a **pie chart** for data that include categoric or discrete variables.

Other variables are **continuous** variables. Their values can be any number. Height, temperature, and time are continuous variables.

If the variables you change and measure are *both* continuous variables, display the results on a **line graph** or **scatter graph**.

Pie charts and bar charts

Student	Time spent on poster (minutes)	Time spent on homework (hours)
Deepak	24	2.5
Jamila	54	4.5
Kasim	12	2.0

Deepak, Jamila, and Kasim research and design a poster together.

The table shows the time that each of them spends. The pie chart drawn from the results helps you to see who did the most work on their project.

Pie charts are useful for showing fractions of a whole. When you want to show data that do not add together, a bar chart is better.

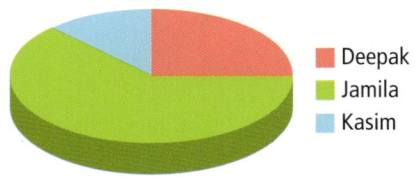

Deepak
Jamila
Kasim

▲ *Time spent on homework.*

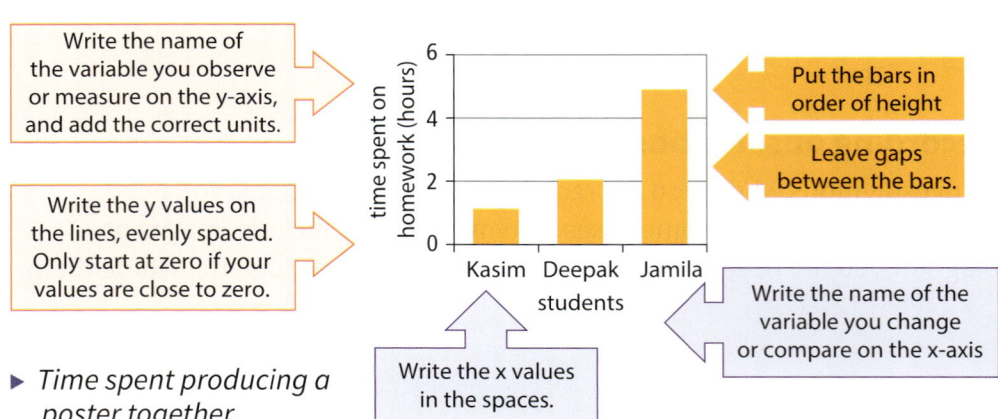

Write the name of the variable you observe or measure on the y-axis, and add the correct units.

Write the y values on the lines, evenly spaced. Only start at zero if your values are close to zero.

Put the bars in order of height

Leave gaps between the bars.

Write the name of the variable you change or compare on the x-axis

Write the x values in the spaces.

▶ *Time spent producing a poster together.*

Line graphs and scatter graphs

Drawing a line graph

A line graph makes it easier to see the link between two continuous variables – the **independent variable** and the **dependent variable**.

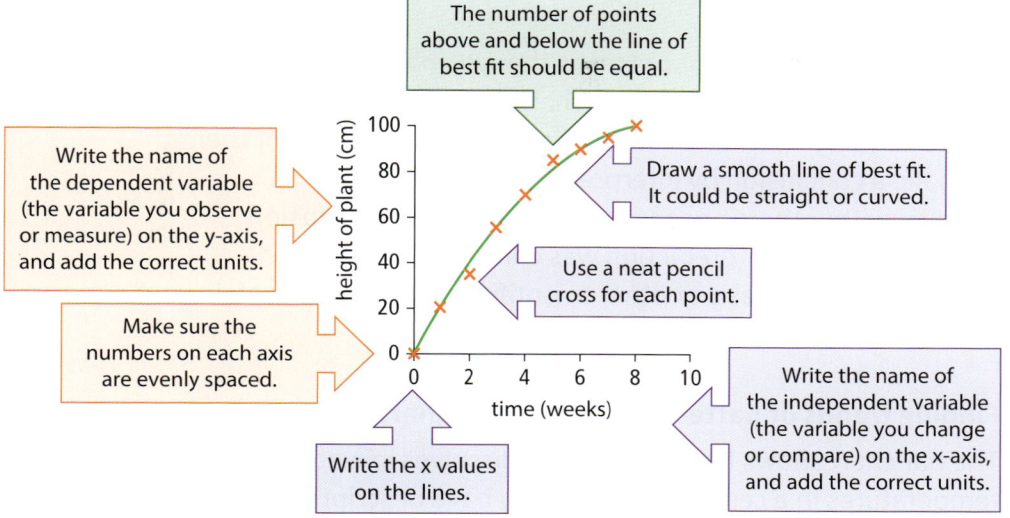

The number of points above and below the line of best fit should be equal.

Write the name of the dependent variable (the variable you observe or measure) on the y-axis, and add the correct units.

Draw a smooth line of best fit. It could be straight or curved.

Use a neat pencil cross for each point.

Make sure the numbers on each axis are evenly spaced.

Write the x values on the lines.

Write the name of the independent variable (the variable you change or compare) on the x-axis, and add the correct units.

▲ *Use a line graph for continuous variables when you think there is a link between them.*

Drawing a scatter graph

A scatter graph shows whether there is a **correlation** between two continuous variables. In the graph below, all the points lie close to a straight line. That means there is a correlation between them. If there is no correlation between the variables, then the points would be scattered all over the graph.

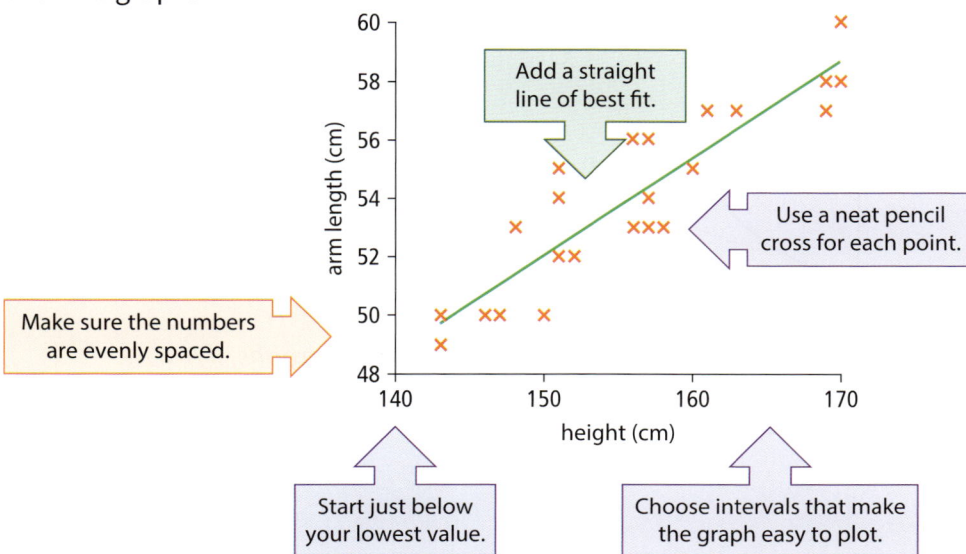

Add a straight line of best fit.

Use a neat pencil cross for each point.

Make sure the numbers are evenly spaced.

Start just below your lowest value.

Choose intervals that make the graph easy to plot.

▲ *A scatter graph will show you if there is a correlation between two continuous variables.*

If you collect continuous data in a fair test investigation, you are trying to find out how one variable affects the other. You will usually plot a line graph.

In other investigations, you may be trying to see if there is a relationship between two variables. You will usually plot a scatter graph.

A line graph shows the link between two variables. You should draw a **line of best fit**. This is a line that goes through as many points as possible with roughly equal numbers of points either side of the line.

A correlation does not mean that one variable affects the other one. Something else could make them both increase or decrease at the same time. For example, if you plotted the number of ice creams sold in a town each day against the number of people going to the town swimming pool that day, you would see a correlation. This does not mean that getting wet makes people eat ice cream, or that eating ice cream makes people go swimming. It probably means that on hot days more people want to go swimming and to eat ice cream.

Objectives

- Describe how to do an analysis of an investigation

- Describe the relationship shown by different lines of best fit on graphs

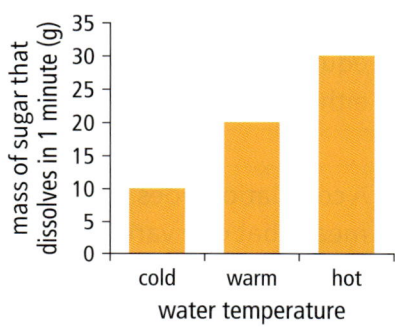

▲ *Mass of sugar which dissolves in water at different temperatures.*

Analysis

Analysing the evidence

When you analyse the evidence that you have collected (yourself or from secondary sources) you should:

- describe the **trends** or **patterns** that you have worked out from the display of your data (a graph, chart, or other display)
- identify any anomalous results, and suggest reasons for them
- make a conclusion by interpreting the results
- say whether there are any limitations to your conclusion
- say whether your prediction was correct
- use your hypothesis or other scientific knowledge to explain your conclusion.

Finding trends or patterns in graphs and charts

The bar chart shows how much sugar dissolves in water at different temperatures in a certain time. We only have descriptions of the temperature, not numbers, so the results are categoric.

You can describe the trend by saying:

'As the temperature of the water increases, the mass of sugar that dissolves increases.'

This is sometimes called the **relationship** between the variables.

Line or scatter graphs show relationships between continuous variables. When you have plotted the points on a line or scatter graph, draw a line of best fit.

In the graphs below the line of best fit is shown, but not the points.

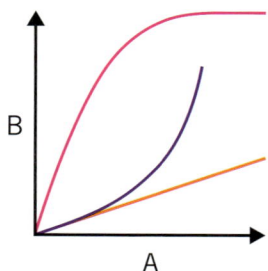

▲ *In these graphs, if A increases then B increases*

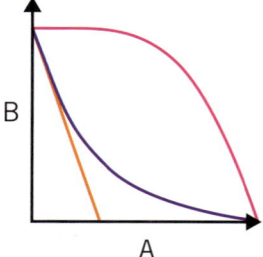

▲ *In these graphs, if A increases then B decreases*

Line or scatter graphs will be different if there is no relationship or correlation between the variables.

You may get a horizontal line in a fair test investigation if changing the independent variable has no effect on the dependent variable.

Identifying anomalous results

When you recorded your data, you may have identified points that did not fit the pattern. You may have ignored them, or repeated the measurement.

- You should identify these points now.
- You might see points that are not close to your line of best fit. They too are anomalous results.
- You should think of possible reasons why they might have occurred.

Writing your conclusion

A conclusion states what you have found out. You should also think about the **limitations** to your conclusion. Your conclusion may be limited if:

- you had lots of anomalous results
- the line of best fit is not clear
- your data is limited in terms of the range of variables that you investigated
- your data was limited in terms of the number of results that you collected.

Checking your prediction

When you have found the pattern, you need to check the prediction you made and say whether it was correct. You should look carefully at the extent to which the evidence (data and observations) supports or refutes (disproves) your prediction.

Explaining your conclusion

Finally, *suggest scientific reasons* for any relationship or correlation differences that you have found. You could refer back to your hypothesis, or other scientific knowledge.

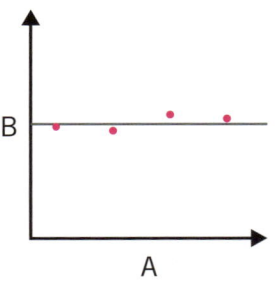

▲ *In this graph, if A increases B does not change.*

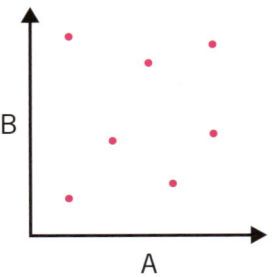

▲ *In this graph, there is no correlation between A and B.*

> If points are scattered everywhere on a graph of two variables, it shows that they do not affect each other.

I think that the powdered sugar dissolved faster than the normal sugar because the pieces were much smaller.

I think that the shoe on the carpet needed a bigger force to move it than the shoe on the wooden floor because there was more friction between the shoe and the carpet.

I think that the plant near the window grew more quickly than the plant away from the window because there was more sunlight there.

▲ *Evaluating means working out what is good and what is not so good.*

Evaluation

After you have collected your data, plotted a graph, written a conclusion, and explained what happened using scientific knowledge, you need to evaluate your investigation.

An evaluation is done in three stages:

- evaluate the *quality of the data*
- evaluate the *methods you used*
- *suggest improvements* to the method.

If you were to do the investigation again, the improvements should make you more **confident** in your conclusion.

Evaluating the data

Are there anomalous results?

When you look at your data tables and graphs, you can see how many anomalous results you had. This is why you do not erase the anomalous results in your results tables.

Anomalous results can limit your conclusions. They can also reduce the confidence that you have in your conclusion.

What is the spread?

The **spread** is between the smallest and the biggest values of repeated measurements. When you repeated the experiment, were the results close together (small spread) or far apart (large spread)? A smaller spread means that you can have more confidence in any conclusion based on your data.

What is the range and number of values?

The **range** of the variables that you have investigated is the difference between the smallest and the largest values. If the range is large, then you can be more confident of your conclusion.

You should collect enough data points to feel confident that you have correctly identified the trend or pattern. Two would not be enough.

Were there systematic or random errors?

There is **uncertainty** in any measurement that you make. This is one of the reasons why there is usually a spread in experimental data.

You should think about possible errors, as well as any anomalous results and the spread, to help you to decide how confident you are in your conclusion. There are two types of error that can affect scientific measurements.

- **Random errors** – these can affect the spread, or cause anomalous results. An example is the temperature of the room suddenly changing because someone opens a door.
- **Systematic errors** – these can make your measurements less accurate. An example is a newtonmeter reading 1 N even when there is nothing attached to it.

Evaluating the methods

When you planned your investigation, you chose the measuring instruments and the methods of using them to collect your data. You should look back at those choices and describe:

I don't think I took the measurements of the volume very accurately because I did not look directly at the scale.

- the extent to which the *equipment* enabled you to collect data that was accurate and precise
- the extent to which the *methods* enabled you to collect data that was accurate and precise

Suggesting improvements

Suggesting improvements is not about making the experiment easier or quicker.

Any improvements that you suggest should be designed to improve the *quality of your data*, because that would mean:

- there are fewer limitations to your conclusion
- you would be more confident in your conclusion.

How do I get better data with the same equipment?

You may improve the quality of the data by:

I think we could have videoed the falling object so we could have played the video to see where it was.

- eliminating systematic errors
- reducing the effect of any random errors
- including a bigger range
- doing more repeat readings
- repeating any measurements that gave anomalous results.

What other equipment would help?

Some types of equipment produce more precise and accurate data than others. You may need to do some research to find out.

- Using light gates to measure speed produces more accurate measurements than clicking a stopwatch by hand.
- Using a measuring cylinder to measure volume produces more accurate measurements than using a beaker.
- Using a balance that measures to more decimal places produces a more precise measurement.

Objectives

- Describe where we get our energy from
- Know the unit of energy

What is energy?

You need energy

You need **energy** from food to walk, run, or ride your bike. Energy from **fuels** is needed for transport and to produce electricity. The idea of energy helps us explain what can happen, but not *why* things happen. For example, fuel allows a car to move, but that doesn't tell you where it will go. But the concept of energy tells you that without fuel the car cannot move.

▲ *It takes a lot of fuel to launch a rocket into space.*

What is the unit of energy?

Energy is *not* an actual substance that moves from one object to another. It is a way of keeping track of a very important quantity, a bit like money.

The unit of energy is the **joule** (**J**). One joule is a very small amount of energy, so we often use **kilojoules** (**kJ**). 1 kJ = 1000 J.

Different foods store different amounts of energy.

▲ *The energy stored in packaged food is usually shown on the nutrition label.*

Food	Energy (kJ) per 100 g of food
banana	340
beans	400
rice	500
cooked chicken	800
chocolate	1500

For a long time a unit called a kilocalorie (kcal) was used for the energy stored in food (1 kcal = 4.2 kJ). People called it a 'calorie' for short and you still often see the kilocalorie content of foods on labels.

▲ *Running or playing can use 3500 kJ per hour.*

▲ *Your body needs 300 kJ per hour to sleep.*

How does your body use energy?

How much energy do you need each day? It depends on what activities you do.

All activities have an energy cost. Keeping your body warm, breathing, moving, and talking all need energy. Children need energy to grow bigger bones, muscles, and brains.

About three-quarters of the energy that you need every day is for processes in your body like breathing. You then need more energy for all the other activities that you do such as walking, running, or lifting things.

Activity	Energy (kJ) for each minute of activity
sitting	6
standing	7
washing, dressing	15
walking slowly	13
cycling	25
playing football	59
swimming	73

▲ *Your brain uses energy to learn.*

🔭 Science in context

The more active you are, the more energy you need. Athletes need lots of energy. People who take on the challenge of walking to the North or South Pole need even more energy, because as well as walking and carrying their food they need lots of energy to keep warm.

Energy balance

An adult should take in only as much energy as they need for the activities that they do. If they take in more energy than they need, their body stores it as fat for future use. If they eat less than they need, then the body will use energy from its store of fat and they will lose weight.

Energy in fuels

Food is *not* the only energy store – fuels such as **coal**, **oil**, or wood also provide us with stored energy that we can use.

We can burn wood or coal to heat a room or to cook food. If we use an electric kettle, then the energy needed to boil the water is transferred by electricity. The energy used to generate this electricity may have been stored in a fuel such as coal or oil.

▲ *One kilogram of wood stores a similar amount of energy to 1 kg of chocolate.*

📖 Key points

- We use the energy stored in food for all our activities and to stay alive.
- Energy is measured in joules or kilojoules.
- The energy in the food someone eats should equal the energy they need.

❓ Questions

1. Name three fuels.
2. Give two reasons why your body needs energy when you are asleep.
3. Calculate the number of joules in 200 kJ.
4. Explain why it is important for young children to take in *more* energy than they need for the activities they do each day.
5. Calculate the number of minutes that you would need to cycle to use up the energy in 100 g of chocolate.

Asking questions: Energy

Objectives

- Recognise that there are many ways to find answers to questions in science
- Understand how to decide if a question can be answered with a fair test investigation

Asking questions

Chima is interested in the energy stored in fuels and food. He asks his friends what they would like to know.

How much fuel will Africa need in the future?

▲ Sanaa

Which city uses the most coal?

▲ Mosi

Should we use fossil fuels to cook our food?

▲ Subira

Which fuel is the best?

▲ Kojo

What types of food do chimpanzees eat?

▲ Thulani

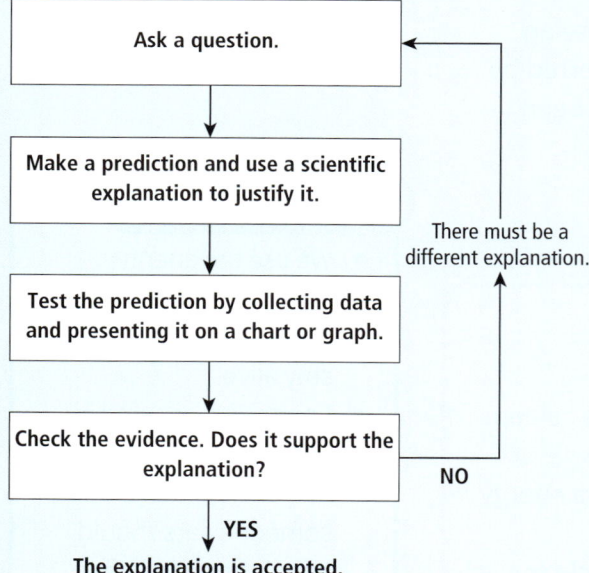

Ask a question.

↓

Make a prediction and use a scientific explanation to justify it.

↓

Test the prediction by collecting data and presenting it on a chart or graph.

↓

Check the evidence. Does it support the explanation?

There must be a different explanation.

NO

↓ YES

The explanation is accepted.

Chima looks at all the questions. He wonders how the questions can be answered.

Answering different types of question

There are lots of different ways of finding answers to questions using science. But not every question can be answered using science. This flow chart shows how some scientists think about answering a question.

To answer some questions you need to collect data. There are lots of different ways of collecting data.

What is a fair test investigation?

You can collect data by doing a **fair test** investigation. This is an experiment in which you change one **variable** (something that can be changed) and keep all the other variables the same. You set up equipment to take measurements to answer the question. If you make sure everything stays the same except the variable you are interested in, then nothing else can affect your measurements and it is a fair test.

▲ *Lumusi*

Lumusi can answer her question by doing a fair test investigation. She can use different fuels to heat some water, and time how long it takes the water to reach a certain temperature. Then she can use the data to find out which fuel heats water the fastest.

She will need to control:

- how much water is used
- the start and end temperatures of the water.

This flow chart shows how she can do her investigation.

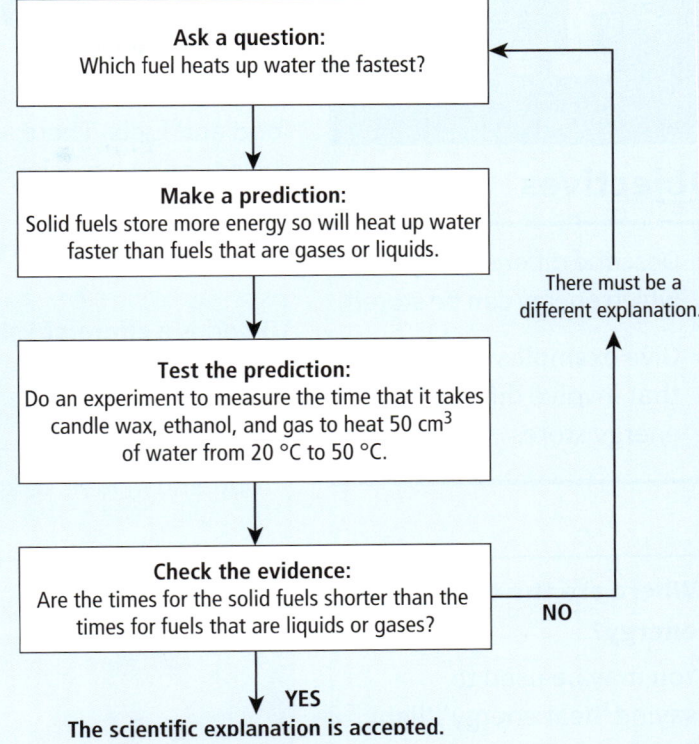

Scientists may do other types of practical work, for example to find out *if* something happens, or how it has changed over time.

Questions

 1. Can you answer the question 'Which fuel is best?' with a practical investigation? Explain your answer.

2. Here are some data that Lumusi collected in her investigation.

Fuel	Volume of water (cm³)	Start temperature (°C)	End temperature (°C)	Time to reach end temperature (s)
ethanol (liquid)	25	20	30	85
natural gas	25	20	30	45
candle wax (solid)	25	20	30	108

a. Write which fuel stores the most energy. Explain how you know.

b. Does the evidence support Lumusi's hypothesis from the flow chart above? Explain your answer.

TWS 3. Explain why scientists repeat their investigations.

Key points

- Scientists ask a question and think creatively about how to collect data.
- Scientists do investigations to collect data to answer a question.
- In a fair test investigation they change one variable at a time and take measurements.

1.3

Energy stores and transfers

When we move, cook food, or turn a light on, we transfer stored energy from food and fuels. There are different ways of calculating energy. We describe the situation associated with a calculation as an **energy store**.

Everyday energy stores

Objectives

- Describe different ways in which energy can be stored
- Give examples of situations that involve different energy stores

Food is a **chemical store** of energy. We say that energy is stored chemically in food and fuels. We use the energy stored chemically from food and in fuels to keep warm.

Food and fuels are *not* the only stores of energy. A change of position or shape can store energy.

A mother lifts her baby up from the floor. Energy is transferred to a **gravitational potential store**, so this energy is stored gravitationally. The position of the baby or **gravity** store. has changed.

If you sit on a bed or a sofa the springs inside it are compressed. Energy is transferred to an **elastic store**, so this energy is stored elastically. Elastic bands and other stretchy materials can also store energy elastically.

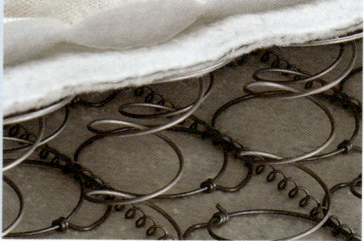

All moving objects store energy kinetically. A fast-moving car has more energy in its **kinetic store** than a slow-moving car.

When fuels burn they heat up the air around them. Energy is transferred to the **thermal store** of the air, so this energy is stored thermally. This is what is sometimes call 'heat'. There is more energy in the thermal store of a hot object than of a cold object.

Where are the types of energy?

You may be used to saying 'heat energy', 'light energy', or 'electrical energy'.

Energy does *not* exist in different forms or types. It is a quantity that we calculate. It can be stored or transferred.

When we say 'kinetic energy' we mean

- energy stored kinetically or
- energy in a kinetic store.

Types of energy transfer

In a torch or flashlight, an **electric current** transfers energy from the battery to the bulb. Electric kettles use an electric current to heat water. Energy is transferred from power stations rather than from batteries.

Energy can be transferred from the Sun or a candle in the form of **light**. Energy is also transferred by **heating**.

Musical instruments transfer energy to your ear as sound. Your vocal chords produce sound when you talk.

An electric current, light, sound, and forces cannot be stored. These are the ways in which energy is transferred.

Energy from the Sun

The Sun produces light and other types of **radiation**. Why does the Sun shine? Inside the Sun a process called **nuclear fusion** is happening all the time. The Sun is a **nuclear store** of energy. Hydrogen atoms combine to form helium, and energy is released.

▲ *When you walk upstairs you exert a force on your body. Energy is transferred when forces act on objects.*

Where are the transformations of energy?

Energy is *not* 'transformed' from one type to another.

Energy is a quantity, so is transferred from one store to another.

Key points

- There are different energy stores including chemical, gravitational potential, elastic, kinetic, thermal, and nuclear.

- Energy is transferred by heating, an electric current, forces, and radiation including light and sound.

Questions

1. Write down two sources of chemical energy.
2. A child picks up a toy from the floor. Name the store that has gained energy.
3. Describe the way energy is stored in a stretched elastic band.
4. A pendulum swings backwards and forwards. Write down two of the energy stores involved.
5. A student thinks that the Sun is a huge ball of fire that provides the energy for the Earth. Would you say that they are correct? Explain your answer.

Energy transfer diagrams and dissipation

Objectives

- Understand how energy transfers are shown in diagrams
- Be able to construct energy transfer diagrams
- Describe what is meant by dissipation

▲ *Sound transfers energy.*

▲ *The elastic in the catapult stores energy.*

Darsh is listening to a battery-powered radio. The radio produces sound. The sound transfers energy stored chemically to the surroundings, which heat up.

Energy transfer diagrams

Energy transfer diagrams are a useful way of showing energy transfers. Here is a diagram for a battery-powered radio. The circles show the *transfer* of energy, and the squares show places where energy can be *stored*.

The diagram shows that the energy is transferred:

- from a chemical to a thermal store
- by an electric current and by sound.

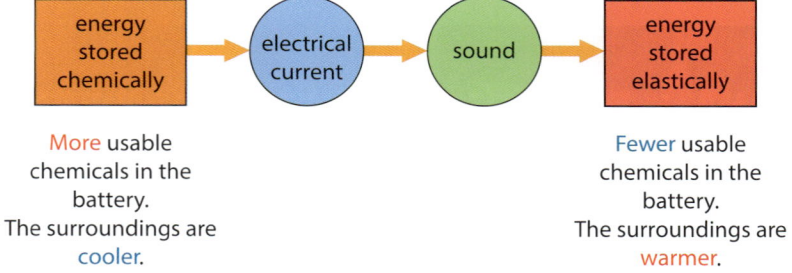

energy stored chemically → electrical current → sound → energy stored elastically

More usable chemicals in the battery. The surroundings are cooler.

Fewer usable chemicals in the battery. The surroundings are warmer.

Transferring energy with forces

There are lots of different devices that transfer energy, such as televisions or catapults.

Dewi is playing with a catapult. She puts a ball in the sling, pulls it back, and lets it go. The ball flies through the air. The food Dewi ate gave her the energy she needs to pull back on the catapult. The energy stored chemically in the food is transferred to energy stored elastically in the catapult. When she lets go this is transferred to the energy of the sling and the ball. The energy transfer diagram is in two stages because you can do calculations of energy stored elastically *and* kinetically.

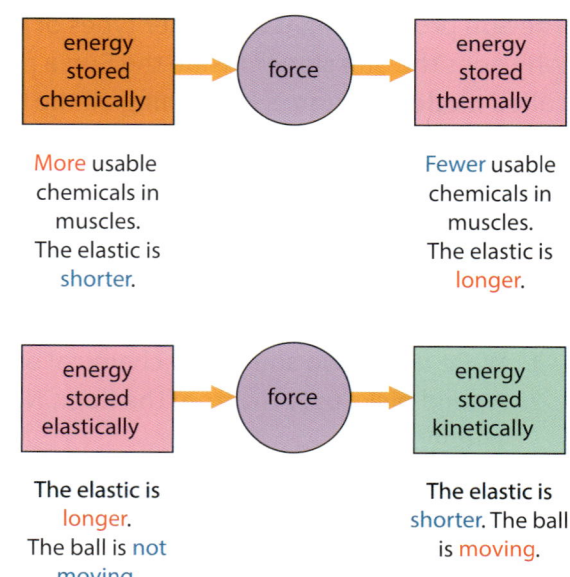

energy stored chemically → force → energy stored thermally

More usable chemicals in muscles. The elastic is shorter.

Fewer usable chemicals in muscles. The elastic is longer.

energy stored elastically → force → energy stored kinetically

The elastic is longer. The ball is not moving.

The elastic is shorter. The ball is moving.

Energy transfers involving gravity

Ikenna is cycling towards a small hill.

- He stops pedalling when he gets to the hill.
- He freewheels up the hill, slowing down.
- He gets to the top of the hill.
- Then he freewheels down the other side, speeding up as he goes.

When Ikenna pedals, energy stored chemically is transferred to energy stored kinetically. The diagram shows what happens next.

There are many examples of this type of energy transfer, for example a child on a playground swing.

As he goes up the hill gravity pulls Ikenna down. Energy is transferred from a kinetic store to a gravity store.

When he goes down the hill the force of gravity transfers energy from the gravity store to a kinetic store.

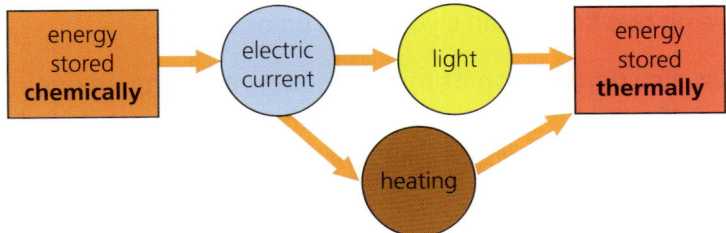

Energy dissipation

Energy is transferred in processes that change something, such as heating food, or moving us from one place to another.

In all processes, some energy is transferred to the surroundings, and heats them up a little bit. This usually happens because:

- things are heated by friction
- sound is produced.

Eventually all energy ends up in the surroundings. In a light bulb, we want energy to be transferred as light, but light bulbs get hot. This heating is *not* useful and energy is wasted or **dissipated**.

energy stored **chemically** → electric current → light → energy stored **thermally**

electric current → heating → energy stored **thermally**

▲ *Energy transfer diagram for a light bulb in a torch.*

▲ *Electrical devices transfer energy using sound and light.*

Questions

1. Draw an energy transfer diagram for:
   a. Dewi walking to school
   b. Dewi climbing the steps of a slide, sliding down, and stopping.
2. Describe what is meant by 'energy is dissipated'.
3. Suggest how energy is dissipated from:
   a. an electric kettle
   b. a car travelling on a road.
4. Suggest why it is hard to describe dissipation when you talk about heating a home.

Key points

- An energy transfer diagram shows the energy changes in a process or device.
- Energy that is dissipated is no longer useful.
- Energy that is dissipated is usually transferred to the surroundings, which get hotter.

Gravitational potential energy and kinetic energy

Objectives

- Describe factors that affect gravitational potential energy and kinetic energy

- Describe situations that involve energy changes between kinetic energy and gravitational potential energy and energy dissipation

▲ *The panda has more energy in a gravity store at the top of the slide than it did on the ground.*

When this panda plays on a slide, energy is transferred between stores.

Gravitational potential energy

Imagine going up to the top of a very tall building. As you climb higher, your gravitational potential energy increases. **Gravitational potential energy** or **GPE** is the energy that something has because of its position. It is another term for 'energy in a gravity store', or 'energy stored gravitationally'. Your GPE depends on your distance from the centre of the Earth. You will gain twice as much GPE if you move up two floors than if you move up one floor.

The GPE an object has also depends on its mass. A small boy going up one floor would gain less GPE than his mother going with him. If his mother had twice the mass of the boy, then she would gain twice as much GPE.

▲ *People on the top floor of this building have more GPE than those lower down.*

Kinetic energy

Kinetic energy is the energy that something has when it is moving. It is another term for 'energy in a kinetic store', or 'energy stored kinetically'. A running lion has lots of kinetic energy.

- If a lion and an antelope are moving at the same speed, then the lion will have more kinetic energy because it has more mass.
- If two lions have the same mass but are running at different speeds, then the faster lion will have more kinetic energy.

▲ *Moving objects have kinetic energy.*

Up and down

Sudarto is playing with a ball. He throws the ball upwards. He watches as the ball slows down, stops for an instant, then speeds up as it falls back down.

- The ball's energy is moving from a kinetic store to a gravitational store on the way up.
- The increase in GPE is equal to the decrease in kinetic energy.
- On the way down, energy is transferred from a gravitational store to a kinetic store.

A very small amount of energy will be transferred to the surrounding because of air resistance. Energy is dissipated.

Back and forth

A **pendulum** is a ball on a string that swings backwards and forwards. To start the pendulum swinging you pull it back and let it go.

- When you pull the pendulum back you also pull it up, so its GPE increases.
- When you let it go it falls, and energy is transferred to a kinetic store.
- As the ball swings up again energy is transferred to a gravity store.

A child's father pushes her on a swing, which behaves like a pendulum. If he stops pushing she will *not* swing so high each time. Friction is making parts of the swing warm up, and air resistance is making the air heat up. Some kinetic energy is being transferred into the thermal store of the surroundings. In other words, energy is dissipated.

A rollercoaster

Engineers who design rollercoasters think very carefully about kinetic energy and GPE. They need to know how much energy is needed for passengers to reach the top of the hills.

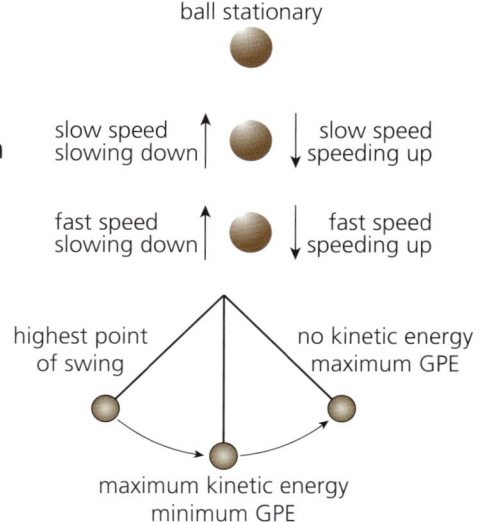

ball stationary

slow speed slowing down | slow speed speeding up

fast speed slowing down | fast speed speeding up

highest point of swing | no kinetic energy maximum GPE

maximum kinetic energy minimum GPE

▲ *A rollercoaster lifts people up high so they can move fast on the way down.*

Questions

1. Describe the difference between kinetic energy and GPE.
2. A man and his son run to the top of a hill and stop. The mass of the man is bigger than the mass of the boy.
   a. Write down who has more kinetic energy while they are both running at the same speed. Explain your answer.
   b. Would it be possible for them to have the same amount of kinetic energy? Explain your answer.
   c. When they are at the top of the hill, who has more GPE? Explain your answer.
3. Suggest how energy is dissipated on a rollercoaster ride.
4. A girl drops a stone down a well and listens for the splash. When the stone is at the top of the well it has 20 J more gravitational potential energy than it has at the bottom.
   a. Calculate how much GPE the stone has halfway down the well.
   b. Write down how much KE it has halfway down the well. Explain your answer.
   c. Write down one assumption you have made.

Key points

- GPE is the energy that something has because of its position.
- Kinetic energy is the energy something has because of motion.
- When things move up and down, energy is transferred between gravitational and kinetic stores. Energy is dissipated by friction and air resistance.

Objectives

- Identify whether a given hypothesis is testable
- Decide how many measurements to take
- Present results clearly

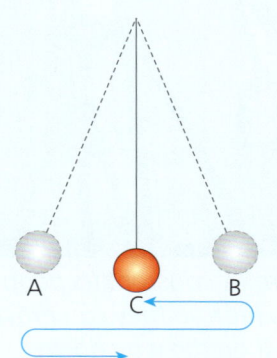

▲ *The period of a pendulum is the time from the centre, out to one side, back through the centre, out to the other side, and back to the centre.*

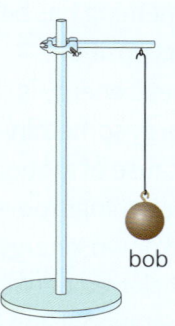

▲ *A simple pendulum.*

Planning: Pendulum motion

Backwards and forwards

A **pendulum** is an object that swings backwards and forwards, like a child on a swing, or a car on a theme park ride.

The time that it takes a pendulum to swing backwards and forwards once is called the **period**.

The pendulum is stationary at points A and B. It is moving fastest in the middle of the swing, at point C. The period is the time that it takes to move from A to B *and back again*.

If you wanted to measure the period of a pendulum, you would measure the time with a stopclock or stopwatch. It is very difficult to measure short lengths of time accurately, so it is important to make repeat measurements.

Asking a scientific question

Aditi and Diya have made a pendulum from some string with a small mass on the end, called a bob.

They decide to plan an investigation. They know that the first thing they need to do is to think what question they want to investigate.

I wonder what would happen if the bob's mass was bigger?

I wonder what would happen if the string was longer?

They decide to investigate what factors affect the period of the pendulum. This is the list of variables that they think might affect the period:

- the length of the pendulum
- the mass of the bob
- the distance that you pull the pendulum bob back.

Producing a testable hypothesis

Aditi and Diya discuss the energy stores and transfers associated with the pendulum. They produce this hypothesis: *How long the pendulum takes to swing depends on the length of the string*. The longer the string, the less difference there will be between the height of the bob at the middle and the end of its swing. A bob on a short string will therefore have more GPE when pulled back for the same distance, so it will have more kinetic energy and go faster.

This is a testable hypothesis because you can collect data to see whether a prediction based on the hypothesis is correct.

Aditi's plan

I will change the length of the string. I will start with 10 cm and go up by 5 cm until I get to 50 cm.
I will time how long the pendulum takes to swing 10 times and divide the result by 10 to get the period.
I will keep the mass of the bob the same and pull it back the same amount each time.
I will put my results in a table like this:

Length	Time

Diya's plan

I am going to change the length of the pendulum. I have tried out the experiment.
I am going to use a pendulum with lengths between 30 cm and 100 cm.
Each time I will time how long it takes to swing once.
I will do this three times for each length and find the average time.
I will put my results in a table like this:

Length (cm)	Time (s)	Time (s)	Time (s)	Average time (s)

Each student has identified the variable that they are going to change and the variable that they are going to measure. They have explained how they will control the other variables.

Key points

- A hypothesis is testable if you can collect data to check a prediction.
- You should always repeat measurements to reduce the effect of errors.
- You should present your results in a table, with a heading and any units at the top of each column.

Questions

1. Describe the best way to display results in an investigation.
2. Write down the type of investigation that Aditi and Diya are doing.
3. Explain why the hypothesis that they produced was testable.
4. Suggest a prediction that they could make based on their hypothesis.
5. Describe what Diya did to make her results more accurate.
6. Describe one way in which Aditi's plan is better than Diya's plan.

1.7 Elastic potential energy

Objectives

- Explain how energy can be stored elastically
- Describe what is meant by 'elastic'
- Describe a situation in which elastic potential energy is dissipated

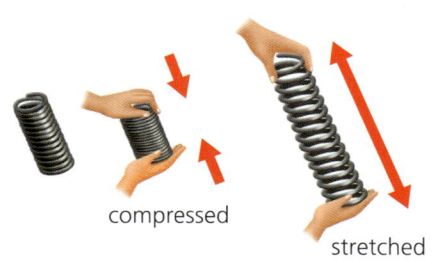

▲ *Springs can be compressed or stretched.*

What is elastic potential energy?

Bows and arrows have been used for thousands of years. The archer pulls back on the bowstring and points the arrow at the target. The **elastic potential energy** (**EPE**) increases because the shape of the bow has changed. 'EPE' is another term for 'energy in an elastic store', or 'energy stored elastically'.

▲ *A bow used in ancient Assyria more than 2000 years ago.*

- An elastic band stretches when you pull it.
- A mattress contains springs that compress when you lie on them. The band and the springs store EPE because their shape has changed.

Bungee jumping

Grace is doing a bungee jump. She first goes up on a high bridge. She attaches a strong elastic cord to her legs.

She jumps and falls, getting faster, and then slows down as the elastic stretches. She stops for an instant, then bounces back up again.

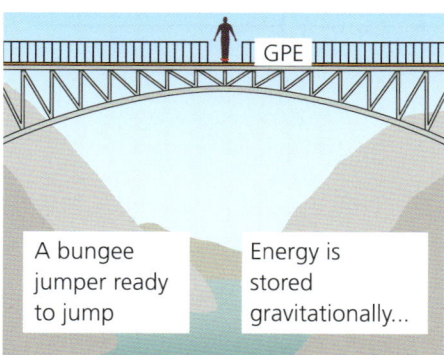

GPE

A bungee jumper ready to jump

Energy is stored gravitationally...

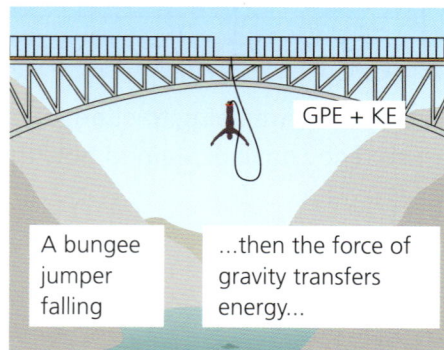

GPE + KE

A bungee jumper falling

...then the force of gravity transfers energy...

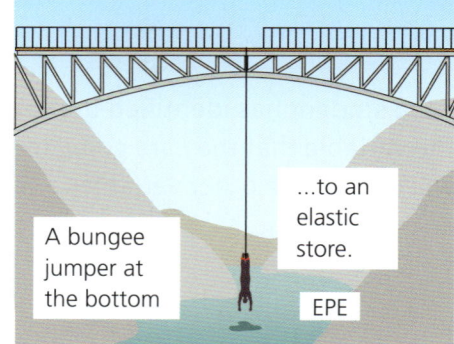

...to an elastic store.

A bungee jumper at the bottom

EPE

Storing energy in materials

▲ *Objects store energy when they deform.*

When materials are stretched or compressed their shape changes. We say that they **deform**. When they deform they store EPE. Materials that return to their original shape after being deformed are called **elastic**.

- When a ball hits the ground it deforms.
- Its kinetic energy is transferred to EPE when it changes shape.
- It changes back to its original shape and bounces back up.
- The EPE that has been stored is transferred to kinetic energy.
- Eventually all its kinetic energy is transferred to GPE and it reaches the top of its bounce.

A ball will *not* bounce back up to original height. This is because not all the kinetic energy that the ball had before the bounce is returned to the ball after the bounce.

Some energy is transferred by sound when the ball lands. When the material deforms its particles change position. This heats the ball a bit. The energy in the thermal store of the surroundings also increases slightly. If some energy is transferred to the surroundings, then the ball will have less kinetic energy after it bounces than before. Energy is dissipated.

The ball will bounce to a lower height each time until it stops. All of the GPE from the start has now been transferred to the thermal store of the surroundings.

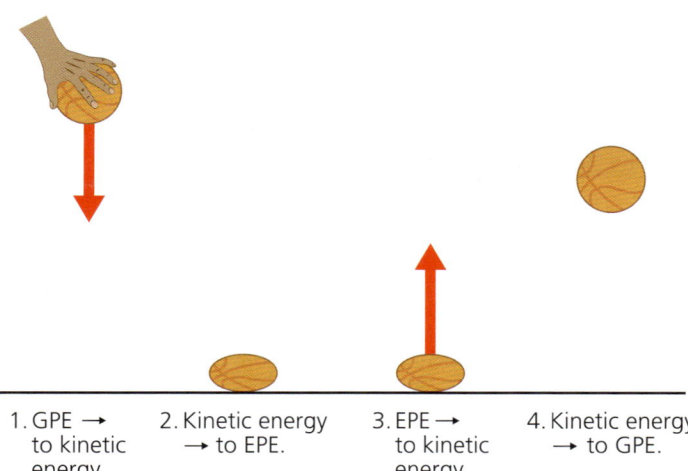

1. GPE → to kinetic energy. 2. Kinetic energy → to EPE. 3. EPE → to kinetic energy. 4. Kinetic energy → to GPE.

EPE stored in your body

A tendon is a band of tissue that connects a muscle to a bone. One example is the Achilles tendon in your calf. When you walk, this tendon stretches and stores energy and then releases it again. Some animals, such as kangaroos and frogs, can make huge leaps because of EPE stored in their tendons.

> **Science in context**
>
> Scientists model the human body using computers. Paralympic athletes use prosthetic legs that are designed using models of elastic potential energy.

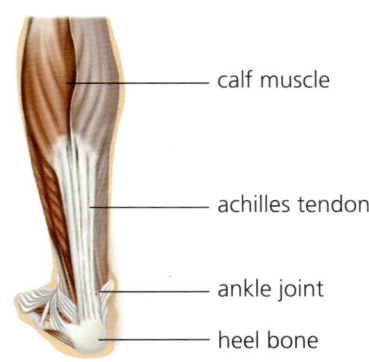

calf muscle

achilles tendon

ankle joint

heel bone

▲ *The Achilles tendon is a store of EPE.*

> **Questions**
>
> 1. A student is playing with an elastic band.
>   a. Describe the changes to the way in which energy is stored as she stretches the elastic band and lets it go.
>   b. Explain in terms of energy why the band will go further if she pulls it back more.
> 2. Look at the pictures of the bungee jumper. She will bounce up and down and eventually come to a stop.
>   a. Draw an energy transfer diagram for the process from the moment she jumps off the bridge until she is briefly stationary at the top of her first bounce.
>   b. Explain where all the energy goes eventually.
> 3. Explain why a ball will never bounce higher than the height from which you drop it.

> **Key points**
>
> - When materials deform they store EPE.
> - An elastic object returns to its original shape when the force is removed.
> - Some energy is dissipated when materials deform.

Review
1.8

1. **a.** Explain why people who do different jobs or sports need different diets. **[1]**

    b. List these activities in order of energy required. Explain why they are in the order that you have chosen. **[2]**

    cycling   sitting   walking slowly

2. List these amounts of energy in order from smallest to biggest. **[1]**

    20 J   20 kJ   2000 kJ   0.2 kJ   2000 J

3. **a.** Give one reason why a question might *not* be a scientific question. **[1]**

    b. Choose the scientific question from these examples. Explain your choice.

    A Which is the best type of bread?

    B How much energy do different types of bread contain?

    C Should I have bread for breakfast? **[1]**

4. Match each word or phrase to its definition. Write out the numbers and the letters.

    Definitions:

    1 the energy that something has because of its position

    2 energy transferred so it is no longer useful

    3 energy stored in food or fuels

    4 the energy that something has because it is moving

    Words and phrases:

    A dissipation

    B energy stored kinetically

    C energy stored gravitationally

    D energy stored chemically **[4]**

5. Which of these statements is *not* correct?

    A Energy is stored chemically in fuels.

    B Energy stored thermally is what some people call heat.

    C GPE depends on the mass of the object.

    D Kinetic energy does not depend on the mass of the object. **[1]**

6. A boy is riding his bike. Complete the sentences using the words below. You may need to use each word once, more than once, or not at all.

    thermal   chemical   heat   light   GPE   kinetic

    a. The food that he ate for breakfast is a store of _____ energy. **[1]**

    b. The useful energy is _____ energy. **[1]**

    c. The wasted energy is _____ energy. **[1]**

    d. As he moves up a hill and down again energy is transferred between _____ and _____ energy stores. **[2]**

    e. The energy stored _____ in the battery decreases when he uses his lights. **[1]**

7. A car transfers energy in the fuel to kinetic energy so the car moves.

    a. Name two ways in which energy is wasted. **[1]**

    b. Draw an energy transfer diagram for the car. **[1]**

8. **a.** Write down the two things on which the kinetic energy of an object depends. **[1]**

    b. Describe a situation in which your kinetic energy increases. **[1]**

    c. A student says: 'If your kinetic energy is increasing that means that your gravitational potential energy is decreasing'. Is this true? Explain your answer. **[1]**

9. List the following in order of amount of GPE, starting with the one that you think has the least.

   A Josie lying in bed

   B Josie about to jump out of a plane to do a sky dive

   C Josie lying on the floor

   D Josie on the top diving board   [1]

10. Here are some pictures of a girl jumping on a trampoline. Choose the picture or pictures that match each statement below.

A just about to hit the mat   B at the top   C at the bottom   D just about to leave the mat

   a. Here the girl has the most GPE.   [1]

   b. Here the girl has the most kinetic energy.   [1]

   c. Here there is the most EPE stored in the trampoline.   [1]

11. Chinonye is timing his friends on a swing. He measures the time it takes for 10 swings.

   a. The time for 10 swings is 6 seconds.

      Calculate the period of the swing.   [1]

   b. Explain why Chimonye should measure the time for several swings and divide by the number of swings.   [1]

Chinonye measures the time for different children to complete 10 swings and calculates the periods. Here are his results:

Mass (kg)	Period (s)
35	1.2
40	1.3
45	1.1
60	1.2

   c. Write a conclusion that gives the link between the mass and the period.   [1]

12. Some students are thinking about questions to ask. For each question write down:

   a. Is it a question that science can answer?   [3]

   b. If so, how could you collect data to answer it? If not, explain why not.   [3]

   These are the questions:

   A Which elastic band is strongest?

   B Do girls or boys have a better memory?

   C How should we grow enough food for everybody in the future?

13. A student has completed an investigation about heating water with liquid fuels. This is her prediction: 'I predict that the more fuel I use the faster the water will heat up.'

   This is a graph of her results:

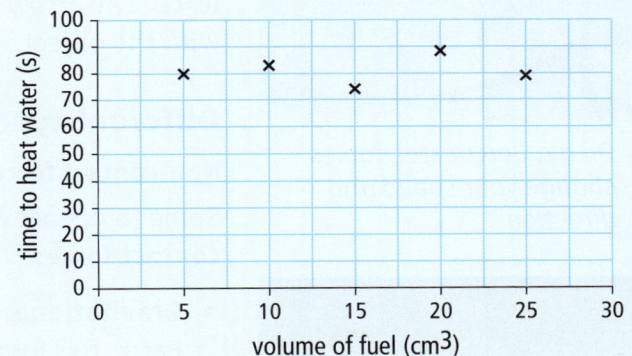

   a. Suggest what the student has missed out in her prediction.   [1]

   b. Do the results support her prediction? Explain your answer.   [1]

   c. Describe a method that the student could have used to get these results.   [1]

14. You are planning an investigation. Which of these things do you *not* do?   [1]

   A Write down what you have found out.

   B Work out how to take precise and accurate results.

   C Decide which results to record.

   TWS

   D Describe how to work safely.

Introduction to forces

Objectives

- Describe how to represent forces on diagrams
- Describe the difference between contact and non-contact forces
- Describe how to measure forces

▲ On a rollercoaster, forces change your speed and direction.

▲ A plastic comb attracts pieces of paper with an electrostatic force.

What is a force?

When a tennis ball hits the ground, a **force** changes its shape, speed, and direction. A force is a push or pull that can change the shape of an object, or change the way that it moves.

You cannot see forces but you can see what they do. If something starts to move, or speeds up, a force is acting on it. Forces can also slow things down or stop them moving. If an object is already moving, a force can also change the direction of motion.

Force arrows

You can show the force acting on an object by drawing an arrow. The length of the arrow shows the size of the force. The direction of the arrow shows the direction of the force. The arrow is drawn in contact with the object.

 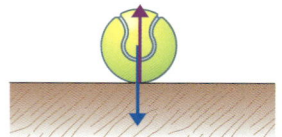

Different types of force

Non-contact forces

Some forces act when even when objects are *not* touching. They are **non-contact forces**.

- **Gravitational forces** act between any objects that have mass. On Earth, this force is called the object's **weight**. This force acts towards the centre of the Earth. It always pulls you 'down', wherever you are on the Earth. The force of gravity attracts you to the Earth. It also attracts the Earth to you!
- **Electrostatic forces** act between objects that are electrically charged. For example, rubbing plastic objects can charge them with static electricity. Once charged they can **attract** or **repel** other charged objects.
- **Magnetic forces** act between magnets, and between magnets and magnetic materials such as iron, steel, or nickel.

▲ A magnetic force attracts iron filings to a magnet.

Contact forces

Some forces , called **contact forces**, only act when objects are in contact (or touching).

- **Friction** acts when solid surfaces are in contact. The force of friction acts to stop objects moving.
- **Air resistance** and **water resistance** act on objects moving through air and water. They are types of **drag**. The moving object collides with the particles in the air or the water, and the collisions slow it down.
- **Thrust** pushes a car or plane forwards.
- **Upthrust** is the upward force on an object that is in contact with a fluid. It is the force pushing up on any **floating** object.
- **Tension** acts when objects are pulled, like the force in a rope that you pull.
- **Reaction** is the force that pushes you up in reaction to your weight pushing something downwards, for example when you sit on a chair. It is sometimes called the **normal** force.

▲ *A hot air balloon rises because of the upthrust from the air around it.*

Measuring forces

You can measure the size of a force. A device for measuring forces is called a **forcemeter**, such as a **spring balance**. Forces are measured in units called **newtons** (N), so force meters are also called **newtonmeters.**

Questions

1. **a.** Name three contact forces.
   b. Name three non-contact forces.
2. A magnet can exert a force of attraction or a force of repulsion on another magnet. Write down which of these forces is a push and which is a pull.
3. A car is travelling along a road. List three of the forces acting on it.
4. Explain why people on the other side of the Earth don't fall off.

Key points

- Weight, electrostatic force, magnetic force, friction, air resistance, water resistance, thrust, upthrust, normal force, and tension are all types of force.
- Forces can change the direction of a moving object, speed it up, or slow it down.
- Forces are measured in newtons using a spring balance.

2.2

Objectives

- Describe the difference between weight and mass
- Describe how the force of gravity arises
- Describe the factors that affect the force of gravity between two objects

▲ *The force of gravity always acts towards the centre of the planet.*

▲ *This scale reads 300 g (mass) but should show a reading in newtons (weight).*

Gravity

The force of gravity

There is a force of attraction between all objects that have mass, called the force of **gravity**. This force is pulling you down towards the centre of the Earth. But forces come in pairs – you are also pulling the Earth with a force of gravity.

The size of the force of gravity between two objects depends on:

- the mass of the two objects
- how far apart they are.

The force of the Earth on you

The force of you on the Earth

▲ *There is a force of gravity between you and the Earth.*

▲ *Relative sizes of Jupiter and the Earth: a more massive planet exerts a bigger force on objects with the same mass.*

Weight and mass

The force of the Earth's gravity on an object is called its **weight**. When you use scales or a forcemeter you are measuring the force of gravity.

- Weight is a force.
- Weight is measured in **newtons** (N).
- **Mass** is the amount of **matter** in an object.
- Mass is measured in **kilograms** (kg).

Another way of thinking about mass is to do with motion. An object with a small mass will accelerate faster when you apply a force to it than something with a big mass.

It's easy to confuse weight and mass. Your weight is a force and should be measured in newtons, but often scales measure your 'weight' in kilograms! Scientifically this is wrong.

The force of gravity depends on the mass of the objects involved, so it is different on different planets. The more massive the planet, the larger the force of gravity.

On the Moon there is no atmosphere (air) and astronauts have to wear spacesuits to be able to breathe. Some people think this means there is no gravity there. This is *not* true. There is a force of gravity on the Moon, but it is about one-sixth as strong as on the Earth.

Gravitational field strength

There is a **gravitational field** around all objects. This is a region where a mass experiences a force.

On the Earth the **gravitational field strength** is 10 newtons per kilogram (N/kg).

Weight (N) = mass (kg) × gravitational field strength (N/kg)

When people say that gravity is different on different planets, they mean that the gravitational field strength is different.

Gravitational field strength decreases as you move away from a planet.

In physics the idea of a 'field' is very important. Fields are linked to forces. There are other types of field, such as magnetic fields (see page 156).

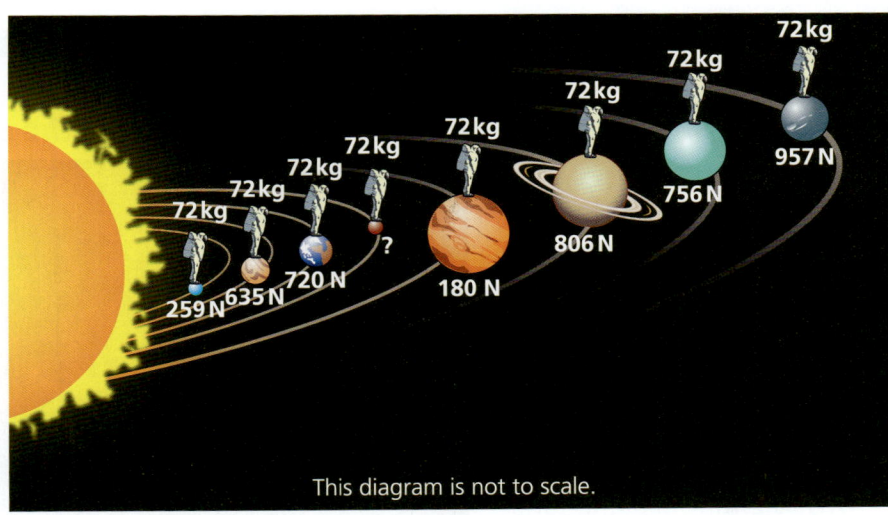

This diagram is not to scale.

▲ *The weight of the astronaut changes, but his mass stays the same.*

▲ *The gravitational field strength on the Moon is 1.6 N/kg.*

see page 156

Questions

1. **a.** Describe what is meant by 'the force of gravity'.
   b. Write down two things that affect the force of gravity.
2. **a.** Describe the difference between weight and mass.
   b. Is the mass of an astronaut on Mars bigger than, smaller than, or the same as their mass on Earth? Explain your answer.
3. A student says that objects get pulled down because the Earth is like a big magnet. Explain how a gravitational force is *and is not* like a magnetic force.
4. A baby has a mass of 4 kg. Calculate its weight on Earth.
5. An astronaut has a weight of 370 N on Mars where the gravitational field strength is 3.7 N/kg.
   a. Calculate the mass of the astronaut.
   b. Write down the mass of the astronaut on Earth.

Key points

- The force of gravity is an attractive force that acts between any objects with mass.
- The force of gravity acting on an object on Earth is called its weight.
- Weight is a force, and is measured in newtons.
- Mass is the amount of matter, and is measured in kilograms.
- The force of gravity is different on different planets or on the Moon.

Objectives

- Describe how scientific knowledge about gravity has developed over time

- Describe some reasons why scientific explanations change

▲ *This Indian observatory was built nearly 500 years ago.*

How 'old' is gravity?

Asking questions about the Earth

◄ *Many people thought about the force holding the Moon in orbit before Isaac Newton.*

Have you ever wondered why we don't fall off the surface of the Earth? Or what pulls us back to the Earth when we jump? Investigations often start with questions like this.

Nearly 1000 years ago the Indian mathematician and astronomer Bhaskaracharya was head of an observatory in Ujjain. He studied the movements of the planets, the Moon, and the Sun. He wondered why the Moon went around the Earth, and didn't get nearer or further away.

Bhaskaracharya noticed that when you drop something, it falls towards the surface of the Earth. He realised that all objects exert a force on other objects – the force that we now call gravity. He also realised that if the Earth attracted small objects towards it, then it would also attract big objects like the Moon. This is the force, he thought, that keeps the Earth, planets, and Moon in orbit.

Evidence from observation

Bhaskaracharya couldn't do any experiments to test his idea. He made observations of the world around him, and used the evidence from his observations to come up with an explanation.

In Europe 500 years later, Sir Isaac Newton thought about the same questions. He knew that a moving object would only change direction if a force acted on it, and the Moon had to keep changing direction to stay in orbit around the Earth. Newton realised that a force must be acting on the Moon to make it do this. Like Bhaskaracharya, he saw that objects fall towards the Earth and wondered if the same force that made them fall kept the Moon from drifting off into space. He developed a mathematical model to describe the force of gravity.

Newton published his idea, called the Law of Gravitation, in a book. A few years later, Newton's law was used to predict the existence of Neptune. In 1846 the planet was discovered as predicted. This was enough evidence for many people to believe Newton's explanation.

Changing ideas about gravity

When there is lots of evidence to support an idea it is usually accepted by other scientists, but it may *not* be a complete explanation.

Over the last 300 years, much more evidence about gravity has been collected by scientists. By making predictions and seeing if they match observations, scientists such as Albert Einstein and Edwin Hubble built on Newton's work. We now know that the force of gravity is much more complicated than both Bhaskaracharya and Newton predicted.

Scientists never stop asking questions. The more we learn about gravity, the more we can use it. For example, scientists have sent satellites into space using the gravitational forces of the Sun and planets to get there.

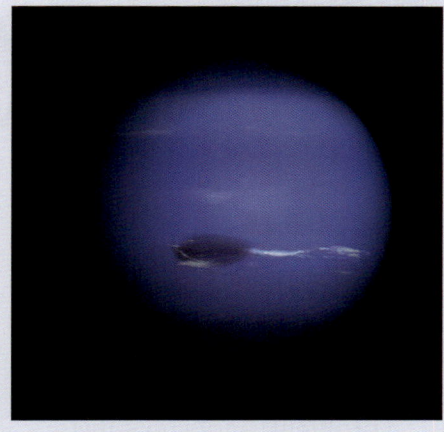

▲ *The discovery of Neptune confirmed Newton's model.*

▲ *The* Voyager *spacecraft has been travelling since 1977, collecting data about the outer planets.*

▲ *In 1969, the first person landed on the Moon and saw a view of the Earth that had never been seen before.*

Questions

TWS 1. What did Bhaskaracharya think held the Moon the same distance from the Earth as it goes around the Earth?

2. Describe how Newton knew that there has to be a force acting on the Moon.

3. Describe one difference between Bhaskaracharya's ideas about gravity and Newton's ideas about gravity.

4. Give one reason why Newton's law of gravitation was accepted.

5. Give one reason why people might *not* have believed Bhaskaracharya or Newton when they said that there is a force on the Earth due to the Sun.

TWS 6. Suggest why our understanding of gravity is *not* yet complete.

Key points

- New evidence from observations can support ideas or models.
- New evidence can also prompt scientists to develop new models.
- The new models are then tested by making predictions.
- People may find some ideas in science difficult to believe.

Air resistance

What is air resistance?

Objectives

- Describe how air resistance is produced
- Describe where it causes a problem, and where it is useful
- Describe what happens in a vacuum

▲ *The air can produce a force that changes your face!*

When an object moves through air, there is a force on it called **air resistance**. Most of the time you don't really notice it. If the air is moving fast, for example if you are walking into a wind, or you are moving very fast through the air, for example if you are cycling fast or leaning out of a car window, then you notice the effect.

An object moving through the air collides with the particles in the air and is effectively pushing the air out of the way. These collisions with air particles provide the resistance. Air resistance is affected by the *speed* of the object moving through the air. Objects moving with a higher speed will push more air out of the way and experience more air resistance.

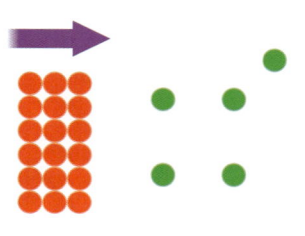

▲ *A solid moves through a gas*

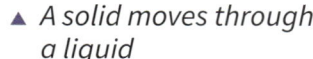

▲ *A solid moves through a liquid*

Reducing air resistance

Air resistance slows things down. Air resistance is less if the area in contact with the air is small. **Streamlining** reduces air resistance by changing the flow of air over a car or plane. Cyclists pull in their arms and crouch forward to reduce the area in contact with the air. They make themselves more streamlined by using special helmets.

Using air resistance

Air resistance can be very useful for slowing things down. A parachute increases the area that is in contact with the air, and therefore increases the air resistance. For example, rocket cars use parachutes to brake.

Air resistance depends on:

- the speed of the object
- the area of the object.

▲ *A parachute slows down a car.*

Science in context

Scientists use wind tunnels to experiment and find the best shape for vehicles. Streamlined cars use less fuel.

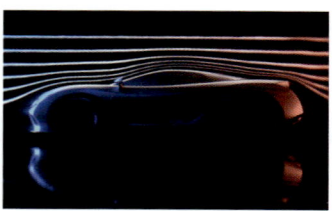

▲ *Smoke shows the path of air in a wind tunnel.*

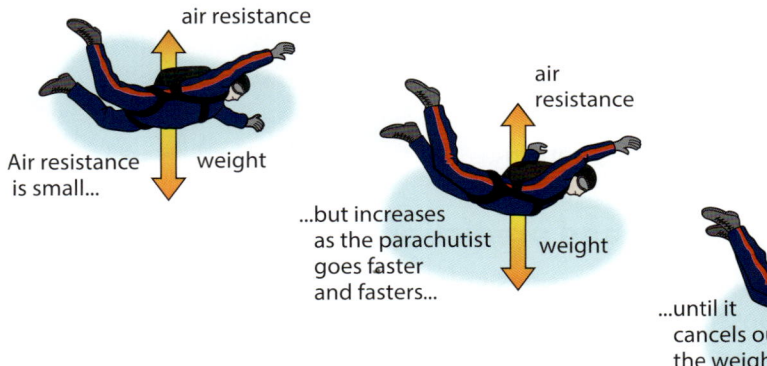

air resistance

Air resistance is small... | weight

air resistance

...but increases as the parachutist goes faster and fasters... | weight

air resistance

...until it cancels out the weight. | weight

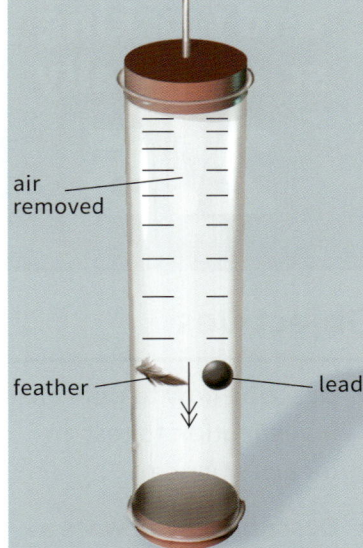

air removed

feather — lead

▲ In a vacuum all objects fall at the same rate.

▲ Without a parachute the parachutist will reach a high, steady speed.

Do heavier things fall faster?

When we see things fall on the Earth, like a feather or a hammer, the heavier object (the hammer) falls faster. This is because of air resistance. Air resistance affects the motion of the feather much more than the motion of the hammer.

If there is no air, there is a **vacuum**, so no air resistance. If there is no air resistance all objects fall at the same rate.

In 1971, an astronaut took a feather and a hammer to the moon, where there is no air, and dropped them together. The feather and the hammer hit the ground at the same time.

Key points

- Air resistance depends on the speed of the object and its area in contact with the air.
- The shape of streamlined objects reduces air or water resistance.
- Parachutes increase air resistance, so can slow down moving objects.
- If there is no air (in a vacuum), then there is no air resistance.
- In a vacuum, all objects fall towards the Earth at the same rate.

Questions

1. Car manufacturers put cars in wind tunnels to help them to design streamlined cars.
   a. Explain what is meant by 'streamlined'.
   b. Write down which object would experience more air resistance – a streamlined car travelling slowly or a lorry travelling fast. Explain your answer.
2. Explain why a tennis ball and a cricket ball dropped together will hit the ground at about the same time, even though the cricket ball is heavier. Assume that they are about the same size.

Objectives

- Describe how to plan an investigation to test an idea in science

- Describe how to write a conclusion

- Describe how to write an evaluation

How does the shape affect the time to hit the bottom?

What are the variables in this investigation?

My indeoendent variables are:
- the shape of the clay
- the mass of clay
- the volume of water in the cylinder
- the temperature of the water

Planning fair tests: Streamlining

Asking questions

Kasini was watching a film about dolphins. The dolphins have to swim fast to catch fish. She wondered what affects how fast things can move through water.

Kasini decided to make different objects out of clay and drop them into a cylinder of water. She could time how long they took to hit the bottom.

Making a hypothesis and prediction

Kasini had learned that engineers design cars and aeroplanes to be as streamlined as possible to reduce drag. She used this information to make a prediction based on a hypothesis.

> I have decided to investigate how the shape of the object affects how long it takes to fall through water.

Making a plan and choosing equipment

This is Kasini's plan.

> I am going to investigate how long it takes different shapes of clay to reach the bottom of the cylinder of water. I will make different shapes from the same amount of clay. These are the shapes I have chosen: cone, cube, sphere, cylinder, cuboid.
>
> I will time how long it takes for the shape to hit the bottom with a stopwatch.
>
> I will write my results in a table.
> This is a list of my equipment:
> - a large measuring cylinder
> - Modelling clay
> - a stopwatch
> - a balance
> - a measuring jug

Writing a conclusion

Kasini completed her investigation and wrote down her results.

She concluded that the evidence did support her hypothesis, and that her prediction was correct.

However, Kasini worried that the data were not very good because it was hard to see when to start and stop the stopwatch.

She asked her friend Nadia. Nadia said that Kasini had not taken enough measurements, so she could not be sure that her measurements were correct. The measurement for the cone shape could be an anomalous result because of a mistake. Nadia also said that there should be a graph and evaluation.

Kasini wrote this for her evaluation:

> To improve this experiment I would find a better method of timing, like using a video camera. I would repeat the experiment several times for each shape. I would look for anomalous results and repeat them. Then I would find the mean. This would make my data more accurate.
>
> I would plot a bar chart of the results because my data are categoric.

It is important that an evaluation shows how to collect better data, *not* make the experiment quicker or easier. Repeating measurements enables you to identify anomalous results.

Shape	Time
cone	0.58
cube	0.65
sphere	0.61
cylinder	0.75
cuboid	0.68

▲ Here is Kasini's results table.

Questions

1. Copy and complete this table to explain why Kasini needed each piece of equipment.

Equipment	Why Kasini needs it
A large measuring cylinder	
Modelling clay	
A stopwatch	
A balance	
A measuring jug	

2. Name the independent, dependent, and control variables in this investigation. Use Kasini's list to help you.
3. What has Kasini missed out of her results table?
4. Explain why repeating the measurements will produce better data.

Key points

- A plan includes your hypothesis and prediction, equipment, and a method.
- A conclusion says what you have found out, whether it matches your prediction, and whether it supports your hypothesis.
- An evaluation includes what you would do to get better data next time.

Round in circles

A force that changes direction

Forces can change the shape, speed, or direction of motion of an object.

Ammon is pushing a trolley. The diagram shows how Ammon applies a force to the trolley.

The trolley will move in a straight line. But what will happen if Eboni pushes from the side? The force will change the direction of the motion. The trolley will start to move in a circle.

Objectives

- Describe what is meant by a centripetal force
- Explain why some objects move in circles
- Explain why some objects move in orbits

▲ *Friction between the tyres and road exerts a force to make a car go around a corner.*

Centripetal force

One of the throwing events in athletics is the hammer throw. The hammer thrower swings the hammer, a big metal ball, around her head on a wire, and then lets it go.

Imagine taking pictures from above as the hammer thrower is swinging the ball. In your first picture, the ball is moving in direction X. In your second picture, it is moving in direction Y. There is a force on the ball from the tension in the wire, which has changed the direction of motion.

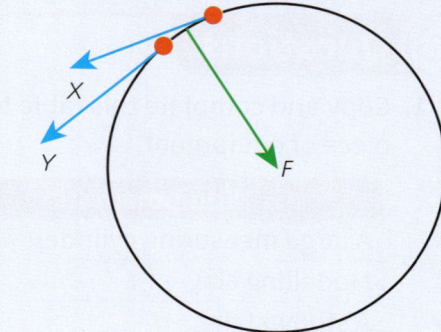

If she lets go of the ball then it will move in a direction that is a **tangent** to the circle.

▲ *The ball moves in a circle because the wire pulls on the ball.*

A force that keeps an object moving in a circle is called a **centripetal force**.

- For the ball on the wire, the centripetal force is provided by the tension in the wire.
- For cars on a curve on the road, the centripetal force is provided by friction between the tyres and the road.

The centripetal force is always directed towards the centre of the circle. This means that the force is always at right angles to the direction of motion of the object.

Centripetal forces in space

The Earth and other planets **orbit** the Sun. The Moon orbits the Earth – it is a natural **satellite**. The Earth also has artificial satellites put into orbit for communication and for studying the surface of the Earth.

▲ *A satellite needs the force of gravity to stay in orbit around the Earth.*

The centripetal force that keeps the Moon and satellites in orbit around the Earth is provided by the force of **gravity**. It is gravity that also keeps the planets and asteroids in orbit around the Sun, and keeps moons in orbit around other planets, such as **Jupiter**.

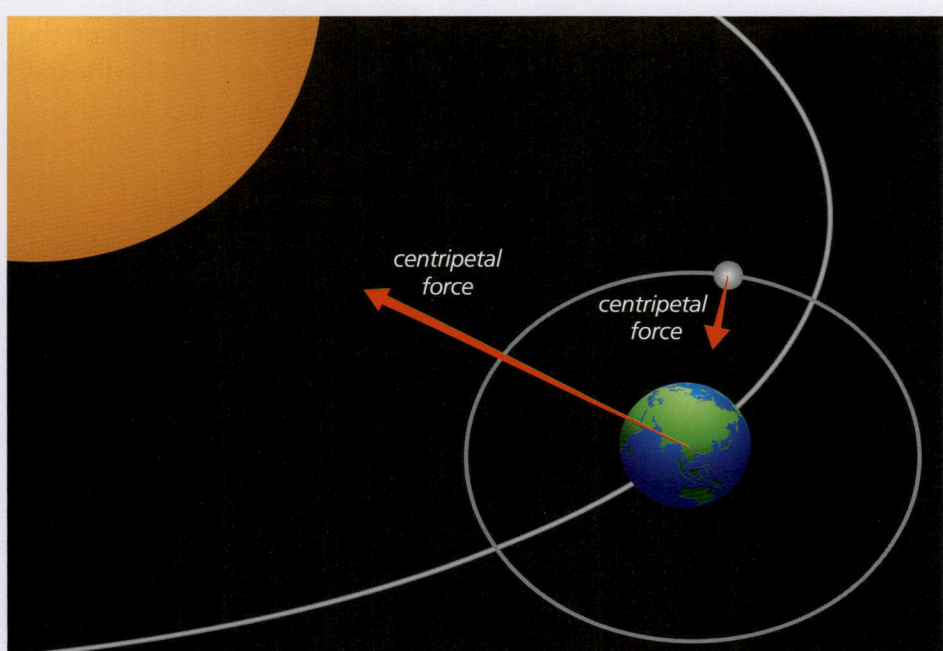

centripetal force

centripetal force

▲ *The force of gravity keeps objects in orbit.*

Science in context

Communication and other satellite systems, like GPS, have changed peoples' lives, usually for the better.

However, there are lots of old satellites and other space 'junk' now orbiting the Earth. They pose a risk to other satellites or when they eventually fall to the Earth!

Questions

1. Write down what is meant by centripetal force.
2. A centripetal force is needed to keep a cyclist moving on an arc of a circle as she goes around a corner. What provides that force?
3. The Moon orbits the Earth.
   a. Name the force that keeps the Moon in orbit around the Earth.
   b. Imagine that this force was suddenly switched off. Describe what would happen to the Moon.
4. A student says that there cannot be any force acting on a satellite if it is moving at a steady speed in orbit. Is she right? Explain your answer.

Key points

- The force that keeps something moving in a circle is the centripetal force.
- The centripetal force can be provided by forces such as gravity, friction, or tension.
- Planets, moons, and satellites stay in orbit because of gravity.

Review

2.7

1. Some forces need contact, and others act at a distance. **[2]**
   Give the letters of the non-contact forces:

   A friction

   B gravity

   C air resistance

   D magnetism

2. Match the words to the definitions. **[2]**

mass	force due to a planet or moon
weight	region in which an object experiences a force
field	amount of matter

3.

   weight

   a. Write down the label on the upward arrow. **[1]**

   b. Describe the other two force arrows that should be added to the diagram. **[1]**

   c. Identify which of the three forces you have described are contact forces and which are non-contact forces. **[1]**

4. Which of these statements about mass and weight is or are correct? There may be more than one. **[1]**

   A Weight is measured in kilograms.

   B Mass is measured in kilograms.

   C Weight is a force.

   D Mass is measured in newtons.

5. On the Earth, a person with a mass of 70 kg has a weight of 700 N. Which of the following statements is or are correct? **[1]**

   A If the person were on a more massive planet their weight will be larger.

   B If their weight on a planet is 300 N, it means that the gravitational field there is stronger than on the Earth.

   C A planet with a weaker gravitational field would make their weight smaller.

   D A planet with a stronger gravitational field would make their mass bigger.

6. Alan Shepherd was the fifth person to walk on the Moon. He took a golf club and ball and hit the ball. The ball went a lot further than it would on the Earth. Give **two** reasons why. **[2]**

7. Scientists are planning a mission to take astronauts to Mars. Suppose an astronaut has a mass of 65 kg on the Earth. The gravitational field strength on Earth is 10 N/kg and on Mars is 3.7 N/kg.

   a. Calculate the astronaut's weight on Earth. **[2]**

   b. Describe and explain what will be the same and what will be different about their weight and mass when they go to Mars. **[2]**

8. a. Explain why ideas about gravity from hundreds of years ago were 'rediscovered' later. **[1]**

   b. Explain why an idea in science might change. **[1]**

9. The table shows some information about Jupiter and one of its moons, Io.

   An astronaut has a mass of 100 kg. Compare their weight on Jupiter and Io. **[3]**

Object	Gravitational field strength (N/kg)
Jupiter	23
Io	1.8

10. A student investigates streamlining.

   a. Define 'streamlined'. **[1]**

   He makes three different shapes using modelling clay and times how long each shape takes to reach the bottom of a tank of water. Here are his results.

Shape	Time 1 (s)	Time 2 (s)	Time 3 (s)	Average time (s)
A	1.2	1.4.	1.3	1.3
B	1.8	0.1	2.0	
C	2.8	2.9	2.7	2.8

 b. Identify the anomalous result in his data. **[1]**

 c. Calculate the average time for shape **B**. **[1]**

   d. List the shapes in order of most to least streamlined. Explain your answer. **[2]**

 e. Give two control variables in this investigation. **[2]**

   f. Describe and explain the type of graph the student needs to plot. **[2]**

   g. Another student says that they have made a shape that takes 4.6 seconds to reach the bottom of a tank.

   Suggest why the first student might not be able to use this result. **[2]**

11. Explain why a streamlined car uses **less** fuel than if it were not streamlined. Use these words in your answer: thrust, air resistance, air particles, force. **[4]**

12. A boy connects a ball to a piece of string and whirls it in a vertical circle, as shown in the diagram below.

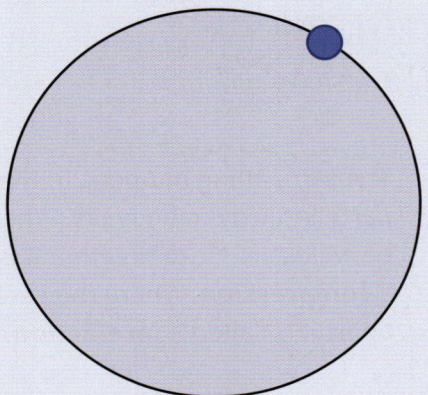

   a. Copy the diagram and add an arrow to show the force keeping the ball moving in the circle. **[1]**

   b. Describe what would happen to the ball if the string suddenly broke. **[1]**

   Jupiter's moons move in a circle around Jupiter.

   c. Name the force keeping them moving in a circle. **[1]**

   d. Describe what would happen to the moons if the force was suddenly removed. **[1]**

13. Copy and complete these sentences about the force of air resistance using these words:

   vacuum   force   particles   same   **[4]**

   When an object moves through the air, air _____ collide with it. This produces a _____ that slows down the object.

   In a _____ there are no air particles, so all objects fall at the _____ rate.

14. Describe a situation in which air resistance is useful. **[2]**

15. When a diving bird is about to enter water, it folds its wings close to its body, so they are not injured when it hits the water. Explain the other effect this action has. **[3]**

Sound waves and how they travel

Making sounds

Objectives

- Describe how sound waves are produced
- Explain how sound waves travel
- Explain why sound waves do not travel through a vacuum

▲ *The skin on a drum vibrates to produce a sound.*

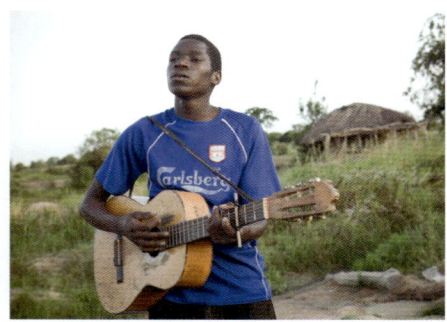

▲ *When you play a guitar you make the strings vibrate.*

When you make music on instruments, or sing, you are making things **vibrate** to produce **sound**. Sound transfers energy from a **source** such as a drum to a **detector** such as your ear.

What is a sound wave?

Phones and televisions contain loudspeakers.

If you put polystyrene beads on a loudspeaker cone, they bounce up and down. This shows that the cone vibrates when it makes a **sound wave**.

▲ *Polystyrene beads bounce on a speaker cone to show that it is vibrating.*

- The vibrating cone makes the air particles next to it vibrate.
- When the layer of air next to the loudspeaker vibrates, it makes the next layer of air particles vibrate.
- The sound wave produced moves through the air to your ear.
- The air itself does *not* travel away from the cone.

You cannot see air particles moving, but you can make a model of a sound wave using a Slinky spring. If you move the end forwards and backwards, the coils move together and then apart as the wave moves along the Slinky. The individual coils do *not* travel to the end, but the wave does.

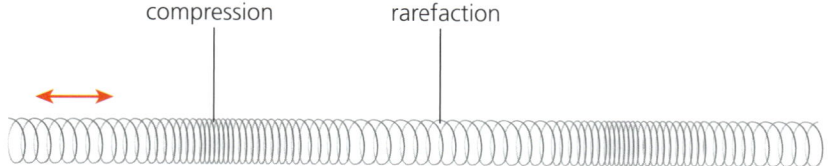

compression   rarefaction

▲ *You can model a sound wave with a Slinky spring.*

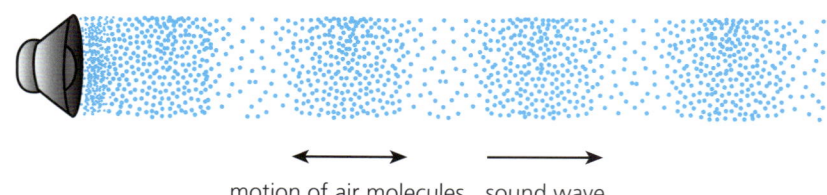

motion of air molecules   sound wave moves this way

The same thing happens to the air particles in a sound wave. Where the air particles are close together it is called a **compression**. Where they are further apart it is called a **rarefaction**.

What can a sound wave travel through?

Sound waves need a material, or **medium**, to travel through, such as air, water, or walls. The vibrating source produces a sound wave that makes the particles in the medium vibrate backwards and forwards as the wave passes. If the sound wave reaches your ear you hear a sound.

You can hear sounds when you are underwater. Animals such as dolphins and whales communicate over very large distances by making and hearing sounds.

If someone is talking loudly in the next room you will probably hear them, because sound travels through solid materials like walls.

▲ *Many animals communicate by sending sound through water.*

What can a sound wave *not* travel through?

A space that is completely empty of everything, including air, is a **vacuum**.

Sound cannot travel through a vacuum because there are no particles to transfer the vibration from the source to the detector.

The speed of sound

Sound waves travel at different speeds in solids, liquids, and gases. They travel fastest in solids and slowest in gases. This is because the particles in a solid are closer together than they are in a gas, so the vibration is passed on more quickly.

Material	Speed of sound (m/s)
air at sea level	330
water	1500
metal	5000

▲ *Sound travels at different speeds in different media.*

Longitudinal and transverse waves

A wave transfers energy without transferring matter.

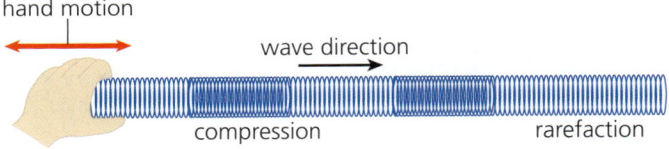

longitudinal wave

- **Longitudinal** waves can be made on a Slinky spring by moving your hand in and out.
- The coils of the Slinky spring move to and fro in the *same* direction as the longitudinal wave is travelling.
- Sound waves and some types of earthquake wave are longitudinal waves.
- **Transverse** waves can be made on a rope by moving your hand up and down instead of in and out.
- The particles of the rope move *at right angles* (90°) to the direction of travel of the wave.
- Stadium (or 'Mexican') waves made by fans at football matches are transverse waves. The people move up and down as the 'wave' moves along.

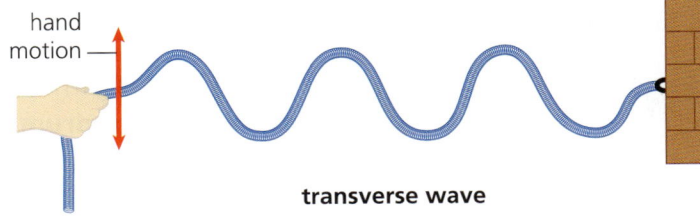

transverse wave

Questions

1. Your vocal cords and a loudspeaker can both produce sound. Name three other sources of sound waves.
2. Explain why sound cannot travel through a vacuum.
3. Does sound travel faster or slower in air than in water? Explain your answer.

Key points

- Sound waves are made by vibrating objects.
- Vibrations are transferred by particles vibrating.
- Sound travels fastest in solids and slowest in gases.
- Sound cannot travel through a vacuum.

Detecting sounds

Your **ear** and a **microphone** both detect sounds.

Objective

- Describe how the ear detects sound
- Explain how your hearing can be damaged
- Describe how a microphone works

▲ A microphone detects sound...

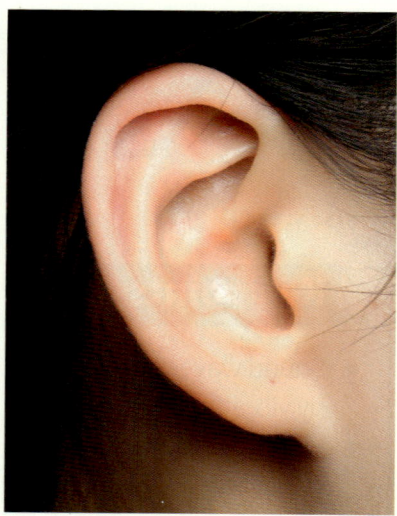

▲ ...and so does your ear.

The human ear

You can only see a small part of your ear, called the **pinna**, when you look at it in the mirror. Your ear has parts *inside* that detect sounds and send signals to your brain.

Having two ears helps you work out where sounds come from. Sounds reach your ears at different times, so you can locate the source of the sound.

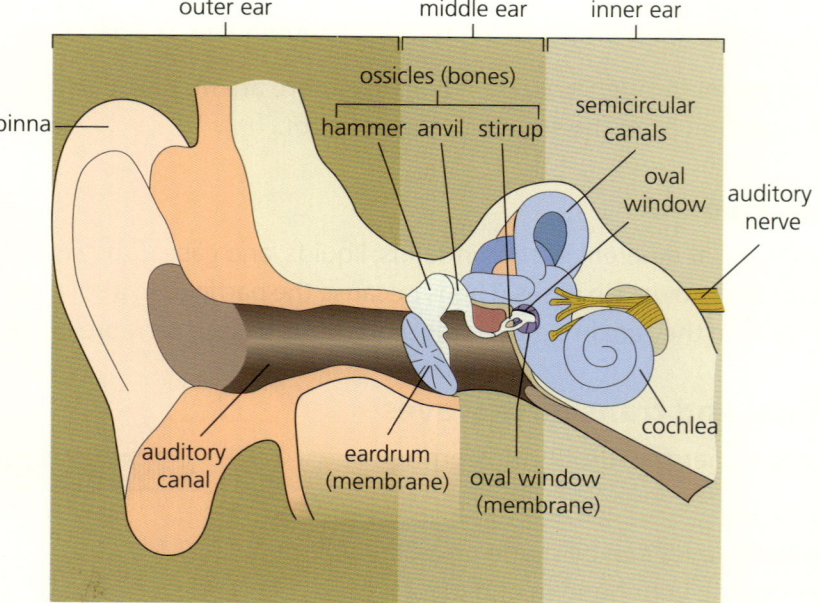

▲ The ear has parts inside that allow you to hear.

The **outer ear** gathers the sound wave and directs it down the **auditory canal** to the **eardrum**. Once there it makes the eardrum vibrate, and this makes the **ossicles** vibrate.

The ossicles make up the **middle ear**. They are the smallest bones in your body. They pass the **vibration** on to the **oval window**, and then the **inner ear**.

The inner ear is made up of the **semicircular canals** and the **cochlea**.

- The semicircular canals help you to balance.
- The cochlea is curled like a snail shell, and contains fluid.
- When the oval window vibrates, it transmits the vibration to the fluid.
- The vibrating fluid makes hairs in the cochlea vibrate.
- The hairs are connected to sound-detecting cells.
- The cells release chemicals that produce a **signal** that travels down the auditory nerve to your brain.
- Your brain processes the signal and you hear the sound.

You can find out more about specialised cells like sound-detecting cells in your *Cambridge Lower Secondary Complete Biology* Student book.

Hearing loss

Your ear contains delicate structures including the tiny ossicle bones and thin membranes of the oval window and eardrum. There are many ways in which these structures can be damaged, affecting your hearing.

- Loud sounds can destroy the sound-detecting cells in the cochlea, causing permanent hearing loss.
- Pressure changes can **perforate** (make a hole in) your eardrum. The hole will usually heal in a few weeks or months.
- Ears naturally produce wax to clean themselves, but too much can affect your hearing. It is very easy to remove excess wax.
- If you have an ear infection, fluid can be produced around the small bones. This interferes with the transfer of the sound wave from the outer ear to the inner ear.
- Head injuries can affect the auditory nerve, which will affect your ability to hear properly.

Older people do *not* hear high-pitched sounds as well as younger people. Some people wear hearing aids to improve their hearing.

Microphones and loudspeakers

A **microphone** is a type of **transducer** – it converts a sound wave into an **electrical signal**. The human ear is the body's microphone.

The microphone contains a diaphragm, which is a flexible plate. Sound waves make the diaphragm move backwards and forwards, like an eardrum. This movement produces an electrical signal that can be amplified and sent to a loudspeaker, or recorded.

A **loudspeaker** is a another type of transducer – it converts the electrical signal into a sound wave. The electrical signal makes the cone vibrate, so the air particles move backwards and forwards to make a sound wave.

▲ *Loud sounds, like the music at concerts, can damage the eardrum.*

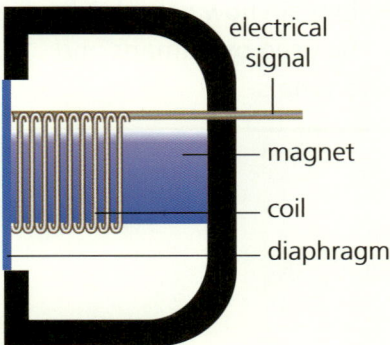

electrical signal

sound waves

magnet

coil

diaphragm

▲ *Like your ear, a microphone produces an electrical signal from a sound wave.*

Questions

1. **a.** Explain why it can be dangerous to put a sharp object into your ear.
   b. Suggest which part of your ear is affected by ear wax.
2. **a.** Write down one similarity and one difference between the ear and the microphone.
   b. Write down which part of a microphone is like the eardrum.
3. A student says 'a microphone is a bit like a loudspeaker in reverse'. Do you agree? Explain your answer.

Key points

- The membranes and bones of your middle ear transfer the vibration of a sound wave from your outer to your inner ear.
- Your inner ear converts the sound wave into an electrical signal that goes to your brain.
- Loud sounds can damage your hearing.
- A microphone produces an electrical signal that allows us to make recordings of sound.

Echoes

What is an echo?

If you stood in a very big cave and shouted, you would hear an **echo** of your voice. The sound wave spreads out and **reflects** off surfaces like the walls of the cave. Sound travels relatively slowly, so there is a time delay between your shout and the echo you hear when the sound wave is reflected from the cave walls .

▲ *Caves have walls that reflect sounds.*

Objective

- Describe how echoes are formed
- Explain how echoes can be used by humans and animals

Problems with echoes

Echoes can be a nuisance inside large rooms such as cinemas and theatres. The sound waves reflect off all the surfaces, and echoes can last for several seconds, muddying the sound. This is called **reverberation**. Cinema walls and ceilings are covered with soft, sound-absorbing materials to absorb echoes.

Using sound

Humans and animals make use of sound that we cannot hear, called **ultrasound** (sound waves with a frequency higher than 20 000 Hz).

Sonar

The word '**sonar**' comes from **so**und **n**avigation **a**nd **r**anging. Sonar is used by submarines for navigating underwater, and by ships for measuring the depth of the water. The ship has an underwater loudspeaker (a **transmitter**) that produces pulses of ultrasound. Beams of ultrasound waves are more focused than beams of audible sound, and they will *not* be confused with sounds made by people, animals, or other boats.

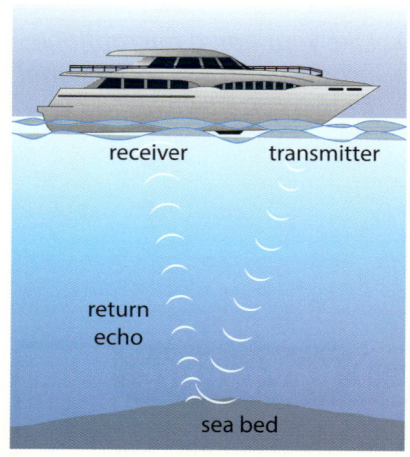

▲ *Sound can be used to work out the depth of the sea...*

The ultrasound waves travel through the water and reflect off objects such as rocks or the sea bed. The echoes returning to the ship are detected by underwater microphones (the **receiver**). The sonar device uses the time of the echo being received to work out the depth of the water.

For example, a boat detects an echo from the sea bed after 1 second. The speed of sound in water is 1500 m/s. How deep is the sea?

Distance = speed × time ÷ 2 ◄——— It takes 1 second for the wave to reach the sea bed and come back again, so the time is divided by 2.

$$= 1500 \times \frac{1}{2}$$

$$= 750 \text{ m}.$$

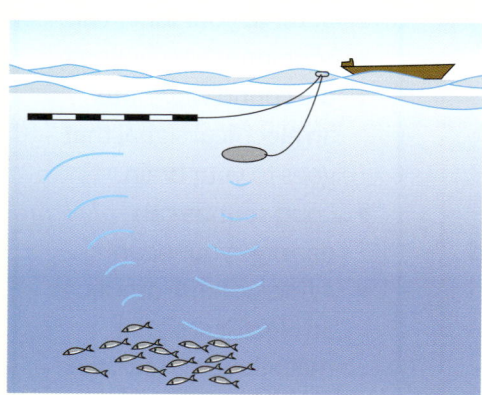

▲ *...and the location of fish.*

Fishermen can use sonar to find fish. Finding the distance to an object in this way is called **echolocation**.

Animal echolation

Animals such as whales and dolphins use echolation more than they use sight. It enables them to find food or recognise their family members even if it is dark or the water is *not* very clear.

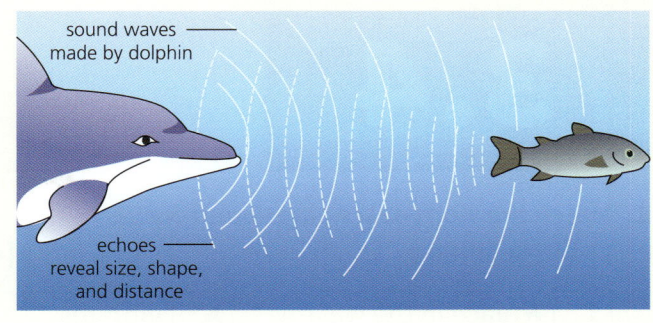

- The dolphin makes a series of clicking sounds.
- The sound reflects off a fish.
- The dolphin detects the echoes and uses the time to work out the location of the fish.

Sound travels much faster in water than in air. Dolphins and whales can communicate over very long distances.

Ultrasound scanning

We cannot hear ultrasound, but it is very useful. Pregnant women should not have X-rays because X-rays could harm both the fetus and the mother. Instead doctors use ultrasound to check the development of the fetus.

The transducer transmits ultrasound waves into the woman. Ultrasound waves are reflected by the boundaries between soft tissue as well as from hard surfaces such as bone. The transducer detects the echo and uses the time delay to calculate where the boundary is. By taking lots of pictures it is possible to build up a three-dimensional picture of the baby.

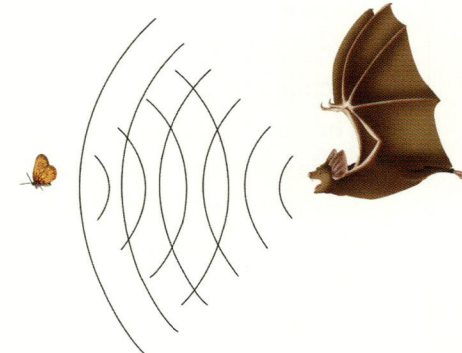

▲ Bats find the distance to objects using echolocation.

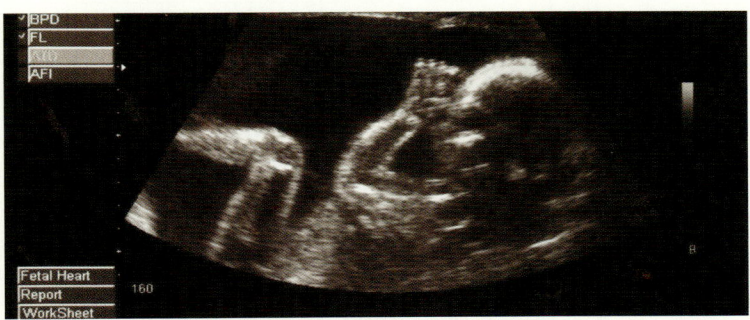

▲ An ultrasound image of a fetus.

Questions

1. **a.** Explain why you need to use half the time taken for an echo to be detected when you are doing calculations of distance.
   b. Sound travels at 330 m/s in air. It takes 1.5 s to detect the echo from a building. Calculate the distance to the building.
   (*Hint*: Remember that the time is for the journey there *and* back).
2. Explain why dolphins can find food at greater distances than bats.
3. A woman is having an ultrasound scan. Ultrasound travels at 1500 m/s inside the body. The ultrasound transmitter sends out a pulse of ultrasound waves, which are reflected from a tissue surface 7.5 cm away. Calculate the length of time before the echo is received. (*Hint*: Be careful with units!)

Key points

- An echo is a reflection of sound.
- Humans and animals use echoes to locate and identify objects or prey.
- Ultrasound is very-high-frequency sound that can be used to make an image of a fetus.

1. Match the word describing sound waves to the definition. [4]

1	A place in the wave where the particles are close together is called a…	A	vibrates
2	The solid, liquid, or gas that the sound wave moves through is called the…	B	rarefaction
3	A place in the wave where the particles are further apart is called a…	C	compression
4	A sound wave is produced by something that…	D	medium

2. Think about what happen when you talk. Are these statements true or false? [3]

   a. The particles in air move away from your mouth as you talk.

   b. The sound wave makes the particles in the air move backwards and forwards as you talk.

   c. The sound wave makes the particles move up and down as you talk.

3. The speed of sound in steel is 5000 m/s, in water is 1500 m/s, but in air is only 330 m/s.

   Use these words or phrases to complete the sentences below. Some words or phrases are not needed. [3]

   closer together   further apart   less   more

   a. The speed of sound in water is greater than the speed of sound in air because in water the particles are _____ than they are in air.

   b. The speed of sound in air is smaller than the speed of sound in steel because in steel the particles are _____ than they are in air.

   c. Sound made under water will take _____ time to travel a certain distance than sound made in air.

4. A sound wave travels through water and bounces off the sea bed.

   Choose the correct words in bold. [2]

   a. The sound wave is **attached to/reflected by** the sea bed.

   b. The sound wave travels **faster/slower** in air than it does in water.

5. Six different materials, A, B, C, D, E, and F are listed with the speed of sound in each of them.

	Speed of sound (m/s)
A	4000
B	1207
C	1497
D	259
E	6420
F	435

   a. Give the letters of the two materials that are probably liquids. [2]

   b. Give the letters of the two materials that are probably gases. [2]

   c. Give the letters of the two materials that are probably solids. [2]

   d. Give the letter of the material in which the particles are closest together. Explain your answer. [2]

   e. Give the letter of the material in which the particles are furthest apart. [1]

6. A bat sends out a pulse of ultrasound. The pulse reflects from the insect. The bat is 1.5 m from the insect. The speed of ultrasound in air = 330 m/s.

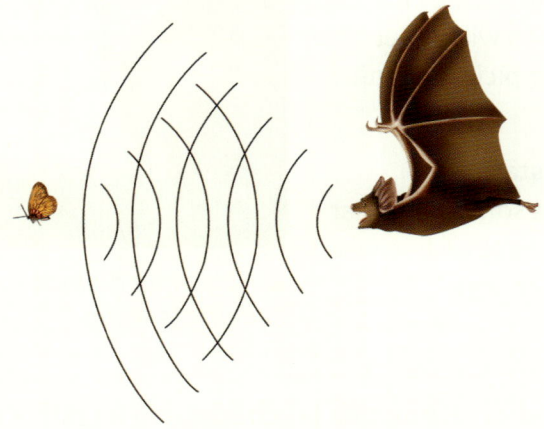

a. Calculate the total distance travelled by the pulse. **[1]**

b. Calculate the time between the emission and detection of the pulse. **[2]** Use this equation: time = distance/speed

7. A fisherman uses sonar to find a shoal of fish. A pulse of sound is sent out and the reflection of the sound is detected 0.4 seconds later.

a. Calculate the time it takes the sound to travel from the boat to the fish. **[1]**

b. The speed of sound in water is 1500 m/s. Calculate the distance to the fish. **[2]**

c. Sometimes fishermen will detect another echo after they detect the echo from the fish. Why? **[1]**

8. Adami is watching a thunderstorm. She counts 4 seconds between seeing the lightning and hearing the thunder. The speed of sound in air is 330 m/s.

a. Calculate the distance to the storm. **[2]**

b. Suggest one assumption Adami has made to do this calculation. **[1]**

9. a. Compare a compression and a rarefaction of a sound wave.

b. Describe what happens to the air particles near your mouth when you talk.

10. Here is a diagram of the ear.

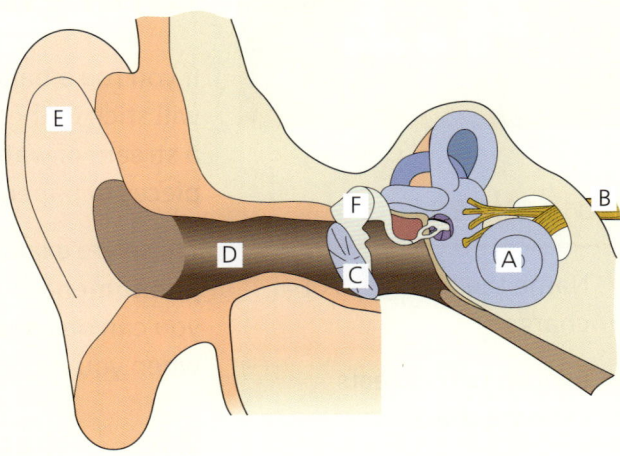

a. Give the name of each part of the ear labelled. **[6]**

b. Give the letters of the parts that make up the inner ear. **[1]**

c. Give the letter of the part that is filled with fluid. **[1]**

11. A friend calls your name. Describe:

a. how the sound wave is produced **[1]**

b. how it travels to your ear **[1]**

c. how it produces a signal that reaches your brain. **[4]**

12. Copy and complete the table. **[4]**

Part of the ear	What it does
pinna	
eardrum	
	pass the vibration from the eardrum to the cochlea
cochlea	

13. A dolphin makes a clicking sound and detects the echo from a fish that is 250 m away. Calculate the length of time before the dolphin detects the echo. The speed of sound in water is 1500 m/s. **[3]**

Charging up

If you rub a balloon against your clothes it will stick to the wall. This balloon will deflect a stream of water from a tap, or pick up small pieces of paper. Why?

These are examples of **electrostatic phenomena**. A phenomenon is something that you can observe. These things happen because when you rub the balloon you **charge** it.

▲ *A charged balloon deflects water.*

Objectives

- Name the two types of charge
- Explain why objects become charged
- Explain the difference between conductors and insulators

What is charge?

There are two types of electric charge, **positive (+) charge** and **negative (–) charge**. Charge is a property of a particle or object, like mass.

When you bring charges together, **electrostatic attraction** or **repulsion** occurs.

- A **positive** charge will repel another **positive** charge.
- A **negative** charge will repel another **negative** charge.
- A **positive** charge will attract a **negative** charge.

▲ *Like charges repel, unlike charges attract.*

When you charge a balloon by rubbing it, it can repel the negative charge in scraps of paper, or running water, or a wall. This leaves a positive charge, which produces attraction.

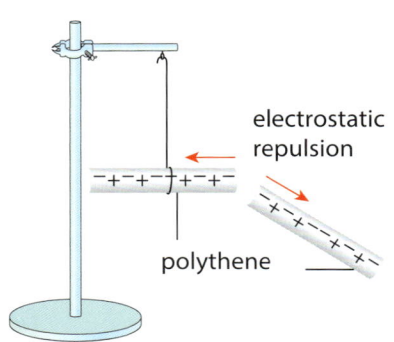

▲ *Two negatively charged polythene rods repel one another.*

Charging and discharging

Kunala rubs a polythene rod with a cloth, and then hangs it up by a piece of string. He rubs another polythene rod and brings it close to the first rod.

- The first rod is repelled by the second rod.
- Both rods have become negatively charged.
- The net charge on both rods is the *same*.

Then he rubs a Perspex (clear plastic) rod and brings it close to the polythene rod.

- This time the polythene rod is *attracted* to the Perspex rod.
- The net charge on the Perspex rod is positive.
- The rods have unlike charges and so they attract.

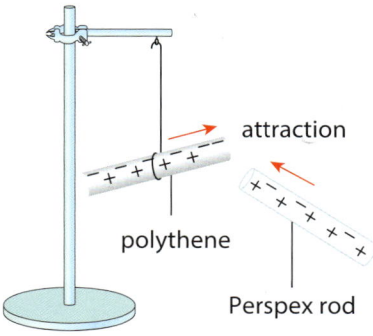

▲ *Two rods with unlike charge attract each other.*

Where does charge come from?

Everything is made up of atoms. In a simple model of an atom, **protons** and **neutrons** make up the central **nucleus**. The **electrons** orbit the nucleus.

In an atom the number of electrons, which are negatively charged, is always the same as the number of protons, which are positively charged, so an atom has no charge overall. We say that it is **neutral**. There is no *net* charge.

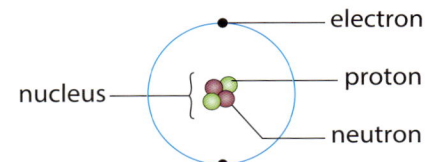
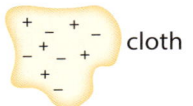

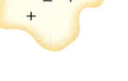

▲ *Atoms are neutral.*

Charging objects

If you move electrons out of one material into another, the materials become charged.

- When you rub a polythene rod with a cloth, electrons are transferred from the cloth to the rod.
- Electrons have a negative charge.
- The *rod* now has a net negative charge because there is *not* enough positive charge to **neutralise** the negative charge.
- The *cloth* has a net positive charge because there is *not* enough negative charge to neutralise the positive charge.

Not all rods are the same. When you rub a Perspex rod electrons are transferred *from* the rod *to* the cloth.

- The Perspex rod has a net positive charge.
- The cloth has a net negative charge.

It is important to realise that not *all* the electrons are transferred, just some. The total amount of charge in the rod and the cloth is always the same. Charge cannot be created or destroyed, just like energy or mass.

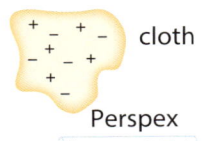
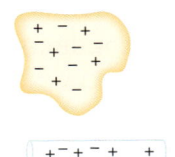

▲ *The total amount of charge is the same before and after.*

Moving charges

There is a region around any charge where another charge experiences a force. This region is called an **electric field**.

Static means stationary or standing still. Rods repel and attract each other because the polythene and Perspex are insulators. In an **insulator** the electrons stay on the rod once you have rubbed it.

If you rub a *metal* rod with a cloth it will also become charged, but the metal is a **conductor**. Electrons can move in it. Any extra electrons will move through the metal to your hand, and then through you to earth.

 Science in context

In 600 BCE a Greek scientist found he could pick up leaves with a piece of amber that he had rubbed. The Greek word for amber is *elektron*.

Questions

1. Explain why an atom is neutral.
2. A teacher rubs a Perspex rod with a cloth. It becomes charged.
   a. Write down the net charge on the cloth and on the rod.
   b. Would the rod attract or repel a negatively charged rod? Explain your answer.
3. Explain why a plastic rod can become charged if you rub it, but a metal rod cannot.
4. Explain why when you transfer charge to a person who is insulated from the ground, their hair moves away from their body.

Key points

- There are two types of charge: positive and negative.
- An object that has equal positive and negative charges is neutral.
- If electrons move from one object to another, each object will become charged.
- In a conductor the electrons are free to move, but in an insulator they are not.

4.2

Electric circuits

Objectives

- Describe how to draw components in circuit diagrams

- Explain how to test whether something conducts electricity

Circuit components

Televisions, movie projectors, and loudspeakers run on electricity. Hospitals use machines that run on electricity to keep people alive. They all contain **electric circuits** that use electrical **components**.

You can make a very simple circuit using a lamp, a battery, some wire, and a switch. This is the circuit that you find in a torch.

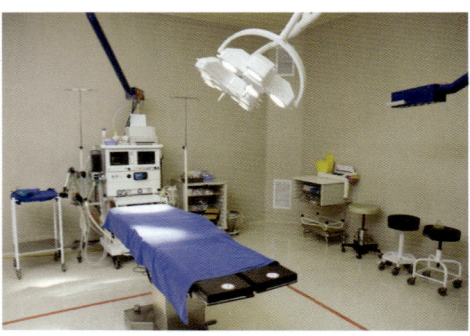

▲ *The anaesthesia machine in an operating theatre needs electricity to work.*

Circuit symbols and diagrams

It would take a long time to draw pictures of the components in a circuit. Instead, we use **circuit symbols** to show which components are in a circuit. You can join circuit symbols in a **circuit diagram** to show how the components in your circuit are connected.

A **cell** is something that many people refer to as a battery. In physics a **battery** contains two or more cells.

A cell has a positive and a negative **terminal**. When you connect cells together in a battery, you must make

Component	Symbol
wire	——
cell	—|⊢—
battery of cells	—|⊢-|⊢—
lamp	—⊗—
open switch	—o o—
closed switch	—o—o—
buzzer	⏝
motor	—Ⓜ—

sure that you do *not* connect two positive terminals (or two negative terminals) together or it will not work. On the circuit symbol the long line represents the positive terminal, and the short line represents the negative terminal.

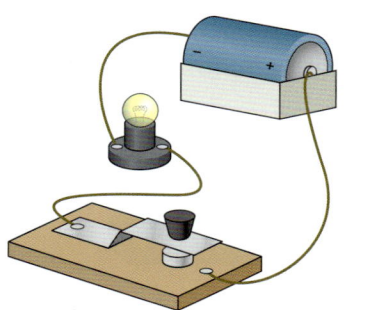

▲ *This simple circuit behaves like a torch....*

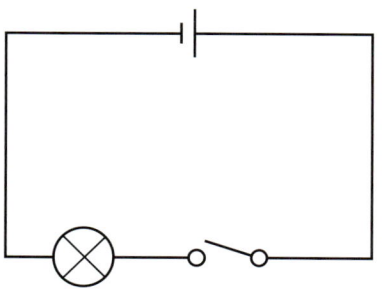

▲ *... and this is a circuit diagram for that torch circuit.*

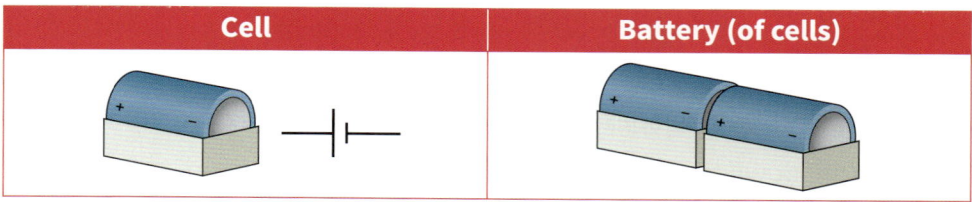

Cell	Battery (of cells)

When will circuits work?

In the circuit diagram for a torch (on the left), the switch is open.

The lamp is *not* lit. The switch is a break in the circuit and you need a complete circuit for the lamp to work. If you close the switch the circuit will be complete and the lamp will light up.

Conductors and insulators

You can use a circuit like the torch circuit to find out whether objects conduct electricity or not.

- An object or a material that conducts electricity easily is a conductor.
- An object or a material that does *not* conduct electricity easily is an insulator.

Mawar and Harta are investigating conductors and insulators. They set up a circuit to work out which things are conductors and which are insulators.

Conductors	Insulators
metal spoon	paper
coin	plastic spoon
nail	wood
aluminium foil	string

I will put different materials in the gap. If the bulb lights up that means the material is a conductor.

It seems that all the conductors are made of metal.

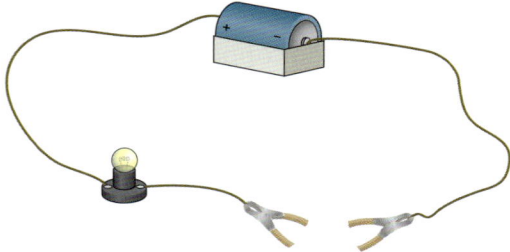

▲ *A circuit for investigating which objects conduct electricity.*

Some people think that *only* metals conduct electricity, but that is *not* correct. The material in the centre of a pencil is graphite, a form of carbon. The graphite conducts electricity but the wood around it does *not*.

The wires for electrical appliances are made of copper, a metal. Bare wires would be very dangerous because you could get an electric shock if you touched them. The wire is covered in plastic, which is an insulator.

⟨?⟩ Questions

1. Look at the electric circuits below.

   A   B   C

   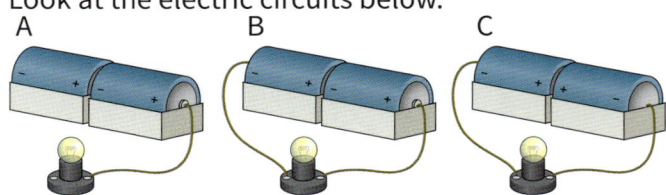

   a. Draw a circuit diagram for each circuit using the correct symbols.
   b. In which circuit will the lamp be lit? Explain your answer.
   c. What do you need to do to make the other lamps light?
2. Explain why the wires that you use in circuits are covered in plastic.

📖 Key points

- We use circuit symbols for components to draw circuit diagrams.
- A complete circuit is needed for components in the circuit to work.
- Conductors allow electricity to flow easily through them, but insulators do not.

4.3 Electric current

When people talk about 'electricity' in a wire they are usually talking about electric **current**. You cannot see a current, so what is going on in the wire?

Objectives

- Describe what is meant by an electric current

- Describe a series circuit

- Describe how to measure current in a series circuit

- Describe how changing the components in a series circuit affects the current

What is electric current?

An electric current is the flow of **charge**. In a circuit made of metal wires the flowing charge is caused by the movement of **electrons**. Electrons are charged particles.

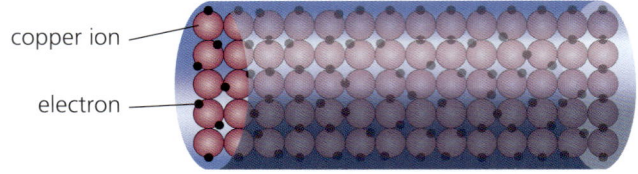

copper ion

electron

▲ A copper wire contains copper ions surrounded by electrons.

Inside a copper wire there are copper **ions**. Ions are atoms that have lost electrons. Some of the electrons on the outside of the copper atoms are *not* strongly bound to the atom and can leave the atoms and move around. You might hear people talk about a 'sea of electrons' inside the metal wire.

When you connect a battery to a circuit, it provides a push to make the electrons move. The moving electrons make an electric current. The current is the amount of charge flowing per second.

How do we measure current?

Current is measured in **amperes** (A), amps for short. Small currents are measured in **milliamps** (mA). 1 mA = 0.001 A, or one-thousandth of an amp.

You measure current with a meter called an **ammeter**.

The circuit symbol for an ammeter is:

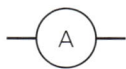

Here is a drawing of a circuit with an ammeter, and a circuit diagram to go with it.

▲ The current is shown by a needle on a scale...

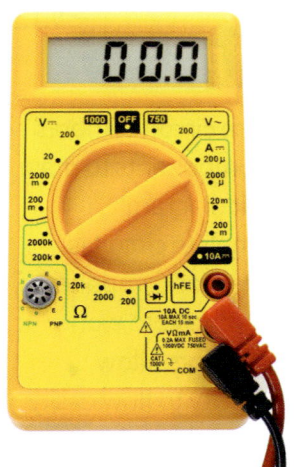

▲ ... or on a digital display.

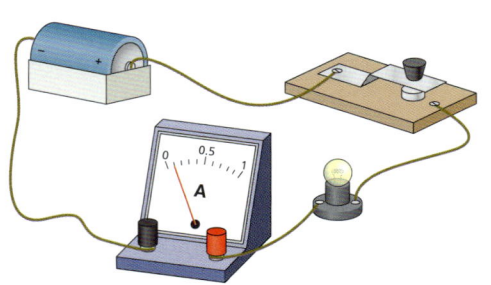

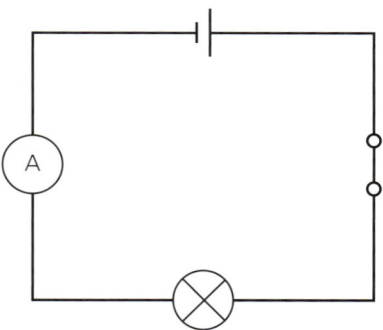

▲ How to measure current.

What happens when you change the number of lamps?

You change the current in a **series circuit** (a circuit with components joined one after another into a loop) by changing the number of components, or the type of components. You can also change the current by changing the number of cells.

Nanda connects up a circuit with one cell, one lamp, an ammeter, and a switch. When he presses the switch the lamp shines with normal brightness.

Then he adds another identical lamp. Two things happen.

- The lamps are now dimmer than normal.
- The reading on the ammeter is less than what it was with one lamp.

Each component, such as a lamp, provides a **resistance** to the flow of charge. If there are two lamps instead of one, there is more resistance so less charge flows per second. The current is smaller.

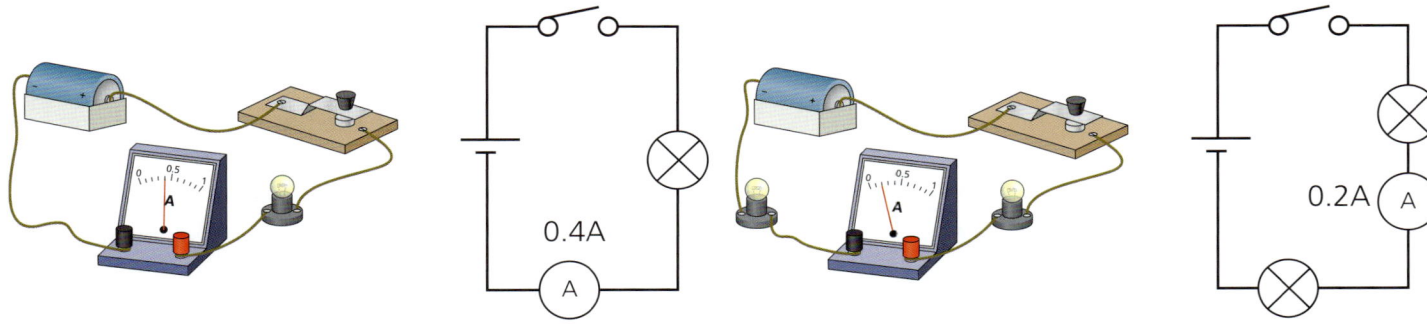

▲ Adding more lamps reduces the current.

What happens when you change the number of cells?

Nanda goes back to his first circuit of a cell, a switch, one lamp, and an ammeter. He adds a second cell.

The lamp is brighter, and the reading on the ammeter is more than what it was before. The current is bigger because two cells push electrons through the circuit more strongly. More charge will pass the lamp each second, and there is more energy per charge, so the lamp gives out more light.

Questions

1. Describe the difference between current and charge.
2. Circuits X and Y are series circuits that each contain *one* lamp.
   - One of the circuits has more cells than the other.
   - In circuit X *twice* as much charge flows through a lamp per second as in circuit Y.
   a. In which circuit is the current bigger? Explain your answer.
   b. Write down the circuit that has more cells. Explain your answer.
   c. There are two cells in circuit X. How many are there in circuit Y?

📖 Key points

- Current is the flow of charge (electrons) per second.
- Current is measured in amperes (A) with an ammeter.
- A series circuit has only one loop.
- A switch controls all the components in a series circuit.
- In a series circuit the current decreases if you add more lamps.
- In a series circuit the current increases if you add more cells.

Modelling electric circuits

Objectives

- Describe a model of an electric circuit
- Describe strengths and limitations of the model

Using models

You cannot see a current flowing in a wire. This can make it difficult to understand what is happening in an electric circuit. To make it easier to work out what is happening, and to predict what might happen in different circuits, scientists use different models. A model is a helpful way of thinking about something that you cannot see, or that is very big, or very small.

You need to remember that all **models** have limits. This means that there may be things that they do *not* explain very clearly, or at all.

The rope model

One way of modelling what is happening in a circuit is to use a piece of rope. The rope is tied to make a continuous loop. One person is the 'battery' and another person is the 'lamp'.

- Person X moves the rope around the circle by pulling it through his hands. This is like the battery pushing the charges around the circuit.
- Person Y grips the rope gently and can feel it moving through. His hands get warm. This is like what happens in a lamp. The filament of a lamp gets so hot that it gives out light.

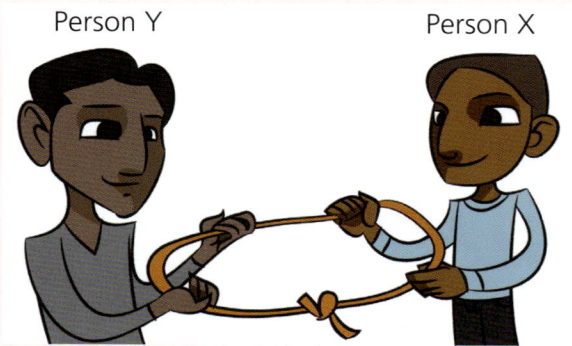

Person Y   Person X

▲ *The moving rope is a model for charge in a wire.*

The rope	… is like …	the charges (electrons) in the wire.
Person X pulling the rope	… is like …	the cell or battery pushing the charge.
The amount of rope going through his hand per second	… is like …	the current (the charge flowing per second).
Person Y gripping the rope	… is like …	a component, like a bulb, in the circuit.

The rope model is a good model because it helps us to visualise what is happening. It also helps us to make predictions. For example, a second lamp is like a second person holding onto the rope: there will be more resistance to its movement.

The rope model shows…	So in an electric circuit…
Person X does *not* produce the rope.	The charges are *not* produced by the battery. They are already in the wires and components.
The same amount of rope passes through any point every second.	The current is the same everywhere in a series circuit.
All parts of the rope start to move as soon as person X starts passing it through.	The current starts to flow as soon as the battery is connected.
The rope transfers energy from person X to person Y.	The current transfers energy from the battery to the components in the circuit.

The factory model and the people model

Another model for an electric circuit uses the idea of a factory that makes things that are sold in a shop, like loaves of bread.

In this model the factory represents the battery, the trucks are like the charges, and the shop is like the lamp.

There are lots of other ways in which you can make a model like this.

- A pot of sweets is like the battery.
- People carrying the sweets are like the charges flowing in the circuit.
- Another person eating the sweets is like the lamp.
- The sweets represent the energy that is being transferred from the battery to the lamp.

However, like all models, this one has limits. Lamps in a circuit light up immediately. This model does *not* explain that very well. It does *not* help you to predict what happens when there is more than one lamp.

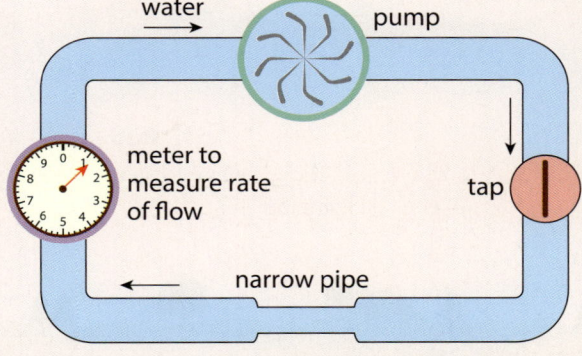

▲ *In this model trucks represent charge, and the bread is like the energy.*

The water circuit model

In this model:

- the pipes are like the wires in a circuit
- the water is like the charges
- the pump pushes the water around the pipe
- the tap can stop the flow of water (like a switch can stop the current in a circuit)
- the meter measures the flow like an ammeter in a circuit.

However, some people think that the insulation around a copper wire is holding the electrons in, like the pipe holding the water in. This is *not* correct. Water can leak out of pipes, but electrons cannot leak out of wires.

▲ *In this circuit water is a model for charge.*

Questions

1. For the water model:
   a. describe one thing you could do to make the water move faster
   b. explain what 'faster water' represents in an electric circuit.
2. In the sweet model, describe how you would represent an ammeter.
3. In each of the three models for an electric circuit, what represents a flat battery (a battery that doesn't work)?
4. A student is wondering how to put a switch in the rope model and the truck model. What would you say to her?

Key points

- Models helps us to work out what is happening in a circuit.
- Models can be used to predict what might happen if you change something in a circuit.
- All models have limitations.

Dangers of electricity

There are lightning storms all over the world every day. Lightning is an example of an electrostatic phenomenon.

Objectives

- Describe how electricity can dangerous

- Explain how the risk of damage or injury from electricity can be reduced

Sparks and current

If enough charge builds up due to friction, then a **spark** will form.
A spark is charge moving through the air, heating it up. We see light and hear a sound. The moving charge is an electric current.

This is what happens in a thundercloud.

▲ *A current transfers a lot of energy.*

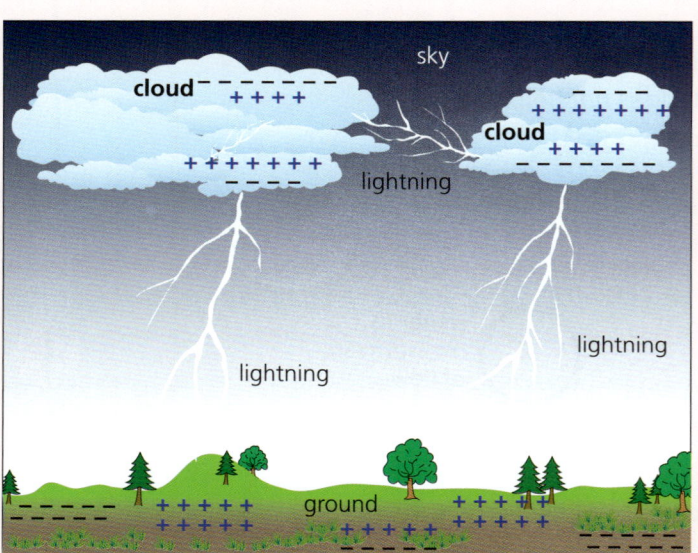

- Inside the cloud the air moves around, like rubbing a rod with a cloth.
- The movement forms a region that is positively charged and a region that is negatively charged.
- Negative charges at the bottom of the thundercloud can make a huge spark as they travel to Earth.
- We see that spark as lightning.
- Sometimes a spark is produced when charge moves between positive and negative regions inside the cloud.

Air does not usually allow a current to flow through it. It takes a big difference in charge to make the air conduct electricity so that a current flows.

Electric shocks

Lightning can be very dangerous. If it strikes a person it can break their bones, cause burns, or even kill them. The charge travels through the person to earth. The electric current can damage their heart or even stop their heart beating.

Doctors in hospitals can use an electric current to try to restart a patient's heart if it stops beating. This can save their life.

Everyday shocks

Sometimes people feel a small shock when they touch a car door. This is because the car has become charged by friction as it has been moving along. You might sometimes hear the crackle of sparks as you remove a piece of clothing, or feel a small shock when you touch a door handle.

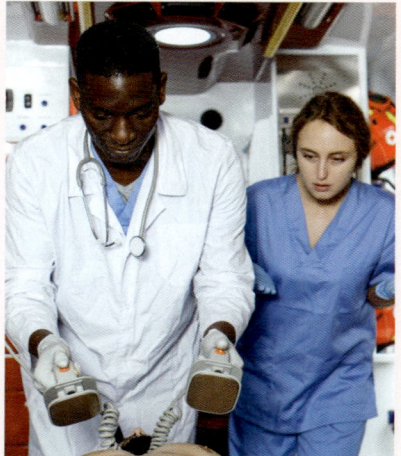

▲ *A defibrillator uses charge to restart someone's heart.*

Reducing risk

Risk is a combination of the probability of something happening and the consequence if it did. We cannot change the probability of lightning hitting a building, but we can change what happens if it does. If you want to reduce the risk of damage by an electric current, you can **earth** the object. This means connecting it to the ground with a conductor.

Lightning conductors

Many tall buildings have a **lightning conductor** to save them from damage by lightning. A lightning conductor is a thick strip of metal, such as copper, running from the top of the building down the wall to a copper plate buried underground. If lightning strikes the building, electrons flow down the strip and into the ground. The metal conducts electricity much better than the building. This is called **earthing**.

Fuel in cars and planes

When cars or planes are refuelled, the friction of the fuel moving along the pipe can cause a charge to build up. This could cause sparks, which would be dangerous because the fuel could catch fire.

Fuel pipes are earthed by attaching a metal wire to the pipe. Any charge that builds up will flow down the wire to earth. This reduces the risk of a spark, and an explosion.

Electronics

Engineers who make circuits for TVs or phones need to reduce the risk of damage to the delicate electrical components. They often wear special wristbands connected to a metal wire that goes to the ground. Any build-up of charge flows down the wire to earth rather than through the delicate components.

▲ *A lightning conductor earths this chimney.*

▲ *A wristband connected to earth.*

Questions

1. **a.** Describe what is meant by 'earthing'.
   b. Describe what lightning is.
2. Explain how an electric current can be dangerous to the human body.
3. Explain how the engineer's wristband can protect delicate electrical components.
4. A student notices that she gets a shock when she touches a metal door handle. She has been charged by walking across a carpet. Would she get a shock if the handle was made of plastic? Explain your answer.

Key points

- Sparks happen when the air conducts electricity.
- Lightning is a big spark that can cause damage to people and buildings.
- Lightning conductors reduce the risk of damage to buildings.
- Earthing reduces the risk of sparks and shocks.

1. Explain what is meant by the following words.

   a. neutral [1]

   b. charged [1]

   c. charge [1]

   d. conductor [1]

   e. insulator [1]

   f. earthed [1]

2. Copy and complete the sentences below. [6]

   a. There are two types of electric charge: _____ charge and _____ charge.

   b. When you rub a polythene rod with a cloth you transfer _____ from the cloth to the rod.

   c. Two charged polythene rods would _____ if you brought them close together.

   d. A charged polythene rod would _____ a rod that had a positive charge.

   e. An electric field is a region in which charges experience a _____.

3. A student rubs a balloon on his jumper and sticks it to the wall. Explain in terms of electrons why the balloon sticks to the wall. [3]

4. Read the sentences below. Which circuit component or type of circuit does each sentence describe?

   a. This is the energy source for an electric circuit. [1]

   b. In this circuit all the components are in a single loop. [1]

   c. This enables you to turn a component on or off. [1]

   d. It measures the charge flowing per second. [1]

5. Copy and complete the following sentences. [6]

   Current is the amount of _____ flowing per _____. In a metal wire, charged particles called _____ move when you connect a battery. You can use a meter called an _____ to measure current. Current is measured in units of _____, which have the symbol _____.

6. a. Draw a circuit diagram to show how you could use a switch to turn a battery-powered motor on and off. [2]

   b. Describe what happens when you close the switch. [1]

7. You can model circuits with a rope loop, or as a series of trucks delivering food to a supermarket.

   TWS

   Copy and complete the table to show how each part of the model goes with each part of a circuit with a battery, lamp, switch, and ammeter. [6]

Component/ feature	Rope model	Supermarket model
battery		
lamp		
switch		
current		
ammeter		
charge		

   TWS

8. Look at these circuits. They have identical cells.

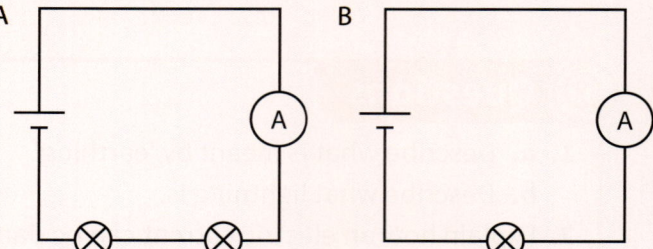

   a. Compare circuit A and circuit B in terms of:

   i. the brightness of the lamps [1]

   ii. the reading on the ammeter. [1]

   b. Describe the effect of adding another cell to circuit B. [2]

9. Look at the circuit diagram below.

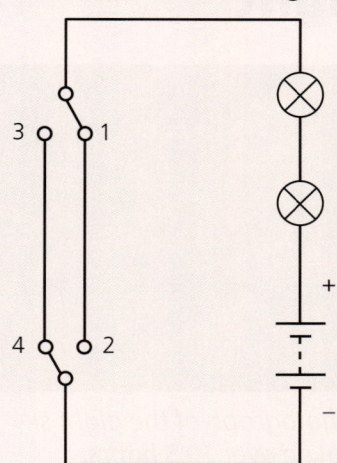

The circuit contains two special switches. The switch at the top can be flipped between connections 1 and 3. The switch at the bottom can be flipped between connections 2 and 4.

a. Copy and complete the table below to show what happens to the light bulbs when the switches are flipped. **[4]**

Position of switch on left	Position of switch on right	Are the bulbs on or off?
1	2	
1	4	
3	2	
3	4	

b. This circuit can be used to control the lights at the top and bottom of a staircase. Explain how it works. **[2]**

10. Copy and complete these sentences using the words in question 1. You may need to use words more than once.

a. In a thunderstorm air moves to produce regions that are _____. **[1]**

b. The air is usually an _____ but if there is a large amount of _____ in the cloud it can become a _____. **[3]**

c. When this happens _____ flows between clouds, or between clouds and the ground, and we see lightning. **[1]**

d. A lightning _____ is a piece of metal attached to a building. If it is struck by lightning the _____ flows through the metal. We say that the building is _____. **[3]**

11. a. In which circuit below A, B, or C, will the current be biggest?

b. In which circuit, A, B or C, will the lamps be dimmest?

A
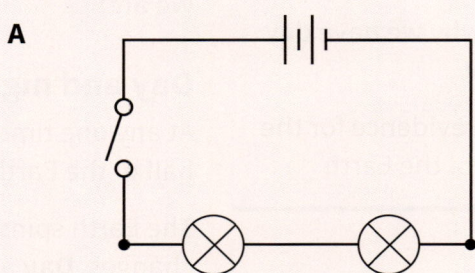

B
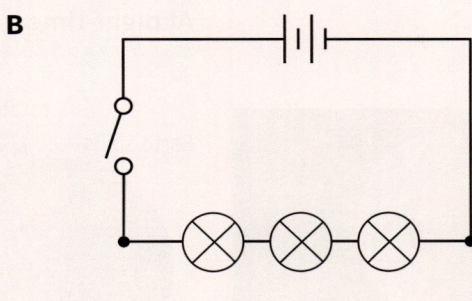

C
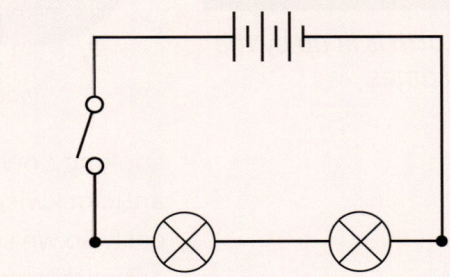

Our planet: Day and night

- Explain why the Sun and stars appear to move across the sky

- Explain why we have day and night

- Describe evidence for the rotation of the Earth

Every day you stand on a planet that seems to be staying still. You see the Sun rise, move across the sky to its highest point at noon, then onwards till it sets. At night the stars seem to move in the same way.

The Sun and stars are *not* moving. We are!

▲ *This photograph of the night sky was taken over 10.5 hours.*

Day and night

At any one time, the light from the Sun lights up half of the Earth. The other half of the Earth is in **shadow**. We can see that from space.

The Earth spins on its **axis**. This means that the area that is in the light changes. **Day** and **night** are caused by the Earth spinning every 24 hours. During the day the part of the Earth that you live on is moving in the light. At night-time it has moved into the dark.

▲ *Half the Earth is lit up by the Sun at all times.*

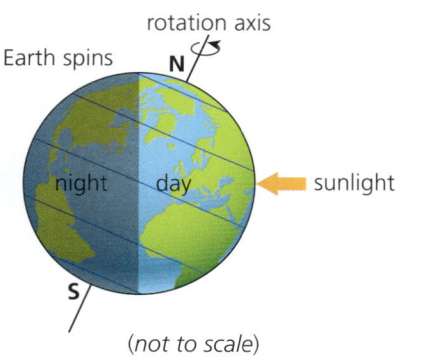

(not to scale)

Looking down from above the North **Pole** of the Earth, the Earth is spinning anticlockwise. This means that the Earth rotates from the west towards the east. So we see the Sun rising in the east and setting in the west. The Sun appears to move across the sky because the Earth is spinning.

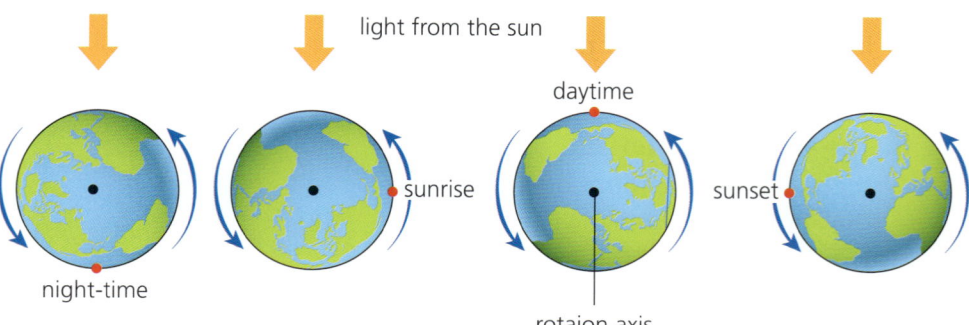

▲ *You experience night when your part of Earth moves out of the light into shadow.*

Is that correct?

I think that it is night-time when the Sun goes behind the clouds.

I think that the Sun goes around the Earth. It is night-time when the Sun is on the other side of the Earth.

▲ A long, heavy pendulum shows that the Earth is spinning.

These ideas are *not* correct.

Some people think that day and night can be explained by the Earth moving around the Sun each day, and that it is night-time when the other side of the Earth is facing the Sun. In this model the Earth is *not* spinning.

It is true that the Earth moves around the Sun, but it takes one **year**, *not* one day!

 Science in context

The spinning Earth

How can we show that the Earth is spinning? One very important experiment to show this was first done in Paris, France, in 1851.

A scientist called Leon Foucault made a pendulum of a very long piece of wire, 67 m long, and a big ball of metal with a mass of 28 kg. He hung the pendulum in a very tall building so that it was just a metre off the floor. There was hardly any friction at the point where the pendulum joined the ceiling. When he started the pendulum swinging it would swing for a very long time because it was so heavy. Over time the direction of the pendulum swing seemed to change against a scale on the floor.

The only explanation was that the pendulum was swinging in exactly the same way, but the Earth was spinning beneath it.

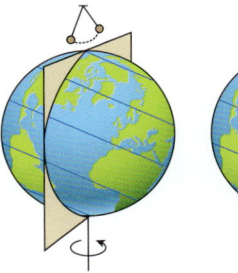

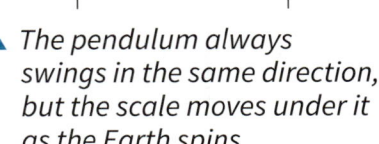

▲ The pendulum always swings in the same direction, but the scale moves under it as the Earth spins.

Key points

- We have day and night because the Earth spins on its axis.
- The Earth spins once in 24 hours.
- The Sun appears to move across the sky because the Earth is spinning.
- You can show the Earth is spinning using a long pendulum.

Questions

1. Copy and complete the sentences.

   The Sun rises and sets each day. The Sun rises in the _____ and sets in the _____. The Sun is highest in the sky at _____. There are _____ hours in a day.

2. A student thinks that we have day and night because the Sun is hidden behind clouds at night. Explain why this is incorrect.

3. Explain why Foucault's pendulum needed a very heavy weight on the end.

5.2

Objectives

- Describe and explain how temperature and the height of the Sun in the sky change over the year

- Explain what causes seasons

Our planet: Seasons

Changing day length

Harish lives in Jalandhar, India. It is October and he is watching the Sun setting. It is setting earlier than it did the night before. The days are getting shorter.

Six months later the Sun sets later than it did the night before. The days are getting longer. He wonders why the Sun sets at different times throughout the year.

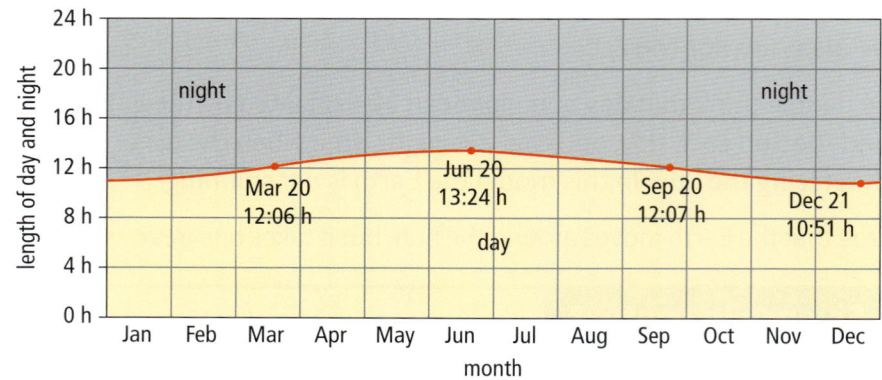

▲ In most places on Earth the length of the day changes over the year.

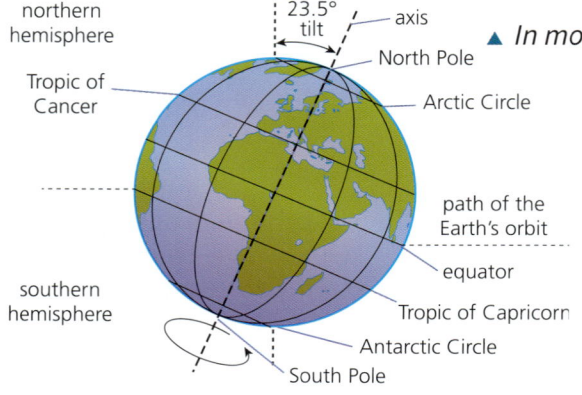

▲ The Earth's axis is tilted by an angle of 23.5° from north.

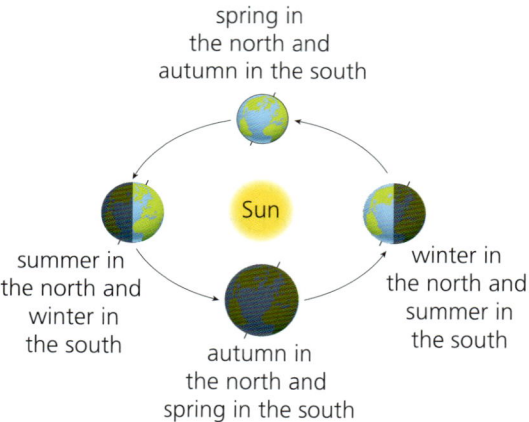

▲ The Earth stays at about the same distance from the Sun throughout the year, but the tilt of the axis changes day length and so produces seasons.

The tilt of the Earth's axis

The Earth orbits the Sun once every **year** or 365 days. We divide the year into **seasons**: spring, summer, autumn, and winter. In some countries near the **poles** there is a big difference in the weather between summer and winter. In summer the days are longer and warmer, and in the winter they are shorter and colder. Other countries near the **equator** do not notice much change.

We can explain the seasons by thinking about the tilt of the Earth's axis. The axis is tilted by an angle of 23.5°. This angle of tilt does not change. As the Earth orbits the Sun, for 6 months the North Pole is tilted towards the Sun, and for 6 months it is tilted away from the Sun.

The Earth's tilted axis explains:

- different day lengths in summer and winter
- different temperatures in summer and winter
- the varying height of the Sun in the sky in summer and winter.

Changing day length

Jalandhar is in the **northern hemisphere**. In summer:

- the North Pole is tilted towards the Sun
- the Sun stays in the sky for longer each day as the Earth spins
- the day is longer than the night.

As Jalandhar goes from summer to winter, sunrise gets later and sunset gets earlier. The days get shorter.

It is winter in the **southern hemisphere** when it is summer in the northern hemisphere. The Sun stays in the sky for less time each day until midwinter.

The tilt of the axis means that in regions in the north inside the Arctic Circle such as Finland, Norway, and Canada, the Sun doesn't set in the summer. These places are known as the 'Land of the Midnight Sun'. The same happens in Antarctica during its summer, but no one lives there to see it. In winter in these places, the Sun does not rise at all. This is called a polar night.

▲ *A series of photographs taken at noon over a year just south of the Arctic circle.*

Changing temperature

It is hotter in the summer than in the winter because:

- the Sun is in the sky for longer
- the rays from the Sun are concentrated over a smaller area.

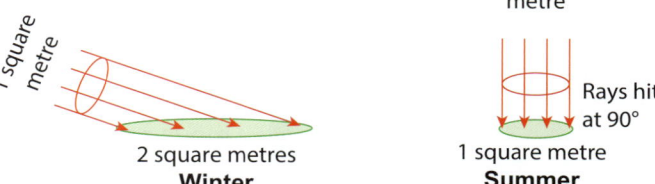

▲ *In the winter the Sun's light is spread over a bigger area than in the summer.*

If the rays hit the surface at 90° then more energy hits the surface per square metre per second. This means:

- the surface heats up faster
- the air temperature increases more quickly.

Some people think that it is hotter in the summer because the Earth is closer to the Sun in the summer. This is *not* true. The Earth's orbit is nearly circular. It is slightly closer to the Sun in January each year.

Changing height of the Sun in the sky

During the day the Sun is highest in the sky at noon. The height the Sun reaches at noon changes over the year. In summer it is higher than it is in winter. This is also because of the Earth's tilted axis.

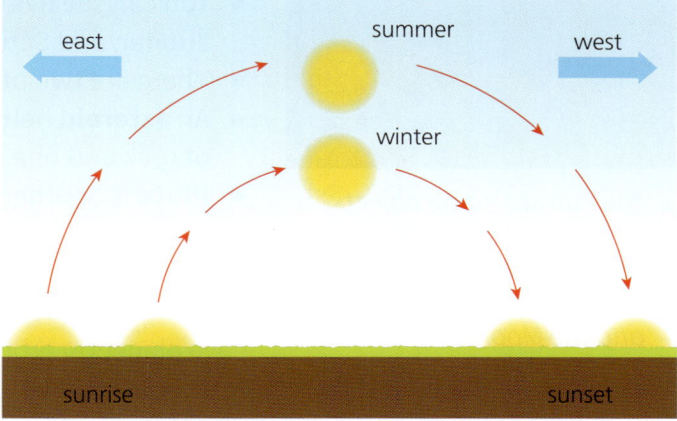

▲ *The tilt of the Earth's axis means that the Sun is higher in the sky in the summer than in the winter.*

Questions

1. Compare the length of a shadow that a fence post makes at noon in the winter and at noon in the summer. Explain your answer.
2. **a.** Explain why the length of a day changes over the year.
   b. Which month has the shortest day in the northern hemisphere, and which month in the southern hemisphere?
3. Describe what would happen to the day length and seasons if the Earth's axis was *not* tilted.

Key points

- The Earth's axis is tilted.
- This is why in summer it is warmer, days are longer, and the Sun is higher in the sky than in winter.

5.3

Objectives

- Name types of objects we can see in the night sky
- Explain how we see different types of object

▲ *You can see some objects in the night sky without a telescope. Here you can see the planet Venus at the top and the Moon at the bottom.*

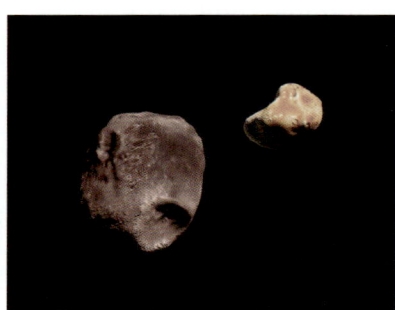

▲ *Phobos and Deimos are moons of Mars. They are not spherical.*

The night sky

Stars

When you look at the night sky you see millions of dots of light. Most of the dots are **stars**. Stars are huge balls of gas that give out light. When you look at a star it might appear to twinkle – the light seems to flicker. This is *not* because the star's light is not constant. It is because the light changes direction as it travels through the Earth's **atmosphere**.

▲ *The bright dots of light in this image are stars.*

Stars are the only objects in the night sky that continuously give out light.

Planets

We always see the stars in the same arrangement. Other objects appear to move against the stars. Ancient astronomers named them **planets**, which means 'wanderers'. Planets are *not* stars – they are objects in **orbit** around stars. They are made of rock or gas and do *not* give out their own light.

- You see planets because light is reflected off them into your eyes.
- You can see five planets without a telescope: **Mercury**, **Venus**, **Mars**, Jupiter, and Saturn.
- There are two other planets: Uranus and Neptune.
- An **asteroid** belt between Mars and Jupiter contains much smaller pieces of rock and one **dwarf planet**.
- **Pluto** is another dwarf planet, much further from the Sun.

These planets and other objects orbiting the Sun make up our Solar System.

Our Sun is *not* the only star to have planets in orbit around it. Astronomers have found lots of planets that orbit other stars. These planets are called **exoplanets**.

Moons

On a clear night you can see the **Moon**. It is made of rock and orbits the Earth. It is the only object beyond the Earth that a person has set foot on.

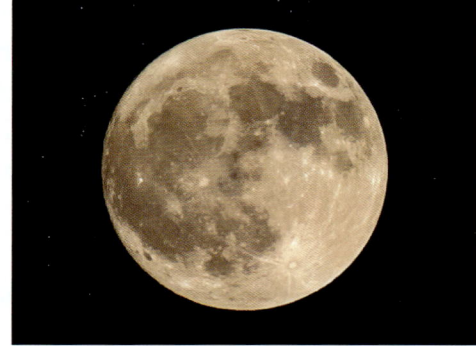

▲ *The Earth has only one natural satellite – the Moon.*

A **moon** is an object that orbits a planet. It is sometimes called a **natural satellite**. Most of the other planets in the Solar System have their own moons, although you cannot see them without a telescope. The Earth has only one moon, while other planets have more. Saturn has over 60 moons. Moons are not always small. Titan, one of Saturn's moons, is bigger than the planet Mercury. The difference is that a moon orbits a planet and a planet orbits a star such as the Sun.

Comets

A **comet** is made of ice and dust like a big, dirty snowball orbiting the Sun. As a comet gets nearer to the Sun it produces a tail that makes a spectacular sight in the sky. The name 'comet' comes from the Greek for 'hairy star'. We see comets because, like planets and moons, they reflect sunlight.

From ancient times people have watched comets, and often thought comets were a bad omen. They seem to appear out of nowhere and then disappear. It wasn't until astronomers started to make measurements that they realised that some comets come back again and again. Comet Encke can be seen every three years. Other comets return after hundreds or thousands of years.

▲ *This comet was photographed through a telescope.*

Meteors and meteorites

Sometimes you see streaks of light in the night sky. People call them 'shooting stars', but they are *not* stars.

- When a particle of dust or a piece of rock called a **meteor** enters the Earth's atmosphere it burns up.
- A piece of rock that survives to reach the ground is called a **meteorite**.

You see a meteor shower when the Earth moves through the dust left by a comet.

Tiny meteorites hit the surface of the Earth all the time. If you go outside about a dozen will land on you every hour.

▲ *You can see the International Space Station moving across the sky because it reflects light.*

Man-made objects

There are thousands of **artificial satellites** in orbit around the Earth. Many are part of the **global positioning system** (**GPS**). This system pinpoints positions of things on Earth using signals from lots of satellites. Sometimes you can see these satellites near dawn or dusk when they reflect light from the Sun. The biggest and brightest of all of these is the **International Space Station**.

Questions

1. Name two planets you can see without a telescope.
2. Some of objects in the night sky are in orbit around the Sun, and some are in orbit around planets.
   a. Name three objects that orbit the Sun.
   b. Name two objects that orbit the Earth.
3. Name three objects can only be seen because they reflect light from the Sun.

📖 Key points

- The objects that we can see without a telescope in the night sky include stars, planets, the Moon, comets, meteors, and artificial satellites.
- Stars emit light, but we can see other objects because they reflect light from the Sun.

Our Solar System

Objectives

- Describe what is in our Solar System
- Describe how the Solar System formed
- Explain why the objects orbit the Sun

We live on a planet that orbits a star called the Sun. You can see some other planets in the night sky with the naked eye. To see other objects that orbit the Sun you need a telescope.

Planets that orbit a star make up a solar system. How did our **Solar System** form, and why do things in it orbit the Sun?

What is in our Solar System?

Our Solar System contains four **inner planets**, an **asteroid belt**, and four **outer planets**. Planets are objects that have cleared their orbits of dust, gas, and rock. There are also **dwarf planets** in the asteroid belt and beyond the outer planets. The outer planets have rings made of rock, dust, and ice.

▲ Our Solar System, with the Sun on the left. The sizes of the planets, but not the distances, are to scale.

	Inner planets	Outer planets
Names	Mercury, Venus, Earth, Mars	Jupiter, Saturn, Uranus, Neptune
Made of...	mostly rock	mostly gas (hydrogen, helium) and/or ice
Size	small	large
Rings?	no	yes
Moons	Mercury 0, Venus 0, Earth 1, Mars 2	Jupiter 79, Saturn 82, Uranus 27, Neptune 14
Seen by...	Reflected light	Reflected light
Life detected?	Only on Earth	No

▲ This table compares the inner and outer planets.

Until 2006 Pluto was called a planet. Then astronomers found other objects like Pluto beyond Neptune, so they started calling Pluto a dwarf planet.

How did the Solar System form?

Our Solar System was formed about 5 billion years ago from a swirling cloud of dust and gas.

Gravity pulled the dust and gas together to make the Sun and all the planets. Some of the material formed rings around the outer planets.

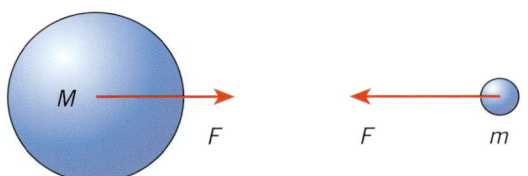

▲ *There is a force of gravity between all objects.*

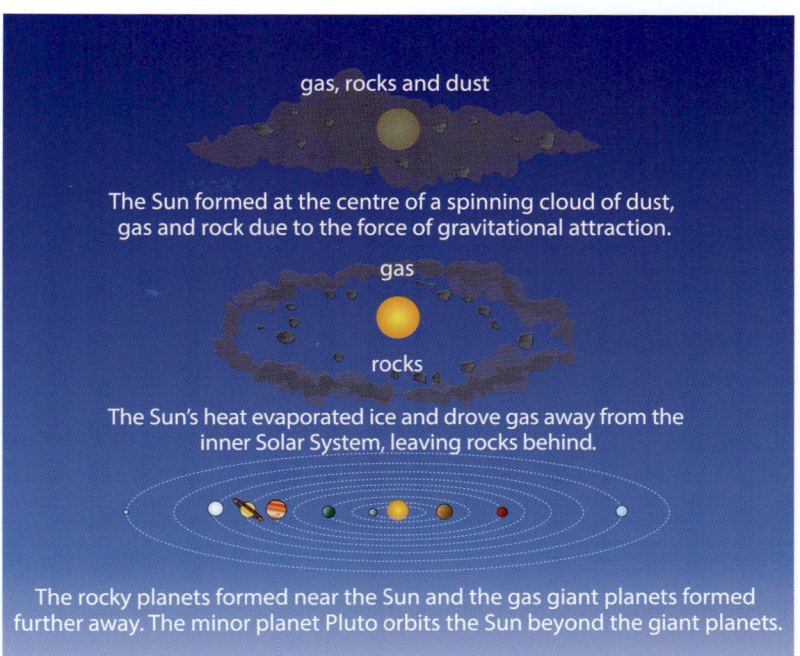

gas, rocks and dust

The Sun formed at the centre of a spinning cloud of dust, gas and rock due to the force of gravitational attraction.

gas

rocks

The Sun's heat evaporated ice and drove gas away from the inner Solar System, leaving rocks behind.

The rocky planets formed near the Sun and the gas giant planets formed further away. The minor planet Pluto orbits the Sun beyond the giant planets.

Why do objects orbit the Sun?

Gravity is a force that acts between all objects with mass. There is a force of gravity between all the planets and the Sun. It is this force that keeps the planets orbiting the Sun.

The Sun pulls on the Earth, and the Earth pulls on the Sun. The mass of the Sun is much larger than the mass of the Earth. The Earth is continuously pulled in, keeping it in orbit. If gravity was 'switched off' the planets would fly off into space.

Objects such as asteroids and comets also orbit the Sun because of the force of gravity.

path of comet

Pluto's orbit

▲ *The orbits of Pluto and comets are different from the orbits of the planets.*

Questions

1. **a.** Describe what a solar system is.
   b. List the planets of the Solar System, starting with Mercury.
   c. Compare the inner planets with the outer planets in terms of:
      i. their moons and rings
      ii. their composition (what they are made of).
2. Describe how the planets of the Solar System formed.
3. Suggest what would happen to the Solar System if gravity was much stronger than it is.

Key points

- The Solar System contains four inner and four outer planets, and an asteroid belt.
- The Solar System formed from a cloud of gas and dust brought together by gravity.
- The planets, asteroids, and comets orbit the Sun because of the force of gravity.

The Moon

Objectives

- Explain why we see phases of the Moon
- Explain why we see solar and lunar eclipses
- Explain why we have tides

Moonlight

You see the Moon in the sky during the day *and* during the night. Moonlight is actually reflected light. The Moon does *not* give out light itself.

The phases of the Moon

The Moon orbits the Earth every 27.3 days. If you took a picture of the Moon each night you would see that its shape changes in a regular way. These are called the **phases of the Moon**.

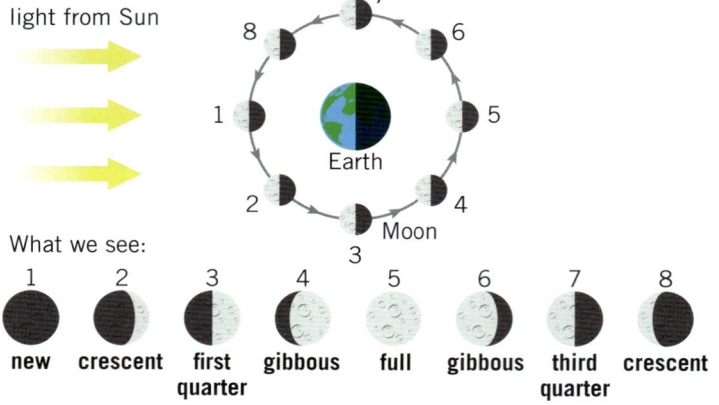

The light from the Sun lights up half of the Moon, just as it lights up half of every planet and moon in the Solar System. The Moon appears to change shape when seen from the Earth because it is moving around the Earth.

▲ *The Moon casts a shadow on the Earth to produce a solar eclipse.*

Eclipses

When an object moves in front of a source of light, a shadow is formed. This can happen when the Moon or the Earth moves in front of the Sun.

Solar eclipses

There is a **solar eclipse** on Earth when the *Moon* blocks the light from the *Sun*.

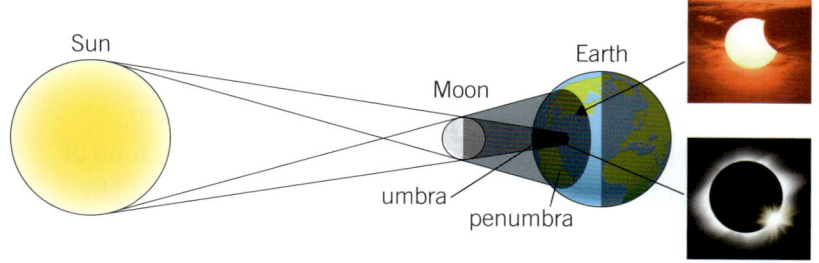

Outside that region, in the **penumbra,** you see a **partial eclipse**.

The region of deep shadow, or **umbra**, is where you see a **total eclipse**.

▲ *A solar eclipse happens when the Moon blocks the light from the Sun.*

We see eclipses because the Moon happens to be the right size and distance away from the Earth to exactly cover up the Sun. This is simply coincidence.

Lunar eclipses

A **lunar eclipse** happens when the Earth comes between the Sun and the Moon. The Moon is in the shadow of the Earth.

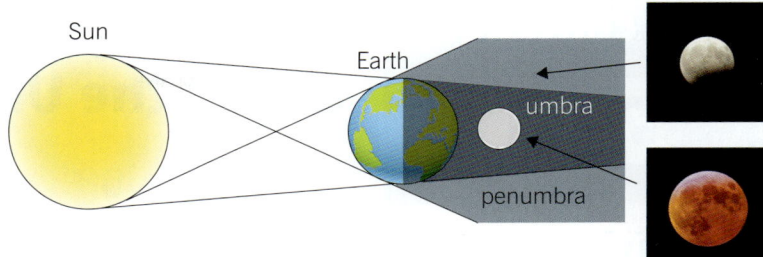

Why are eclipses rare?

You might expect to see a solar eclipse every 27.3 days, because the Moon orbits the Earth in that time. In fact, total eclipses are rare. This is because the orbit of the Moon is tilted slightly. The Earth, Sun, and Moon need to be lined up for an **eclipse** to happen.

If the orbit of the Moon was *not* tilted then the Earth would block out the light and we would never see a full moon.

Why are there tides?

Water in the oceans is free to move. Gravitational attraction of the Moon and Sun produces **tides**.

- There is a bulge of water either side of the Earth due to the force of gravity from the Moon.
- The Earth spins through the bulges, producing two high tides a day.
- When the Sun and Moon are aligned there is extra force on the water, and we get a higher tide, called a **spring tide**.
- When they are *not* aligned we get a lower tide, called a **neap tide**.

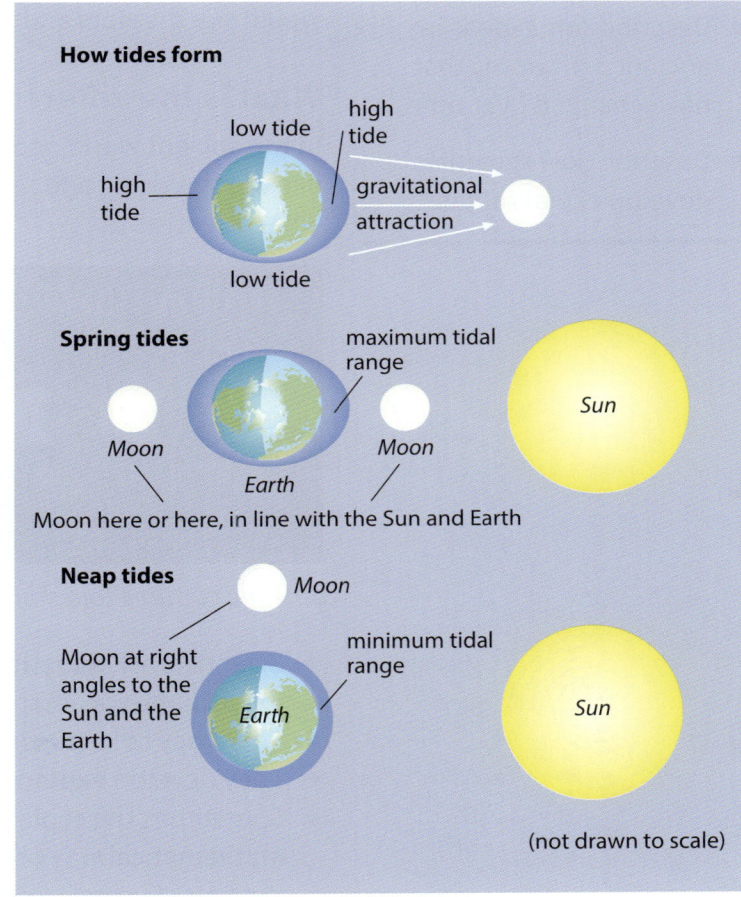

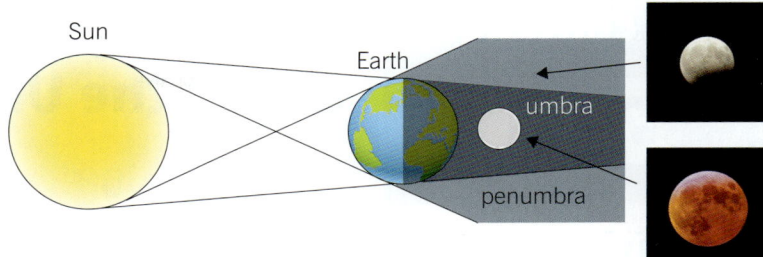

Questions

1. **a.** Write down the time between one full moon and the next.
   b. Write down how much of the Moon is lit by the Sun when there is a new moon.
2. Describe the difference between a solar eclipse and a lunar eclipse.
3. Describe one difference that we would notice if the orbit of the Moon was *not* tilted.
4. Explain the connection between a new moon, a solar eclipse, and a spring tide.

Key points

- Phases of the Moon happen because half the Moon is lit, and it moves around the Earth.
- Eclipses happen when the Earth or the Moon blocks light from the Sun.
- The gravitational force of the Sun and Moon produces tides.

Objectives

- Describe some ideas in ancient astronomy that have changed over time
- Describe how scientists develop explanations

Changing ideas 1: Ancient ideas about the Universe

Past times

Imagine that you were living 2000 years ago. Many things about your everyday life would be different, but if you looked up at the night sky you would see the stars and planets change position just as they do today.

What is the difference between a story and an explanation?

In the ancient world people told stories to explain why things happened, like the sun rising and setting, eclipses, and seasons of the year. This is *not* what scientists do.

▲ *Many cultures told stories about the Sun being eaten during an eclipse.*

- Scientists ask **questions**, for example, 'Why do the stars and planets move in the way that they do?'
- Then they collect **evidence** by making observations or measurements.
- They develop **explanations** to account for the evidence.
- Sometimes the explanations use a **model**. A model is a physical or mathematical way of explaining how something works, like a model of the Solar System.

The invention of numbers and of writing were important for developing explanations. If people could *not* make measurements and write down their results, they could *not* look for patterns.

Astronomy in India

The Indian astronomer Aryabhata was born in Bihar, India, in the year 476. He thought that the motion of the stars was because the Earth was spinning on its axis, *not* because the stars were moving. The Earth's spin also explained day and night.

At the time some people thought that eclipses happened because dragons or demons attack the Sun during an eclipse, or that the Sun is swallowed by gods. Aryabhata showed that his model could predict when lunar and solar eclipses would happen.

▲ *Part of an old observatory in Jaipur, India.*

Another Indian mathematician–astronomer, Brahmagupta, described a law of gravity about a thousand years before Newton. He worked out a method for calculating the positions of planets, and when they rise and set.

Astronomy in Africa

Thousands of years ago the Egyptians made a huge number of observations and worked out a calendar with 365 days. This is *not* the same as saying that the Earth goes around the Sun in 365 days. It was an observation that the seasons change over a year.

One of the oldest astronomical sites known is in southern Egypt at Nabta Playa. Scientists think that more than 7000 years ago people put stones in places that would line up with where certain stars would rise.

People were probably making measurements of the stars many thousands of years ago in West Africa. Scientists think that people put stones at Namoratunga on the west side of Lake Turkana in Kenya in 300 BCE and tilted them to make an astronomical calendar pointing to stars at particular times.

Islamic astronomy

The earliest records of astronomical observations came from Babylonia, a region that is now in Iraq. People there recorded the positions of planets and the dates of eclipses. Their observations were later used by Greek and Indian astronomers.

Ibn Al-Haytham, who was born in Basra in Iraq in 956, thought that the Earth was spinning on its axis. Al-Biruni, a scientist born in Uzbekistan about 1000 years ago, developed a model to explain the phases of the Moon.

One of the big achievements of Islamic science was the construction of many observatories and invention of new instruments, such as the astrolabe. With these instruments astronomers could make very precise measurements of the stars and planets. They made detailed charts of the positions of the stars.

▲ *Part of an ancient observatory in Africa.*

▲ *Astronomers using a range of instruments.*

Questions

1. Explain why the invention of writing was important for the development of scientific explanations.
2. Suggest how the scientific explanation of day and night was developed.
3. **a.** Describe a story people used to tell about eclipses.
   b. Give a scientific explanation for eclipses.
   c. Explain how the story and the explanation are different.

Key points

- Scientists ask questions, collect evidence, and develop explanations.
- Ancient astronomers measured positions of objects in the sky.
- They produced models to explain eclipses, moon phases, and day and night.

Changing ideas 2: The geocentric model

Objectives

- Describe the geocentric model

- Describe evidence for the geocentric model

- Explain why scientific explanations change

You know that the Earth is *not* at the centre of the Universe, but people did not always believe that.

Astronomy in Greece

Plato and Aristotle were Greek astronomers who lived about 2400 years ago. They developed the **geocentric model** of the Universe. 'Geo' means Earth in the Greek language. In this model the Earth is at the centre of the Universe. In their model:

- the Earth did *not* move, but the stars and planets moved around the Earth
- there were seven planets: Mercury, Venus, Mars, Jupiter, Saturn, the Sun, and the Moon, with stars beyond them
- all these objects were spherical, including the Earth
- all the objects moved at a steady speed on spheres around the Earth
- light could travel through the spheres, called 'crystalline spheres'
- the Moon was on the sphere closest to the Earth, and the stars were on the sphere furthest from the Earth.

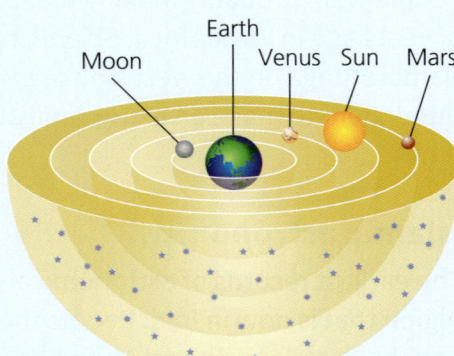

▲ *The geocentric model.*

Moving backwards

The geocentric model explained a lot of astronomical observations, but it did *not* explain one important observation. The planets seemed to wander across the sky. They did not always go in the same direction. This motion is called **retrograde motion**.

Ptolemy was another Greek astronomer who lived about 1800 years ago. He changed the geocentric model to explain the retrograde motion of the planets.

- He added little circles on top of the bigger circles that were the paths of the planets.
- He worked out how big they needed to be to explain the 'backward' motion of the planets.
- He predicted the future positions of the planets.
- He wrote about his model in a book that was used for hundreds of years.

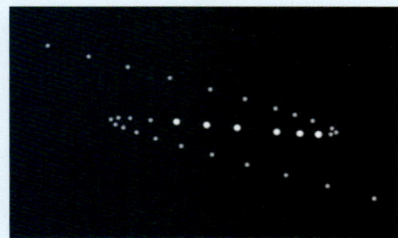

▲ *The position of Mars photographed every week for several months.*

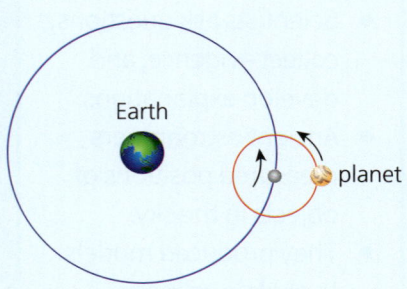

▲ *Little circles are added to orbits to explain retrograde motion.*

Evidence for the geocentric model

The ground that you are standing on does *not* feel as if it is moving. This supported the idea that the Earth was at the centre of the Universe. There were many other ideas.

- Greek philosophers thought that all objects fall towards the centre of the Universe. If you drop an object it goes down, so the Earth had to be at the centre.
- Ptolemy saw that half the stars were above the horizon and half were below the horizon at any time. If the Earth was *not* at the centre, then the numbers of stars above and below the horizon would *not* be the same.
- If the Earth was moving then stars would appear to move between January and July as the Earth moves around the Sun. In fact the stars *do* move, but they are so far away you cannot see that movement without a telescope, so Ptolemy did not know this.

Ptolemy's geocentric model was very successful because it could explain lots of observations and make predictions. It was a popular model because it fitted what people could see and feel. Only when people had evidence from things that they *couldn't* immediately see would the scientific explanation change.

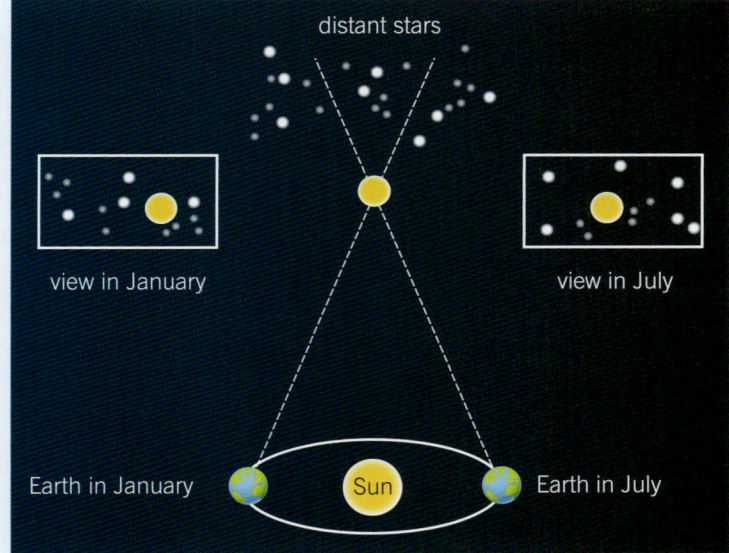

▲ *Ancient astronomers could not detect the apparent movement of stars.*

Questions

1. Complete these sentences:

   In the geocentric model the _____ is at the centre of the Universe. The Moon, Sun, planets, and stars orbit on _____. This model made sense to people because the Earth does *not* appear to _____.

2. Explain why people found it easy to believe the geocentric model.

3. Describe how the geocentric model was changed to explain the observation that the planets seemed to wander backwards and forwards across the sky.

Key points

- In the geocentric model the Moon, planets, Sun, and stars orbit the Earth on crystal spheres.
- This model was easy to believe because the Earth does not appear to be moving, but the Moon, Sun, and stars do appear to move.
- Scientific explanations change when there is new evidence.

Objectives

- Describe the heliocentric model of the Universe

- Describe the evidence for the heliocentric model

- Explain how scientific explanations develop

Changing ideas: Modern ideas about the Universe

People who lived before 1600 had only ever seen five planets. They watched the planets, the Sun, and the Moon move across the sky. They were happy with the explanation that astronomical objects were carried on celestial spheres.

Only new evidence would change people's view of the **Universe**.

That new evidence came with the invention of the **telescope**.

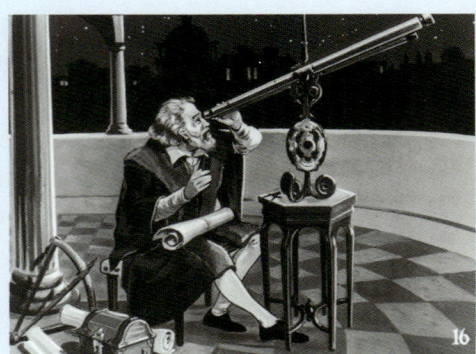
▲ *Galileo and his telescope*

Challenging the geocentric model

Nicholas Copernicus was born in 1473 in Poland, before telescopes were invented. He is famous for developing the **heliocentric** model of the Universe. 'Helio' means 'Sun' in the Greek language. Copernicus promoted the heliocentric model in Europe, but he did *not* invent the idea. The Greek astronomer Aristarchus had produced a model with the Sun at the centre about 1200 years earlier. Al-Biruni suggested that the Sun might be at the centre about 500 years earlier. He thought that it was *not* possible to prove that the Earth was moving.

In the heliocentric model:

- the Sun was at the centre of the Universe
- all the planets were in orbits around the Sun
- the Moon was in orbit around the Earth.

This model was able to explain the motion of the planets much more *simply*. Scientists look for the simplest explanation of their observations. If explanations begin to get very complicated, then they will look for a simpler one.

It was very difficult for Copernicus to discuss his heliocentric model openly. Most people believed the Earth to be at the centre of the Universe at the time. This was an important part of religious beliefs. Copernicus was worried that he would be persecuted if he published his ideas. His book about the heliocentric model was published as he died. It was published in Latin, which was *not* a language that many people could understand.

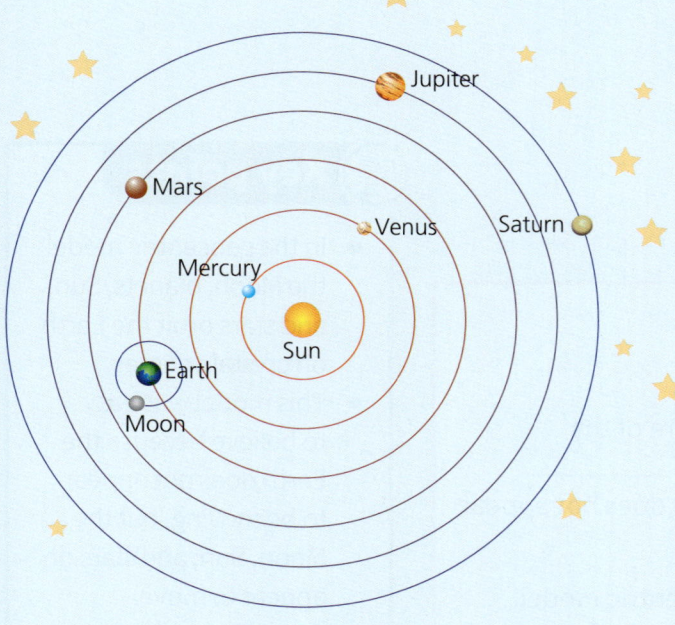
not to scale

▲ *The heliocentric model.*

Galileo and the telescope

The telescope was invented in 1600 in the Netherlands. Galileo Galilei was an Italian mathematician and scientist who made his own telescope in 1609. One night he pointed it at the planet Jupiter and made some important observations.

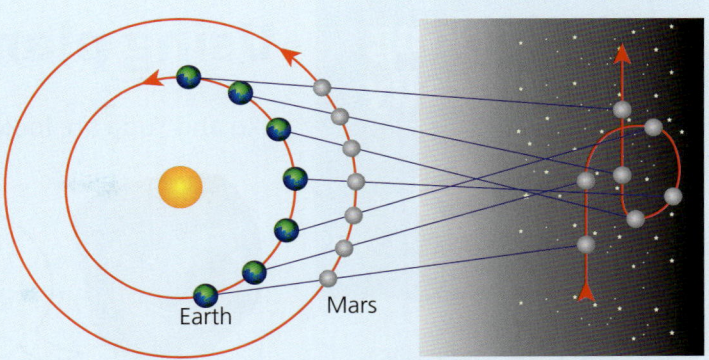

- He saw four objects that seemed to be in orbit around Jupiter, *not* around the Earth.
- These were the four biggest moons of Jupiter.
- These moons are now called the Galilean satellites in his honour.

This was evidence that not all the objects in the Universe orbited the Earth. Galileo thought that planets orbit the Sun just like moons orbit Jupiter.

▲ *The planets show retrograde motion when viewed from the Earth because they are orbiting the Sun at different speeds.*

Galileo also used his telescope to make observations of the Sun. He observed dark spots on the Sun called **sunspots**. The observations showed that astronomical objects such as the Sun were *not* perfect.

Unlike Copernicus, Galileo talked about his model in public. Church leaders in Rome were not happy with Galileo. They tried to make him take back his claim. He refused, and published his ideas in 1632.

- His book showed how the heliocentric model explains the evidence more simply than the geocentric model.
- The book was published in Italian, which a lot of people understood.

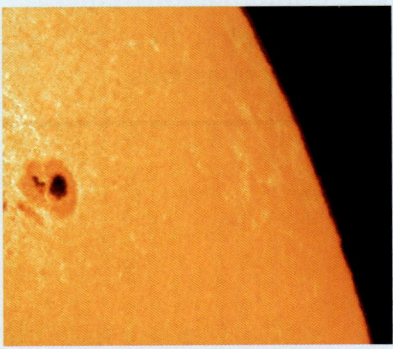

He was kept imprisoned in his house until he died in 1642. After his death, the heliocentric model gradually became accepted.

▲ *A sunspot on the surface of the Sun.*

In fact, the Sun is at the centre of our Solar System, *not* of the Universe. Since Galileo, astronomers have built up a model of the Universe that is much larger than the Solar System and contains galaxies.

How do ideas change?

Scientists might *not* know that an explanation has already been developed by someone else because:

- it may *not* have been published
- it might have been written in a book that hadn't reached them yet
- it might have been written in another language.

Objectives

- Describe the difference between primary and secondary sources of data
- Name some secondary sources
- Use information from secondary sources to answer questions

Using planetary data

Eko and Yuda are looking at pictures of the planets.

I wonder if there is a link between the length of a day and the length of a year on each planet.

I wonder if there are seasons on other planets.

They cannot answer these questions by doing experiments in a laboratory.

Primary and secondary sources

In an experiment you take measurements or make observations. These observations or measurements are called **data**. Scientists record their data in tables and display them using charts or graphs. An experiment is an example of a **primary source** of data in science. There are other primary sources of data, such as fieldwork.

Answering Eko's question

Eko wants to know if the day length on other planets is related to the year length.

- She knows that the Earth takes 24 hours to spin on its axis.
- This is the length of one day on Earth.
- It takes 365 days to orbit the Sun.
- This is the length of one year.

She finds a book and looks up data about other planets.

The book is a **secondary source** of data. Eko did *not* make the measurements herself. She used measurements that other people have made.

It is very important to check data, whether they are primary data or secondary data. This could mean repeating an experiment, or looking at more than one secondary source to check that the data are the same.

These are the data Eko found.

Planet	Time to spin once on its axis (hours)	Time to go once around the Sun (years)
Mercury	1416 (59 days)	0.24
Venus	2 137 440 (244 years)	0.69
Earth	24	1.00
Mars	25	1.88
Jupiter	10	11.86
Saturn	11	29.46
Uranus	17	84.01
Neptune	16	164.80

Eko can use the data to answer her question.

Answering Yuda's question

Yuda knows that seasons on Earth are due to the tilt of the Earth's axis.

- He finds a website about the planets.
- He writes down the angle at which the axis of each planet is tilted.
- He checks on several other websites and finds that the data are very similar on all the sites.

Data are *not* always presented in the form of a table. Yuda found the data on this diagram. He took the numbers from the diagram and made a table for them. When you are using secondary data you may need to display the data in a different way, such as on a chart or a graph.

Planet	Angle of tilt (°)
Mercury	0
Venus	177.4
Earth	23.5
Mars	24.0
Jupiter	3.1
Saturn	26.7
Uranus	97.9
Neptune	28.8

Mercury
0°

Venus
177° (−3°)

Earth
23.5°

Mars
25°

Jupiter
3°

Saturn
27°

Uranus
98° (−82°)

Neptune
30°

not to scale

Questions

 TWS **1.** Look at Yuda's data.
   a. Which planet doesn't have seasons? Explain your answer.
   b. Which planets have seasons that would be similar to Earth's seasons? Explain your answer.
   c. On which planets would there be seasons, but you would hardly notice them? Explain your answer.
   d. On which planet would the seasons be very different from the Earth's seasons? Explain your answer.

2. Look at Eko's data. Write down whether there is a relationship between day length and year length. Explain your answer.

TWS **3.** Explain why you should look at more than one secondary source.

Key points

- Experiments and fieldwork are primary sources of data.
- Secondary sources of information, such as a book or the Internet, contain data that other people have collected.
- You can use primary and secondary data to answer questions.

1. Match each object to its correct definition.

   1 a piece of dust or rock that burns up in the atmosphere

   2 a piece of rock in an orbit between Mars and Jupiter

   3 a meteor that has landed on the ground

   4 a big, dirty snowball that orbits the Sun

   A asteroid

   B comet

   C meteor

   D meteorite   [4]

2. Write the letter of the statement that explains why we have day and night.

   A The Earth orbits the Sun once a day.

   B The Sun orbits the Earth once a day.

   C The Sun spins on its axis once a day.

   D The Earth spins on its axis once a day.   [1]

3. Look at the diagram of the Solar System.

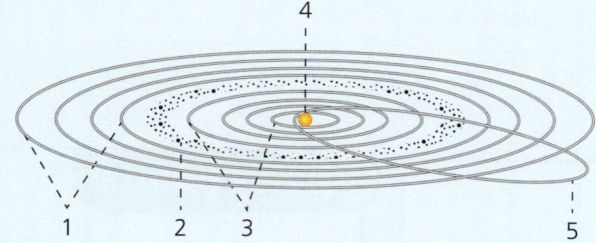

   Write the number that shows where you find:

   a. the inner planets   [1]

   b. the outer planets   [1]

   c. the asteroid belt   [1]

   d. a comet   [1]

   e. a star.   [1]

4. Here is a list of objects that you can see from the Earth, with the naked eye or with a telescope.

   Sun   comet   Jupiter's moon   star (*not* our Sun)
   Moon   planet   Earth   Venus   Mercury   Neptune

   a. Write down the objects that you can see *without* using a telescope.   [1]

   b. Explain why you cannot put the objects in order of size.   [1]

   c. 'Mercury' could form a group with three other words on the list. Which three, and why?   [1]

   d. 'Sun' could form a pair with one other word on the list. Which one, and why?   [1]

5. Copy and complete these sentences. You may use each word once, more than once, or not at all.

   away from   towards   North   South   day   month
       year   equator   axis   shorter   longer

   a. We have seasons because the _____ of the Earth is tilted. The Earth orbits the Sun once every _____.   [2]

   b. In summer in the southern hemisphere the _____ Pole is tilted towards the Sun and the _____ Pole is tilted away. This means that the days in the northern hemisphere are _____ and the days in the southern hemisphere are _____.   [4]

   c. In winter the energy from the Sun is spread out over a _____ area because that part of the Earth's surface is tilting _____ the Sun.   [2]

6. Here is a diagram of the first four planets in the Solar System.

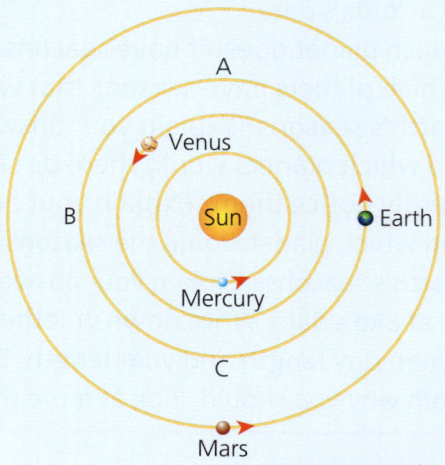

   not to scale

Look at where the Earth is on the diagram. How many months later will the Earth be at:

a. A [1]

b. B [1]

c. C [1]

The distance between the Sun and Mercury is 60 million km. The distance between the Earth and the Sun is 150 million km.

d. Calculate the closest distance that Mercury gets to the Earth. [2]

e. Use the diagram to explain why we see different stars at different times of the year. [1]

7. When studying astronomy you cannot do experiments. Scientists make observations and develop explanations to account for the observations. Here is a list of observations that astronomers made thousands of years ago.

A The ground that we are standing on does not appear to move.

B We see different stars in June and December.

C The Sun moves across the sky.

D Planets appear to wander across the sky.

E The stars move together across the sky without wandering.

a. Write the letter of each observation that fits with a geocentric model of the Universe. (There may be more than one.) [1]

b. Write the letter of each observation that fits with a heliocentric model of the Universe. (There may be more than one.) [1]

c. Write the letter of the observation that is explained more simply by the heliocentric model than by the geocentric model. [1]

8. Below are some statements about how the Solar System formed. List the letters in the correct order.

A The Sun formed.

B A cloud of gas and dust was spinning in space.

C Gravity pulled the dust and gas together.

D The Earth and other planets formed.

E Asteroids are leftover pieces of dust and rock.

9. There are tides twice a day.

a. Draw a diagram of the positions of the Earth, Moon, and Sun during a spring tide. [1]

b. Describe the type of eclipse that would be produced by the positions of the Earth, Moon, and Sun you have described in part **a**. [1]

c. Explain why we do not get eclipses every month. [1]

d. Describe the phase of the Moon when the Earth, Moon, and Sun are in the position you described in part **a**. [1]

10.a. For each statement below write 'primary' or 'secondary' to describe the type of data.

i. An astronomer uses data from a science book to develop an explanation about the atmospheres on different planets. [1]

ii. A student uses a table from a magazine to find out about moons around the planets. [1]

iii. A teacher uses a telescope to observe the craters on the Moon. [1]

b. Which data are likely to be more reliable? Explain your answer. [1]

TWS

TWS

1. A student says 'There is a force of gravity between any objects with mass. There is a force of gravity between me and my chair. The force of gravity between me and my chair makes it hard to get out of my chair!'

   Write down which sentence or sentences is or are correct. [1]

2. A baby gorilla is playing.

   He climbs up a tree and sits on a branch. He pulls some leaves off the tree and they fall to the ground. Then he jumps down from the branch. He sees his mother and they run towards each other.

   a. Name the energy store that gained energy as he climbed the tree. [1]

   b. Describe where this energy comes from. [1]

   c. He dropped a large leaf and a small leaf from the branch. Write down which one hit the ground first. Explain your answer. [2]

   d. Describe the energy transfers from when he jumps off the tree to when he has landed on the ground. [2]

   e. His mother runs towards him at the same speed that he is running towards her. Write down who has more kinetic energy. Explain your answer. [2]

3. Here are some data about different types of music player.

Type of player	Sound energy produced for each 200 J of electrical energy supplied
A	70
B	100
C	80

   a. Calculate how much energy is dissipated by music player A. [1]

   b. Suggest how the energy is dissipated. [1]

   c. Put the music players in order of efficiency starting with the least efficient. [1]

   d. Explain why you have chosen this order. [1]

   e. Describe how the sound travels from the music player to the ear. [1]

   f. A student takes the music player into a large canyon. They hear multiple sounds. Suggest where the sounds come from. [1]

4. A planet has been discovered in orbit around a star that is not our Sun. The star (51 Pegasi) is about the same size and brightness as our Sun. Here is some information about the planet, which is called 51 Pegasi b.

	51 Pegasi b	Earth
Distance from the star (million km)	7.7	150
Time to orbit the sun (days)	4	365
Time to spin once on its axis (days)	4	1
Tilt of the axis (degrees)	79	23.5

   a. Suggest whether the surface temperature of 51 Pegasi b is higher or lower than the temperature on the Earth. Explain your answer. [2]

   b. How long is a day on 51 Pegasi b? [1]

   c. How long is a year on 51 Pegasi b? [1]

d. Does 51 Pegasi b have seasons? Explain your answer. [1]

51 Pegasi is a star in the constellation of Pegasus. It is 51 light years away.

e. Suggest how 51 Pegasi formed. [1]

f. When astronomers look at 51 Pegasi, are they seeing it as it is now? Explain your answer. [1]

5. A teacher is showing the class how you can make a balloon stick to the wall by rubbing it on your clothes.

a. As she rubs the balloon on her clothes the balloon becomes positively charged. Describe what has happened to charge the balloon. [1]

b. The wall is neutral. Explain what that means. [1]

c. As she brings the positively charged balloon towards the wall the outside of the wall becomes negatively charged. Explain why. [1]

d. She leaves the balloon stuck to the wall but it eventually falls off. Why? [1]

e. A student goes home and tries to stick a balloon to the metal case of a refrigerator. Will it work? Explain your answer. [1]

6. a. Here are some statements about the geocentric and the heliocentric models of the Universe. Write down which statements are true and which are false. [4]

TWS

A The geocentric model could only work if you added smaller orbits to the orbits of planets.

B A heliocentric model explains the motion of planets more simply than the geocentric model.

C Nobody before Copernicus had thought of the heliocentric model.

D Copernicus built his heliocentric model using the ideas of others.

E The orbit of the Moon around the Earth was the evidence that Galileo used to convince people that the Earth orbited the Sun.

F The invention of the telescope was very important for the development of the heliocentric model.

G There were no astronomical observatories in India before Copernicus.

b. Astronomers use telescopes to collect data. Is that primary or secondary data? [1]

c. Scientists ask questions, and then use evidence to develop explanations and test them. Think about the development of the heliocentric model. Fill in the table below to show how the question/evidence/explanation system fits the development of the heliocentric model. [3]

	Applied to the development of the heliocentric model
Question	
Evidence	
Explanation	

TWS

Objectives

- Calculate speed
- Explain what is meant by average speed
- Describe the difference between average and instantaneous speed

▲ *Speed limits are described in km/h in India.*

▲ *The speed of a boat changes during a voyage.*

Speed

How fast?

Bicycles, cars, and planes all travel at different **speeds**. Speed is a measure of how fast something is moving. It is the distance moved per second or per hour. The unit of speed is **metres per second** (m/s) or **kilometres per hour** (km/h).

▲ *A fast car can drive at over 300 km/h, or about 80 m/s...*

▲ *... but you can walk at about 5 km/h, or 1 m/s*

You may have heard people describe a cricket or tennis ball as moving at 80 miles per hour, which would be written down as 80 mph. In the UK and the USA speeds are usually measured in mph rather than km/h.

How to calculate speed

To find the speed of a moving object, you measure the time it takes to travel a known distance. The speed is the distance divided by the time:

$$\text{Speed} = \frac{\text{distance}}{\text{time}}$$

A long-distance runner runs part of his race at a **steady speed**. It takes him 20 seconds to run 100 m.

$$\text{Speed} = \frac{\text{distance}}{\text{time}}$$
$$= \frac{100\ \text{m}}{20\ \text{s}}$$
$$= 5\ \text{m/s}$$

It is helpful to write out calculations like this. If you put the units of distance and time in the equation, you will have the correct units for the speed.

Average speed

In September 2017 six female officers of the Indian Navy set off to sail around the world. They made four stops on the way, crossing three oceans. They had to sail through storms with waves as high as nine-storey buildings. They arrived safely in Goa in May 2018. They had sailed 40 707 km in 254 days, which is 6096 hours.

Their speed varied depending on the wind:

- with strong wind they went faster
- in storms, wind could reach hurricane force (over 100 km/h)
- sometimes they were stuck for days with very little wind.

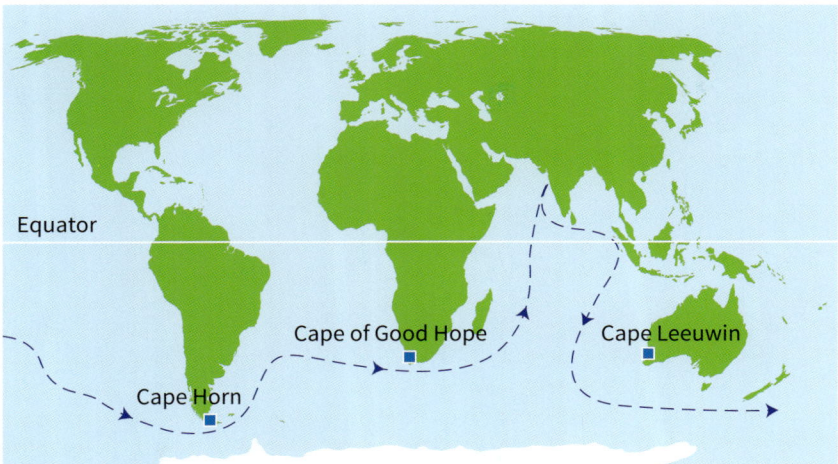

▲ *Many round-the-world sailors take this route.*

You can work out their **average speed** over the whole journey by dividing the *total distance* by the *total time* that it took. This average speed makes it easier to compare how fast different people, or boats, or cars travel.

$$\text{Average speed} = \frac{\text{total distance}}{\text{total time}}$$

$$= \frac{40707 \text{ km}}{6096 \text{ h}}$$

$$= 6.7 \text{ km/h}$$

Instantaneous speed

During a car journey the speed that the driver sees on the speedometer is the speed at that moment in time, called the **instantaneous speed**. That speed is worked out from the size of the wheels on the car. It is *not* the average speed.

	Average speed (m/s)	Average speed (km/h)
walking quickly	1.7	6.1
sprinting	10	36
typical driving speed limit	14	50
category 1 hurricane (typhoon)	33	119
plane cruising speed	255	918

▲ *The table shows some typical average speeds.*

Questions

1. Write down the difference between average and instantaneous speed.
2. **a.** David drives 150 km in 2 hours. Calculate his average speed in km/h.
   b. Fatima runs 200 m in 40 seconds. Calculate her average speed.
3. Preeti runs 100 m in 19 seconds. Nikita runs the same distance in 15 seconds.
   Write down who is faster. Explain your answer.
4. A student says that a speed of 50 m/s is slower than a speed of 140 km/h because the number is smaller. What would you say to them?

Key points

- Speed is distance divided by time, and has units m/s or km/h.
- Average speed is total distance divided by total time.
- Your speed at a particular moment is the instantaneous speed.

Objectives

- Explain the difference between accuracy and precision
- Define reaction time
- Know how to measure time precisely

▲ *A stopwatch.*

Precision and accuracy: What's the difference?

▲ *Short races need precise measurements of time.*

Usain Bolt won an Olympic gold medal in 2012 by running 100 metres in a record time of 9.63 seconds. The silver medal went to, Yohan Blake, with 9.75. Justin Gatlin, who won the bronze medal, took 9.79. When differences are so small, timing must be as accurate and precise as possible.

Precision is shown by the number of **significant figures**. Significant figures are the number of digits in a number. So, 24.3 has 3 significant figures, but 24 has 2 significant figures.

Yohan Blake and Justin Gatlin both had times of 9.8 seconds to 2 significant figures. You need 3 significant figures to say who is faster.

Accuracy tells you how correct a measurement is.

- A measurement can be very precise, but *not* be accurate.
- If Usain Bolt's time had been recorded as 9.73 seconds it would have the *same precision*, but would have been very inaccurate.

What is reaction time?

For a long time the 100 m race was timed with a stopwatch. In the 1970s automatic timing was introduced.

- The starter's gun triggers the start of the timer.
- The athlete breaks a light beam when their body crosses the line.
- A light sensor automatically stops the clock.

When automatic timing was first introduced, the measured times for the 100 m got longer! When a stopwatch was used there was a short delay between the judge hearing the gun and starting the stopwatch.

- This delay is called the **reaction time**.
- It is the time that it takes the brain to process information.
- It is typically about 0.2 seconds.

The judge stood by the finish line. They could anticipate when the athlete would cross the line. This meant that there was *not* the same reaction time when they *stopped* the stopwatch. The times measured were *shorter* than they should be by about 0.2 seconds. Automatic timing is more accurate.

▲ *Starting blocks are connected to the timer.*

Measuring time in the laboratory

It is important to think about reaction time if you are timing an experiment.

- If you are timing something over a *long* period of time, then your reaction time will *not* make much difference.
- If you are timing something that happens *very quickly,* then reaction time will have a big effect.

Dev is doing an experiment about air resistance.

- He drops a ball from a metre off the ground.
- He times how long it takes to hit the ground.
- The reading on his stopwatch is 0.65.
- If Dev's reaction time is 0.2, then the accurate time could be 0.45!

Dev could improve the accuracy of his experiment by using **timing gates or light gates**. These gates work like the finish of the 100 m race. When the ball falls through the light beam of the first gate, the timer is started, and when it falls through the second beam the timer is stopped.

These timers can record times that are accurate and are precise to 3 significant figures.

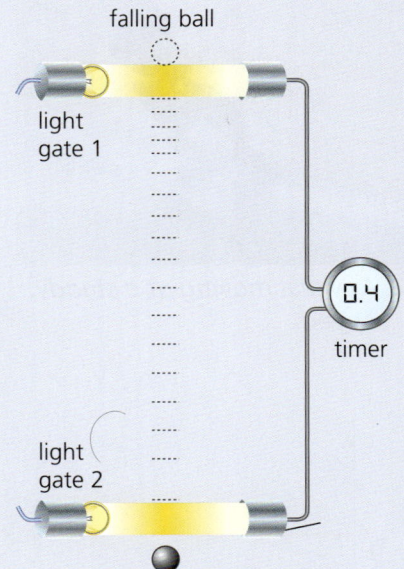

▲ *Light gates automatically start and stop the timer.*

Questions

1. Explain which measurement from the list below. below is the most precise.

   1.3   1.348   1.35   1

2. Explain why timing gates give you a more accurate measurement of the time it takes a ball to fall than a stopwatch does.

3. Explain why automatic timing is more important in a 100 m race than in a marathon (over 42 km, or 26 miles).

4. When sprinters start a race, sensors in their starting blocks detect a change in pressure. If the time difference between the starting gun going off and the change in pressure is too short, a false start is registered. Which of these times could be the time difference that gives a false start: less than 0.2 s, 0.2 s, more than 0.2 s? Explain your answer.

Key points

- Accuracy is how close a measurement is to its true value.
- Precision is the number of significant figures to which a value is given.
- Your reaction time is about 0.2 seconds.
- The reaction time is the time that it takes the brain to process information.

Objectives

- Describe the motion of an object using a distance–time graph

▲ A cyclist moving at a steady speed.

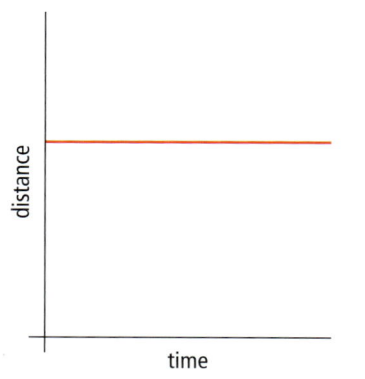

▲ A horizontal line (flat) means the object is stationary.

Distance–time graphs

What is a distance–time graph?

A **distance–time graph** is a way of showing how something is moving.

You can measure the distance a moving object has travelled from its starting point each second. Then you can plot a distance–time graph from those data.

A cyclist is photographed every 2 seconds as he cycles along a track. The lines on the track show the distances. These data are displayed in a table and on a graph.

Time (s)	Distance (m)
0	0
2	12
4	24
6	36
8	48
10	60

▲ You can plot distances and times...

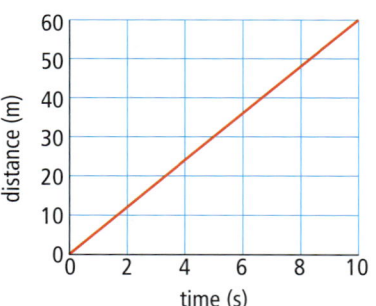

▲ ... on a distance–time graph.

Telling a story

In each second the cyclist moves the same distance – his speed does *not* change. This means the graph is a straight line.

The slope, or **gradient**, of a distance–time graph tells you the *speed*.

In the graph on the right, both objects are moving at a steady speed but the lines have different gradients.

The graph on the left also tells a story, but not a lot is happening! The line is flat, so the slope is zero. This tells you that the speed is zero – the object isn't moving.

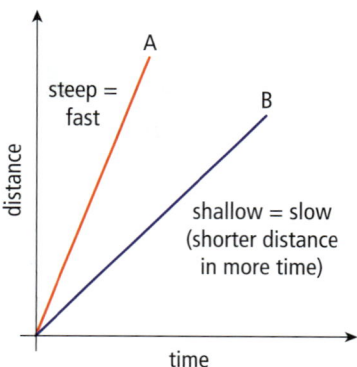

▲ A and B are moving with different steady speeds.

A journey to school

Rana walks to school. She stops to chat, and changes speed along the way.

At the top of the next page is a distance–time graph of Rana's journey to school.

Sections B and D of this graph are flat. This tells you that Rana stopped. In sections A, C, and E, the graph is a straight line that is sloped. She is moving at steady speed.

Rana's graph suddenly changes between flat sections and sloped sections.

But usually when we are walking we don't suddenly change speed – our speed changes continuously. A more realistic graph would have curved lines as Rana's speed changed gradually.

Working out the speed from a distance–time graph

You can calculate the speed of a moving object from its distance–time graph. For example, in section A of the graph Rana walks 1000 m in 10 minutes. Ten minutes converted to seconds is 600 seconds.

$$\text{Average speed} = \frac{\text{distance}}{\text{time}}$$

$$= \frac{1000 \text{ m}}{600 \text{ s}}$$

$$= 1.7 \text{ m/s}$$

This is how to calculate the average speed in section C.

$$\text{Average speed} = \frac{(2400 \text{ m} - 1000 \text{ m})}{12.5 \text{ min} \times 60 \text{ s/min}}$$

$$= 1.9 \text{ m/s}$$

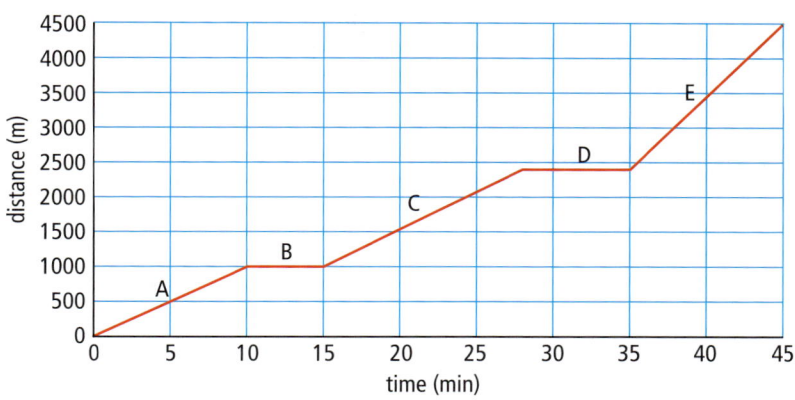

▲ This distance–time graph tells the story of Rani's journey to school.

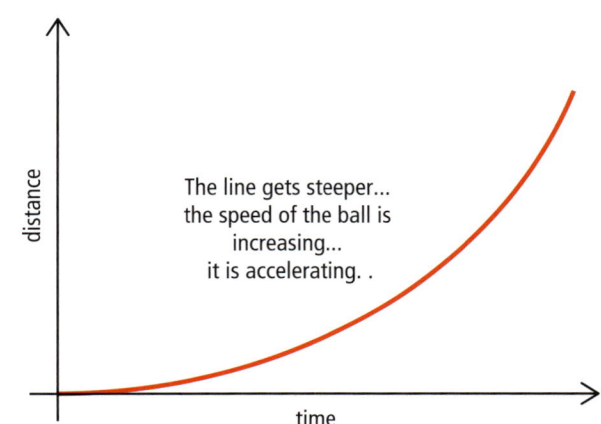

The line gets steeper... the speed of the ball is increasing... it is accelerating. .

Changing speed

When you drop a ball its speed increases as it falls. A distance–time graph for the ball will *not* be a straight line because the speed is *not* constant. It is a curve and the slope of the curve is increasing. This shows that the speed is increasing.

A distance–time graph of something that is slowing down would also be a curve, but the line would become less and less steep. This would show that the speed is decreasing.

📖 Key points

- A distance–time graph shows how the distance moved changes with time.
- A straight, upward-sloping line means a steady speed.
- If the line is horizontal, then the object is not moving.
- If the line is steeper, the object is moving faster.
- If the line is curved, the speed is changing.

❓ Questions

1. Describe the difference between a distance–time graph showing a steady speed and one showing acceleration.
2. In which section of her journey to school is Rana moving fastest? Explain your answer.
3. Calculate Rana's speed for section E of the graph at the top of the page.

Objectives

- Describe how to calculate acceleration
- Explain what is meant by deceleration
- Explain how speed–time graphs tell a story

Acceleration and speed–time graphs

Changing speed

When the speed of something changes, we say that it is **accelerating**. The speed of a ball can change quickly. Acceleration tells you how much the speed changes in one second.

Calculating acceleration

To calculate average **acceleration** you need to know:

- the speed at the start
- the speed at the end
- the time it took for the speed to change.

$$\text{Acceleration} = \frac{\text{change in speed}}{\text{time}}$$

$$= \frac{\text{final speed} - \text{starting speed}}{\text{time}}$$

We measure it in m/s *per second* or metres per second squared (m/s^2).

Sinita is using her scooter to get to work. She sets off from her house and accelerates from 0 to 20 m/s in 5 seconds.

$$\text{Her acceleration} = \frac{\text{final speed} - \text{starting speed}}{\text{time}}$$

$$= \frac{20 \text{ m/s} - 0 \text{ m/s}}{5 \text{ s}}$$

$$= 4 \text{ m/s}^2$$

▲ *Football players make a ball speed up and slow down.*

▲ *A hockey player accelerates a ball.*

Slowing down: decelerating

If the final speed is lower than the starting speed, it means that the object is slowing down or **decelerating**. The acceleration is negative.

Sinita sees a friend. She slows down from 20 m/s to 10 m/s in 2 seconds.

$$\text{Her acceleration} = \frac{20 \text{ m/s} - 0 \text{ m/s}}{2 \text{ s}}$$

$$= \frac{-10 \text{ m/s}}{2 \text{ s}}$$

$$= -5 \text{ m/s}^2$$

We say that her **deceleration** is 5 m/s^2. This is the same as an acceleration of –5 m/s^2.

Speed–time graphs

You can calculate the gradient of a line on a graph.

- The gradient of a line on a *distance–time graph* tells you about the speed of an object.
- The gradient of a line on a *speed–time graph* tells you about its *acceleration*.

In the graph on the right, the speed is increasing by 10 m/s every second.

- The acceleration is 10 m/s^2.
- The acceleration is **constant** – the line is straight and the gradient, or slope, is constant.
- If the acceleration was bigger then the line would be steeper.
- If the speed does *not* change, then the line is horizontal. The acceleration is zero.

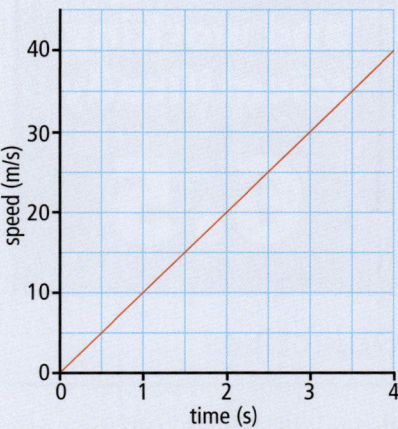

▲ *A straight line shows constant acceleration.*

Terminal velocity

A cyclist is riding a race. He accelerates at the start and gets faster and faster.

As he does so the force of air resistance increases until eventually it balances the forward force. He cannot go any faster – he has reached **terminal velocity**. A speed–time graph tells the story of this journey. At first the line is steep, showing that his speed is increasing very quickly. The line gets less and less steep as his speed changes less quickly. This does *not* mean that he is slowing down. It means that it takes longer for his speed to increase. Eventually he travels at a steady, terminal velocity. The line is horizontal because his speed is no longer changing.

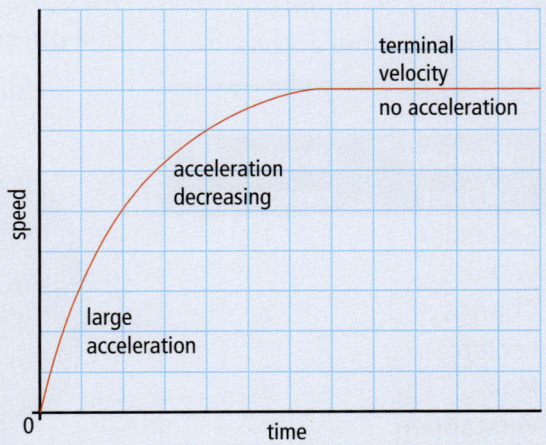

⑦ Questions

1. Explain the difference between speed, acceleration, and deceleration.
2. Copy these sentences, choosing the correct words from each pair.

   The slope of a speed–time graph tells you the **acceleration/distance**. If the line is **straight/horizontal** the speed is not changing. This **is/is not** the same as a speed of zero.

3. A football decelerates from 10 m/s to 0 m/s in 0.1 seconds.
   a. Calculate the acceleration of the ball.
   b. Write down the deceleration of the ball.
4. Use the information on the graph to calculate the acceleration of car A and car B.

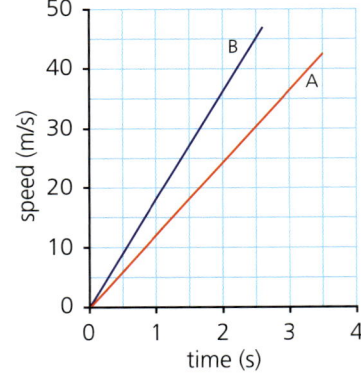

📖 Key points

- Acceleration is change in speed divided by time.
- If an object slows, its acceleration is negative. It is decelerating.
- A straight, sloping line on a speed–time graph shows constant acceleration.
- A horizontal line on a speed–time graph shows constant speed.

Objectives

- Explain which type of graph to plot from different types of data

- Apply ideas about distance–time graphs

Winning driver	Number of races won
Sebastian Vettel	1
Charles Leclerc	2
Max Verstappen	3
Valtteri Bottas	4
Lewis Hamilton	11

Time (s)	Speed (km/h)
0	0
2	100
4	195
6	270
8	310
10	320
12	320
14	320
16	300
18	280
20	300

Presenting data from racing

Formula 1 racing

The first Formula 1 racing driver from India, Narain Karthikeyan, races on race tracks all over the world. Formula 1 cars are designed to reach speeds far greater than ordinary cars.

▲ *Narain Karthikeyan waits to start a race.*

Drawing charts: winners

In the 2019 season there were 21 races in the World Championship. The table shows the number of races won by each driver.

We could show this information in a **bar chart** or a **pie chart**.

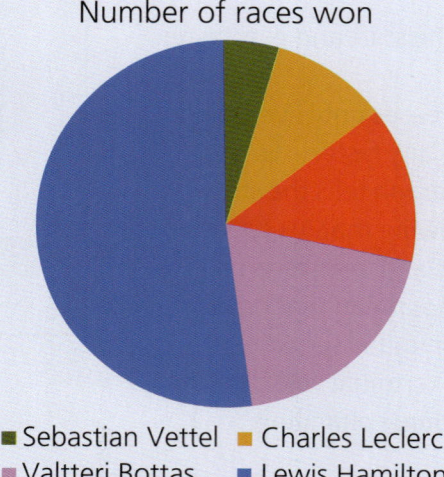

Number of races won

- ■ Sebastian Vettel ■ Charles Leclerc
- ■ Valtteri Bottas ■ Lewis Hamilton
- ■ Max Verstappen

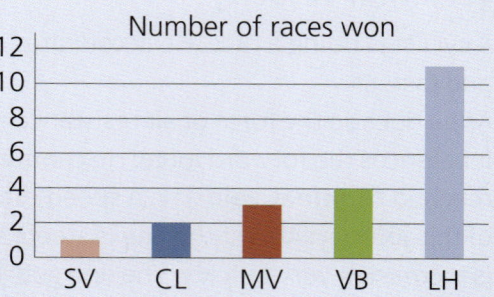

▲ *Bar chart for the same data.*

◄ *Pie chart. It has 21 equal parts, one per race. The slice for each race is 360° ÷ 21 = 17.1°.*

- The name of the driver is a **categoric** variable. Categoric variables are words, *not* numbers.
- The number of races is a **discrete** variable. You can only have whole numbers of races.

If one of the variables in your table is a discrete variable, then you will need to plot a bar chart or pie chart.

A variable that can take any number, like time or distance, is **continuous**.

- The data in both columns must vary continuously to plot a line graph.

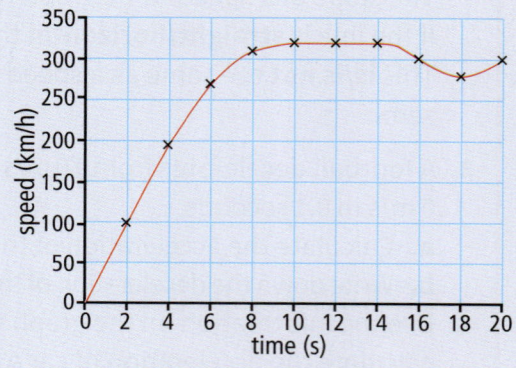

▲ *Speed–time graph for a racing car.*

Drawing line graphs: speed

The speed of a racing car changes as it goes round the track – the car accelerates and decelerates. The speed of a racing car is a continuous variable because it varies all the time.

The table on the previous page shows the speed of a car during the first 20 seconds of a race. Both the table and the graph show that the speed is changing. It is much easier to see *how* the speed is changing from the graph.

- When the line slopes upwards, the car is accelerating.
- When the line slopes downwards, the car is decelerating.
- When the line is flat, the car's speed is constant.

The line is smooth because the speed changes continuously, *not* suddenly.

Time (s)	Distance (m)
0	0
2	28
4	80
6	170
8	330
10	510
12	680
14	860
16	1030
18	1200
20	1360

Drawing line graphs: distance

This table (top right) shows the *distance* that the car travelled in the first 20 seconds of the same race. We can show these data on a line graph too. Like speed, distance changes continuously.

Telling a story

On a distance–time graph for a car:

- when the line is horizontal the car is stationary
- a straight sloping line means that the car is moving at a steady speed
- when the line curves up the car is accelerating.

On a speed–time graph for a car:

- when the line is horizontal the car is moving at a steady speed
- a straight sloping line means that the car is accelerating at a steady rate.

You can calculate the acceleration from a speed–time graph by reading the values of speed and time from the axes.

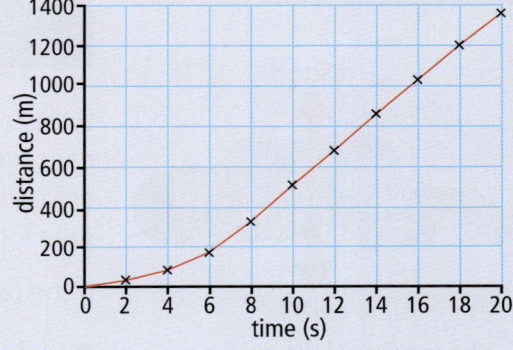

▲ *Distance–time graph for a racing car.*

  Questions

1. Look at the distance–time graph.
   TWS
   a. How far did the car travel in the first 18 seconds?
   b. After what time is the speed constant? Explain your answer.
2. A student wants to plot the number of races that each driver finishes each season. Suggest which type of graph or chart they should use. Explain your answer.
3. The car stops in the pits for 10 seconds. How would this be represented on:
   a. a distance–time graph
   TWS
   b. a speed–time graph?

Key points

- Variables that are names are categoric.
- Variables that can only have certain numerical values, are discrete.
- Variables that can have any numerical value are continuous.
- Pie and bar charts are used when a variable is not continuous.
- Line graphs are used when both variables are continuous.

Balanced and unbalanced forces

You can use forces to explain why an object is moving in the way that it is, or why it is *not* moving. The force of gravity, or weight, is always acting on the diver. Why is she moving in only one of the pictures?

Objectives

- Explain the difference between balanced and unbalanced forces
- Describe the effect of balanced forces
- Describe the effect of unbalanced forces

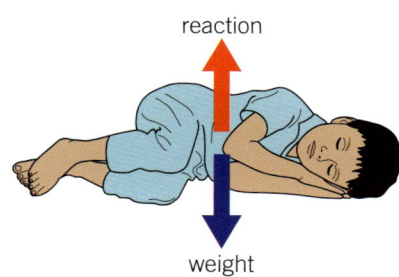

▲ *The boy's weight and the reaction force are balanced. He doesn't move.*

If the forces on an object are the same size but in opposite directions, they cancel out. The forces are **balanced**. It's like a tug-of-war when the teams are equal. The object behaves as if there is *no* force on it. When the diver floats, her weight is balanced by the upthrust of the water.

If the forces on an object are balanced, its motion will *not* change:

- If it is *not* moving it will stay still.
- If it *is* moving it will keep moving at a steady speed.

If the forces on an object are *not* equal and opposite, they are **unbalanced**. If the forces are unbalanced then the motion of the object *will* change:

- If it is *not* moving it will start moving.
- If it *is* moving it will speed up (**accelerate**), slow down (**decelerate**), or change direction.

Resultant forces

If you know the size of each force acting on an object, you can work out the **resultant force**, or **net force**. If the arrows are in the same direction you add the forces. If they are in opposite directions you take away one force from the other.

- If the resultant force is *zero* the forces are balanced.
- If the resultant force is *not zero* the forces are unbalanced.

▲ *The resultant force is 400 N to the right.*

▲ *The resultant force is 800 N to the left.*

▲ *The resultant force is zero.*

Lots of forces

Sometimes more than one pair of forces is acting on an object.

- The weight and the upthrust acting on the boat are balanced. The boat does not move up or down.
- The thrust of the engine is balanced by the air and water resistance. The boat moves forwards with a steady speed.

If an object is stationary or moving at a steady speed, then you know that the forces on it are balanced.

If it is speeding up, slowing down, or changing direction, then the forces acting on it must be unbalanced.

Do you need a force?

Lots of people think that if something is moving, such as a leaf falling at a steady speed, or a football rolling along the ground, then there is a force acting on it. This is not the case. An object can move without a *resultant* force.

The ball...	The forces...	The ball will...
is in contact with the player's foot	are unbalanced	accelerate
leaves the player's foot	are unbalanced	decelerate
is rolling on the ground	are balanced	continue at the same speed

If we could remove friction and air resistance, then the ball would carry on travelling at a steady speed forever. You will learn more about Galileo, who had this idea, on pages 106–107.

▲ *The forces acting on the boat are thrust, water resistance, air resistance, weight, and upthrust.*

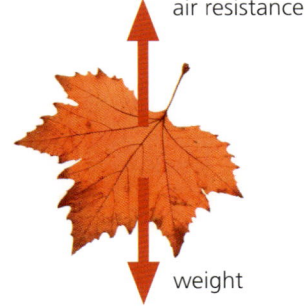

air resistance

weight

▲ *The weight of the leaf and the air resistance are balanced. The leaf falls with a steady speed.*

Questions

1. Copy and complete the table for forces acting on an object. Tick the correct column for each statement.

	Balanced	Unbalanced
there is zero resultant force		
the forces do not cancel out		
the forces cancel out		
there is a resultant force		

2. A cricket player drops a ball, and the forces on it are unbalanced. Describe its motion.

3. Vil says that if a car is moving, then there must be a resultant force acting on it. Alom says there doesn't need to be a resultant force acting on it. Write down who you agree with, and explain your answer.

Key points

- Forces are balanced if they cancel each other out. They are unbalanced if they do not cancel out.
- An object stays still or moves at a steady speed if the forces on it are balanced.
- Unbalanced forces change the speed, direction, or both of a moving object.

6.7

Using forces: Friction

Objectives

- Describe the effect of friction on moving objects
- Describe how to reduce friction
- Describe how friction can be useful

What is friction?

When an object is moving, there will often be a force of **friction** acting on it.

When a child goes down a slide, there is a force of friction between them and the slide. This is because all surfaces, even surfaces that feel very smooth, are uneven. A metal slide looks smooth, but under the microscope you can see how uneven it is.

▲ Friction slows down moving objects.

You need to push things to get them to move.

- If you push a book on a table and it does not move, the forces are balanced.
- The uneven surfaces produce the force of friction that you have to overcome.
- When your force is bigger than friction, the forces are unbalanced.
- Then the book will accelerate.

Reducing friction

People often want to reduce friction, in order to go faster or reduce damage.

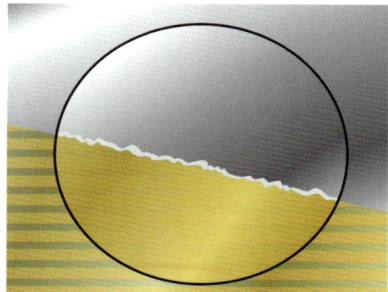

▲ Even smooth surfaces like shiny metal are rough if you look at them under a microscope.

▲ Skiers and snowboarders wax their skis and boards to reduce friction and to go faster.

▲ Cyclists lubricate the chains of their bicycles to reduce friction.

You put oil in a car to reduce friction between engine parts that move. This is called **lubrication**. A layer of oil between two surfaces makes it easier for the surfaces to slide over each other. It reduces the force of friction.

▲ Ball bearings in the wheels slow down surface wear.

There are ball-bearings inside the wheels of a skateboard. They roll over each other as the wheel turns, and reduce friction.

Friction can be useful

Friction is not always a bad thing.

- When you walk, friction acts between your feet and the ground, making it possible for you to move.
- Vehicles too need the force of friction between the tyres and the road to make them move. In icy conditions friction is reduced. Wheels skid because there is not enough friction for them to grip the road.
- Bicycles and cars also need friction to stop. The force of friction acts between their brakes and the wheels. The friction between brake pads or brake blocks and the wheels wears the brakes away and they have to be replaced.

▲ *When there is not much friction, you realise how important friction is!*

Measuring friction

You can use a newtonmeter to measure the force of friction between two surfaces.

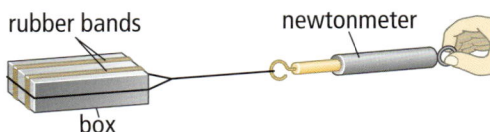

rubber bands   newtonmeter

box

▲ *The newtonmeter reads the force of friction when the box moves at a steady speed.*

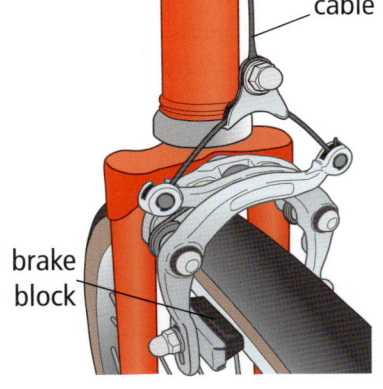

cable

brake block

▲ *Friction between the brake blocks and the wheel helps you to control the speed of a bicycle.*

Questions

1. Name one situation where friction is useful, and one situation where friction is not useful and needs to be reduced.
2. **a.** Explain what causes the force of friction between two surfaces that are sliding over each other.
   b. Explain how lubrication with oil reduces friction.
3. **a.** Ice skates create a thin layer of water between the blade and the ice.
   Suggest how this affects friction and helps the skater.
   b. Suggest why ice skates have a jagged edge at the front.
4. Tyres have treads to remove water from between the tyre and the road.
   Explain why it is important to remove the water.

Key points

- The force of friction slows things down.
- You can reduce friction by lubrication or by using ball-bearings.
- Friction is useful for grip, to start moving, or for braking.

Changing ideas about motion

Objectives

- Describe how explanations about motion were developed

- Explain why ideas take a long time to change

Starting and stopping

Moving objects are slowed down by friction, air resistance, and water resistance.

A moving object still needs a force to balance the forces slowing it down in order to keep it moving. To cycle at a steady speed, the cyclist pushes the pedals around. If he stops pushing on the pedals he will slow down and stop.

▲ *It takes a force to start something moving when it is stationary.*

air resistance and friction balance the driving force

driving force

▲ *The cyclist moves at a steady speed.*

What did people think before?

Over 2400 years ago a Greek philosopher called Aristotle developed ideas about motion. He said that objects move only when they are pushed, and slow down when you remove the force. He did not really understand friction.

His ideas were accepted for a long time.

Asking questions about motion

Galileo lived over 400 years ago. He was interested in experiments about motion. He made lots of observations that did not agree with Aristotle's ideas about motion. He asked this question:

What would happen if there was no friction?

He could not do experiments without friction in a laboratory. There is always some friction between surfaces, even if they are really smooth.

Instead of doing an experiment with equipment, Galileo did a '**thought experiment**'. This can be very useful to work out what might happen in a situation where it is not possible to carry out the experiment.

Galileo's 'thought experiment'

Galileo knew that something rolling down a slope speeds up. He thought that if there was *no* friction and…

…if the slopes were equally steep…	… the ball would reach the same height	
		h same height, h
… if the second slope was longer…	…the ball would reach the same height again…	
		h same height, h

... if the second slope was completely flat	... the ball would continue moving!	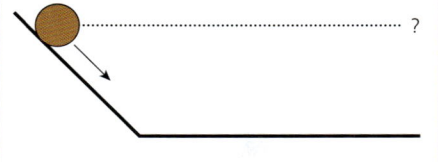

Galileo realised that the ball would keep rolling at a steady speed forever!

He had shown that you do *not* need a force to keep something moving. This is an example of someone using **creative thinking** to solve a problem in physics.

What happened next?

Galileo wrote a book about motion in 1590. The idea that you did not need a force to keep something moving wasn't very popular.

- People saw in the world around them that you did need a force to keep something moving.
- They didn't realise that in most situations the force is needed to cancel out friction or air resistance.
- Without friction and air resistance no force is needed to keep something moving.

It took a long time for his idea to be accepted.

Newton's first law and inertia

About 100 years later, Sir Isaac Newton developed Galileo's work into three laws of motion. **Newton's first law of motion** states that, *unless an external force is applied*, stationary objects will remain stationary, and moving objects will continue to move in a straight line at a steady speed.

Objects resist changes to their motion – this is called **inertia**. This is obvious when you try to move an object with a very large mass.

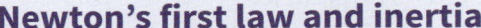

⑦ Questions

1. Imagine that there is no friction for a day. Make a list of things that it would *not* be possible for you do. Which things would still be possible?
2. A boat on a river is moving with a steady speed. The engine is running.
   a. Explain why you need to keep the engine running to move at a steady speed.
   b. Describe what would happen if the engine was turned off.
3. Suggest what would happen if the slopes in Galileo's experiments had been curved rather than straight. Explain your answer.

📖 Key points

- Some questions can't be tested easily using equipment.
- You can use your imagination to do thought experiments.
- Sometimes it takes a long time for an idea to be accepted.

Using forces: Tension and upthrust

Objectives

- Describe what happens when you stretch a spring
- Describe what is meant by elastic limit
- Explain how upthrust is produced
- Explain why things float or sink

▲ A bungee jumper jumps off a platform with a long elastic cord attached to them.

Under tension

Ropes for rock climbing are designed to **stretch** a certain amount, just in case the climber falls. If the rope did not stretch then the climber could be injured. Ropes, cords, and springs all stretch when you pull them.

There is force in the rope that balances the climber's weight. The force in a rope that is being stretched is called **tension**. As you pull on the rope the particles inside are moving apart. The tension force that you feel is the force of attraction pulling the particles back again.

Elastic bands and bungees

A bungee cord is designed to stretch a very long way. It is made of lots of elastic cords all bound together.

- If something is **elastic**, it will go back to its original length when you remove the force. The amount that it stretches is called the **extension**.
- If something does not go back to its original length after stretching, we say it is **plastic**.

Stretching a spring

Springs are used in spring balances, trampolines, and car suspension systems. It is important to use the right spring for the job, so you need to find out how much a spring will stretch when you apply a force to it.

- A bigger force will produce a bigger extension.
- If you double the force you double the extension.
- The extension is **proportional** to the force.
- We say that the spring obeys **Hooke's Law**.

This is why we can use a spring balance to measure forces.

▲ The tension in the rope exerts an upwards force on the climber.

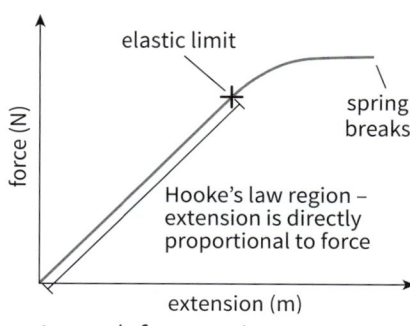

▲ A graph for a spring.

(graph labels: elastic limit, spring breaks, Hooke's law region – extension is directly proportional to force, force (N), extension (m))

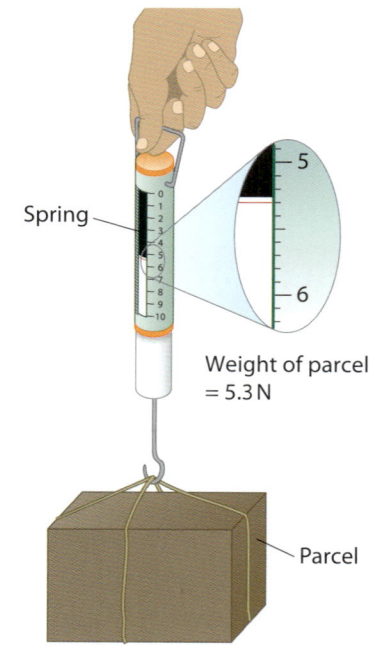

Spring

Weight of parcel = 5.3 N

Parcel

▲ A newtonmeter contains a spring because the extension is proportional to the force.

If you keep loading more and more on, eventually the spring will not return to its original length when you remove the force. It has reached its limit, called the **elastic limit**. The spring cannot spring back and it is **permanently extended**.

Elastic materials will break if the force applied to them is too big.

Why do things float?

◄ *Oil tankers can be 400 m long and have a mass of 500 000 tons fully laden.*

In the water a force called upthrust pushes up on the boat. The upthrust balances the weight so the boat floats.

- Water particles collide with the bottom of the boat.
- They push the boat up.
- A large area produces a big force that can balance a big weight.

Apparent weight

If you weigh an object underwater it appears to weigh less because of the upthrust. It is much easier to pull yourself out of a swimming pool than it is to pull yourself up when you are *not* in water.

In this diagram, the upthrust is 4 N. The water level has risen because the weight has displaced some water. The weight of the displaced water is 4 N. This is **Archimedes' principle**. The *apparent* weight of the object is now 6 N.

Questions

1. A spring is 3 cm long. A student hangs a weight of 2 N on it. It stretches to 4.5 cm long.
   a. Calculate the extension.
   b. Calculate the extension with a weight of 4 N on it.
   c. Calculate the *length* of the spring with a weight of 6 N on it.
2. A floating boat weighs 20 000 N. Write down the magnitude and direction of the upthrust.
3. When people do a bungee jump, they are asked how much they weigh. Explain why.

Key points

- Springs, ropes, and elastic materials have a tension in them when stretched.
- The extension is proportional to the tension, up to the elastic limit.
- Springs in tension pull back.
- Objects float when the upthrust equals the weight.

Objectives

- Describe how to present results in tables
- Describe how to draw lines of best fit and identify anomalous results
- Explain which data points are reliable

How can I find out the elastic limit of a spring?

Presenting data from springs

The elastic limit

Suma is on the trampoline. The label says that the weight limit for the trampoline is 1200 N.

The springs must be below their elastic limit when the weight is 1200 N.

▲ *A trampoline contains many springs.*

Recording results

Suma collects a spring, some weights, and a ruler. She draws a table ready to record all of her repeat measurements. Each column has a heading with a unit.

Weight (N)	Extension (cm)			Mean extension (cm)
	1	2	3	

Suma measures the length of the spring with no weights on it. This is the original length.

- She records zero force and zero extension in her table.
- She adds one 1 N weight to the spring and measures the new length.
- To find the extension she subtracts the original length from the new length, and writes the result in her table.
- She repeats the experiment by adding weights until she reaches 8 N.
- She writes the result in her table.
- Suma calculates the mean extension for each weight.

Weight (N)	Mean extension (cm)
0	0
1	1.1
2	2.0
3	2.5
4	4.2
5	5.4
6	6.8
7	8.5
8	11.0

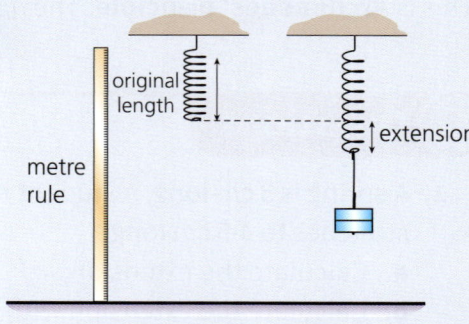

What sort of graph?

Suma knows that to find a pattern in her results, she needs to plot a graph.

- Both weight and extension are continuous variables.
- She plots a line graph of the weight and the mean extension.
- She draws a **line of best fit**.

Her line starts at the origin (0, 0) because when the force is zero the extension is zero.

Her line was straight at the start, but then curved up as the load increased.

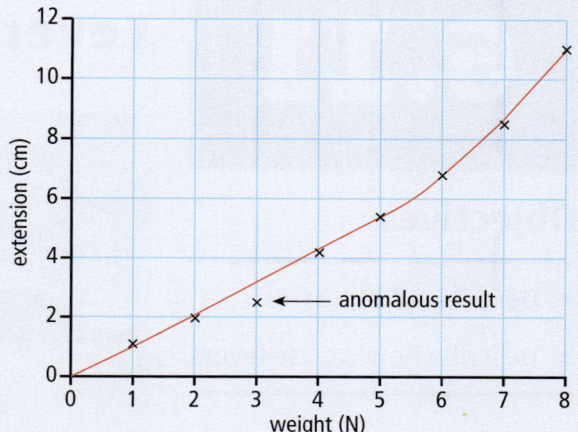

Points that don't fit

One point was far away from the line. This is called an **anomalous result**. Suma probably measured the extension wrongly. Anomalous results:

- should be identified, as shown on the graph
- should be repeated or checked.

Suma applied a force of 6 N to the spring and then removed it. The spring did *not* go back to its original length. She had found the elastic limit!

Are the results reliable?

Suma can be confident in the conclusions based on her data because:

- she repeated each measurement three times
- repeated measurements make the data more accurate and reliable.

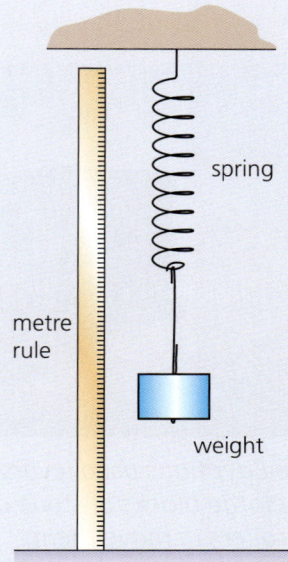

 Questions

1. Describe what you should write at the top of each column of your results table.
2. Describe how to draw a line of best fit.
3. Identify the force that was the weight for Suma's anomalous result.
4. Jamila did a different experiment. She put 2 N weights onto lots of different coloured springs and measured the extension of each spring.
   a. Draw the table that she would need to record her results.
    b. Explain why she could *not* draw a line graph for her results.

 Key points

- A results table shows what you are measuring and the unit.
- A line of best fit goes through most points and is a smooth line or curve.
- Anomalous results are results that do not fit the pattern.
- Repeating measurements makes results more accurate.

6.11

Levers

Objectives

- Describe what a lever is
- Describe how we use levers

▲ *The pyramids at Giza were built over 4500 years ago.*

When the ancient Egyptians built the pyramids there were no cranes or big machines. How did they do it? We know that they carried huge blocks of stone long distances by putting them on rollers and pulling them along, but how did they get them onto the rollers?

They used **levers**. Levers work because some forces produce a **turning effect**.

Turning forces

Forces can change the motion of objects, and they can also make things turn. Every time you close a door you are applying a **turning force.** When you use a wheelbarrow or sit on a see-saw you are applying a turning force.

▲ *The Egyptians used levers to get large blocks of stone on to rollers to move them.*

Levers

A lever is a bar that can turn when you exert a force on one end.

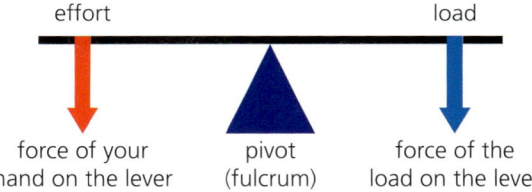

- The force that you exert on the bar is called the **effort**.
- The weight of the object that you lift is called the **load**.
- The bar rotates about a **pivot** (or **fulcrum**).

You can use a lever to lift a block of stone. When you push down on one side of a lever you produce a **turning force**. This will make the other end of the lever move up, lifting the stone.

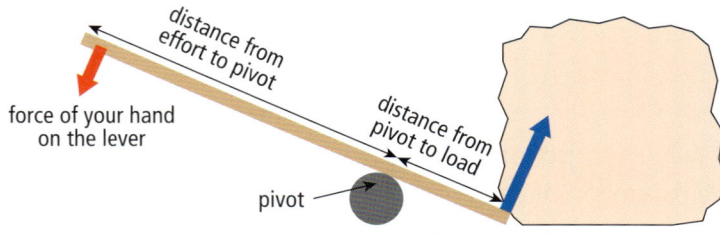

▲ *The lever is a force multiplier.*

A lever works because:

- the distance from the effort to the pivot is bigger than the distance from the load to the pivot

- you push down with a force that is much *smaller* than the weight of the stone to move it.

Science in context

Levers in the world around you

Levers are not always straight bars.

Other examples of force multipliers include using a hammer to remove a nail, or using a wheelbarrow. In a wheelbarrow, the load is *between* the effort and the pivot.

Some levers are **distance multipliers**. You move your hands (or fingers) a smaller distance than the load moves. For example, using tongs to hold a test tube or beaker: if you push the handles together a small amount with your fingers, the ends of the tongs move much further. The distance is bigger, and the force is smaller.

▲ *A wheelbarrow is a force multiplier.*

▲ *A hammer taking out a nail is a force multiplier.*

▲ *Tweezers reduce the force.*

Questions

1. Circle the names of objects that act as force multipliers.

   lever   tongs   wheelbarrow

2. The picture shows how you could make a lever to lift a stone using a plank of wood.

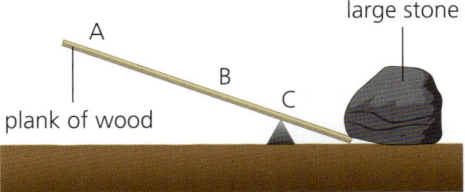

   a. Which letter (A, B, or C) is the pivot?
   b. Which letter (A, B, or C) shows where you should apply the least effort to lift the stone?
   c. Which letter (A, B, or C) shows where you should apply the most effort to lift the stone?

3. Using tongs to hold things in a chemistry lab means that we need a *bigger* force. If this is the case, explain why we use them.

Key points

- Forces can turn an object clockwise or anticlockwise about a pivot.
- A lever can be used as a force multiplier, or a distance multiplier.
- Levers can produce forces that are much bigger than the effort, to lift heavy loads.

6.12 Calculating moments

Objectives

- Identify turning forces, or moments
- Calculate moments
- Use the law of moments to identify equilibrium

▲ *You use a turning force when you open or close a door.*

On the high wire

Walking on a tightrope can be difficult and dangerous. The tightrope walker uses a long pole to help him balance. If he feels that his body is going to tilt one way, he can adjust the pole to tip him the other way. The pole cancels out the turning force to help him balance.

▲ *The pole is used to alter the turning forces.*

What is a moment?

The turning effect of a force is called a **moment**. The size of the moment depends on the force being applied and how far it is from a pivot.

- Moment (Nm) = force (N) × perpendicular distance from the pivot (m)
- A moment is measured in **newton metres** (Nm).

A boy uses a force of 5 N to close a large door. The handle is 1.5 m from the hinge (the pivot). What is the moment of the force?

Moment = force × distance

$$= 5\,N \times 1.5\,m$$

$$= 7.5\,Nm$$

Balancing

When two people balance on a see-saw, we say that it is in **equilibrium**. This means that the **anticlockwise** moments and **clockwise** moments are the same. This is called the **principle of moments**:

If an object is balanced, the anticlockwise and clockwise moments are the same.

Example 1

Sunita and her mother sit on a see-saw (left). Is it balanced?

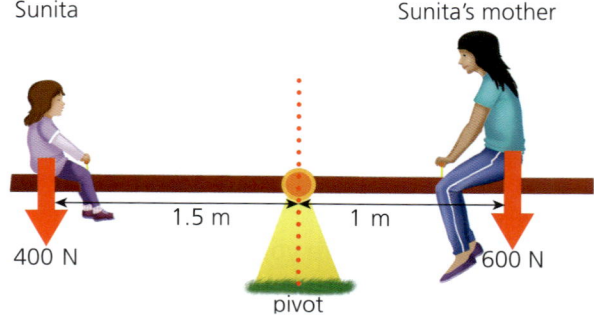

Sunita   Sunita's mother

1.5 m   1 m

400 N   600 N

pivot

Anticlockwise moment = 400 N × 1.5 m	Clockwise moment = 600 N × 1.0 m
= 600 Nm	= 600 Nm

The anticlockwise moment and the clockwise moment are equal, so the see-saw is balanced.

Example 2

Davina is balancing some 10 N weights on a see-saw. Is it balanced?

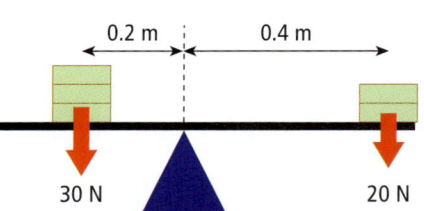

Anticlockwise moment = 30 N × 0.2 m	Clockwise moment = 20 N × 0.4 m
= 6 Nm	= 8 Nm

The anticlockwise moment and the clockwise moment are *not* equal, so the see-saw is not balanced.

Balancing a see-saw

You can use the principle of moments to work out how much weight you need to balance a see-saw or where you need to put the weight.

Problem 1: what's the weight?

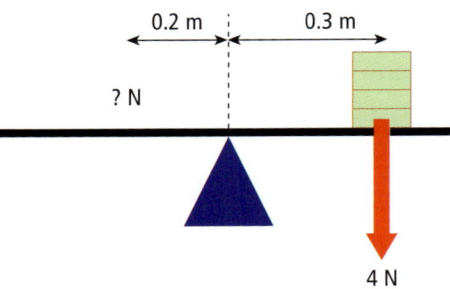

Anticlockwise moment = clockwise moment

$$? N × 0.2 m = 4 N × 0.3 m$$

$$? N × 0.2 m = 1.2 Nm$$

$$? N = \frac{1.2\ Nm}{0.2\ m}$$

Force = 6 N

Problem 2: what's the distance?

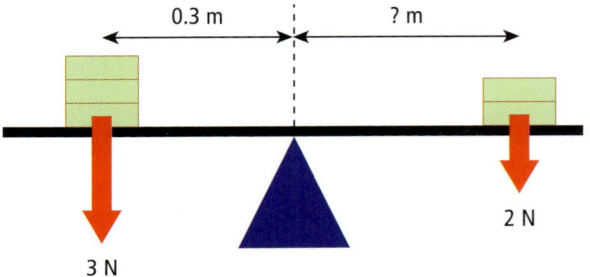

Anticlockwise moment = clockwise moment

$$3 N × 0.3 m = 2 N × ? m$$

$$0.9 Nm = 2 N × ? m$$

$$\frac{0.9\ Nm}{2\ N} = ? m$$

Distance = 0.45 m

Questions

1. Copy and complete the sentences using these words:
   small   big   moments   equilibrium
   On a see-saw, a _____ force acting a short distance from the pivot will be balanced by a _____ force acting a long distance from the pivot. Then the see-saw is in _____.
   This is the principle of _____.

2. Explain why the handle of a door is not close to the hinges.

3. For Davina's experiment in Example 2 above, write down which way the see-saw will turn.

4. Citra has a weight of 450 N. She sits 1.2 m from the pivot of a see-saw. How far away from the pivot must Ari sit on the other side to balance the see-saw if his weight is 600 N?

Key points

- The turning effect of a force is a moment.
- moment = force × distance.
- A see-saw is in equilibrium if the anticlockwise moments balance the clockwise moments. This is the principle of moments.

Objectives

- Define centre of mass (or centre of gravity)
- Explain why some objects topple over easily and others do not

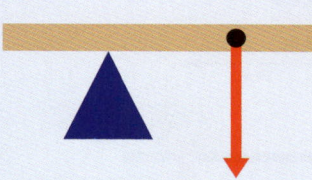

▲ *The plank turns.*

Centre of mass and stability

Centre of mass

If you want to carry a plank of wood on your shoulder, you need to place it so that your shoulder is underneath the *middle* of the plank. There is a point in an object through which all of the weight of the object seems to act, called the **centre of mass** (or **centre of gravity**). If this point is above your shoulder, which is the pivot, there is no turning force acting on the plank of wood.

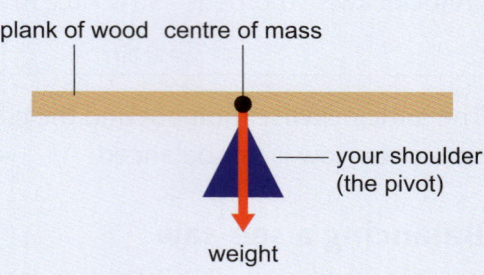

▲ *The plank balances.*

- If you move the centre of the plank away from your shoulder, there will be a turning force on the plank.
- The plank's weight acts through its centre of mass.
- This produces a turning force that turns the plank clockwise.
- To balance it you will need to apply another force with your hand.
- This force will turn the plank anticlockwise and balance its weight.

The centre of mass of a regular shape, such as a plank or a ball, is in the *centre* of the shape because the mass is evenly distributed throughout the object. For an irregular shape such as a broom, the centre of mass of will not be in the centre.

Why do objects topple over?

We can work out why some objects are easier to topple than others by thinking about the position of the centre of mass.

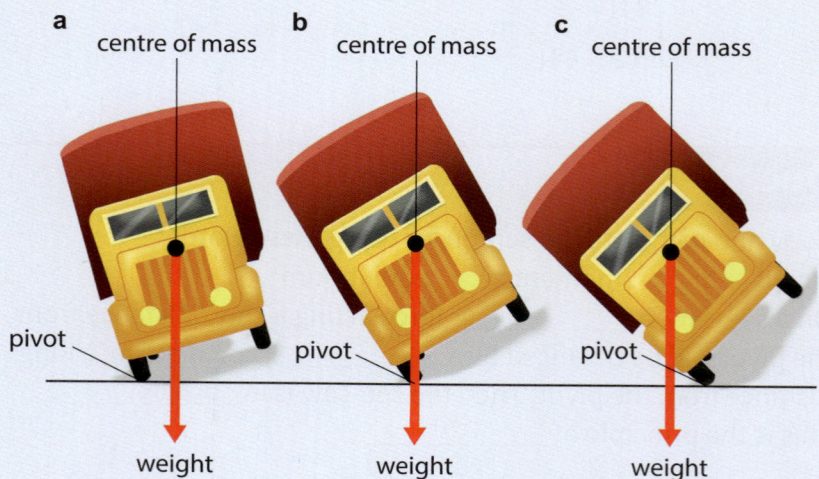

▲ *The acrobats balance so that there is no net turning force acting on the group and the bicycle.*

a As you start to push the toy lorry over, the centre of mass moves toward the pivot. If you stop pushing it at this point, it will fall back.

b When the centre of mass is above the pivot the lorry it will balance, but this is hard to achieve.

c When the centre of mass is the *other* side of the pivot, the lorry will topple over.

Stability

If an object is in a position that makes it difficult to topple over, we say that it is in a **stable** position.

It is harder to topple objects with a low centre of mass than objects with a high centre of mass.

- If the centre of mass is higher, the object will be less stable.
- The object does not need to be tipped as far before the centre of mass moves outside the pivot.
- A car has a lower centre of mass than a truck.
- High winds can blow over a tall vehicle like a truck, but are less likely to topple a car.

centre of mass

Moving your centre of mass

You can demonstrate how your centre of mass can move. If you bend over to pick up a chair, you naturally lean back slightly so that your centre of mass stays over your feet. If you stand against a wall and try and pick up the chair, you will find it impossible. You cannot lean forward and still keep your centre of mass over your feet. You feel you are going to topple over.

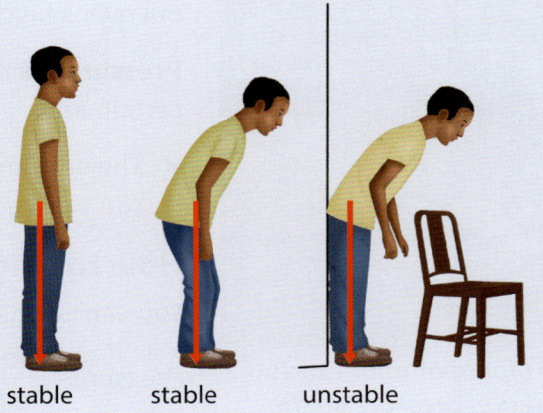

stable   stable   unstable

▶ *In July 2012 the world record for human mattress toppling was set by 1001 volunteers in Shanghai, China. The gaps between the mattresses were fixed so that when one toppled over it knocked the next one over. Each person falls safely because the mattresses are soft.*

Questions

1. Describe where the centre of mass of a cube is.
2. If you lean back in a chair, the two back legs act as a pivot. You can only lean so far back without falling over. Explain why in terms of your centre of mass and a turning force.
3. Sketch a diagram of a brush with a long handle that you use to sweep the floor. Label where you think the centre of mass might be. Explain your answer.
4. Explain why it becomes more and more difficult for the acrobats on the bicycle to keep their balance as more acrobats climb to the top.
5. It is almost impossible to balance a pencil on its point. Explain why.

Key points

- The centre of mass (or centre of gravity) of an object is the point through which the weight of the object seems to act.
- Objects with a high centre of mass are easier to topple than those with a low centre of mass (gravity).

6.14

Pressure

Objectives

- Explain what causes pressure
- Calculate pressure

▲ *The hippopotamus exerts a greater pressure on the ground than the bird.*

You can also use these equations:

Force = pressure × area

$$\text{Area} = \frac{\text{force}}{\text{pressure}}$$

What is pressure?

Earthmovers use wide tracks instead of ordinary tyres so that they can move across mud without sinking in.

An earthmover is very heavy. It has a weight of about a million newtons, equal to about 15 000 people! If you stand on the same muddy ground you might sink. The earthmover does *not* sink because its **weight** is spread out over a bigger **area**.

▲ *Wide tracks reduce the pressure on the ground.*

Pressure is a measure of the force exerted by an object or substance over a certain area.

- The earthmover's tracks have a much bigger area than your feet.
- You exert a greater pressure even though your weight is much smaller.

How to calculate pressure

You can calculate pressure using this equation:

$$\text{Pressure} = \frac{\text{force}}{\text{area}}$$

Force is measured in newtons (N) and area is measured in metres squared (m²), so pressure is measured in newtons per metre squared (N/m²). 1 N/m² is also called 1 **pascal** (Pa).

It is often easier to measure smaller areas in centimetres squared (cm²). If you measure the area in cm² and the force in N, then the pressure is measured in N/cm².

When you do calculations it is very important to look at the units of area. If you write them next to the number in your equation, then you will see which unit of pressure you need to use.

A force of 100 N is spread over an area of 2 m². What is the pressure?

$$
\begin{aligned}
\text{Pressure} &= \frac{\text{force}}{\text{area}} \\
&= \frac{100 \text{ N}}{2 \text{ m}^2} \\
&= 50 \text{ N/m}^2
\end{aligned}
$$

A force of 20 N is spread over an area of 4 cm². What is the pressure?

$$
\begin{aligned}
\text{Pressure} &= \frac{\text{force}}{\text{area}} \\
&= \frac{20 \text{ N}}{4 \text{ cm}^2} \\
&= 5 \text{ N/cm}^2
\end{aligned}
$$

Large and small pressure

Sometimes it is useful to *increase* the pressure.

The studs on a football boot have a small area compared with the area of the foot. This produces *more* pressure, so that the studs sink into the ground and help the footballer to grip the ground and move more easily.

▲ *Football boots have studs on the bottom.*

The same force over a smaller area produces a much bigger pressure.

The weight of a footballer is 800 N.

The area of his two feet is 200 cm².

$$Pressure = \frac{force}{area}$$

$$= \frac{800 \text{ N}}{200 \text{ cm}^2}$$

$$= 4 \text{ N/cm}^2$$

reducing the area increases the pressure →

The total area of the studs is 20 cm².

$$Pressure = \frac{force}{area}$$

$$= \frac{800 \text{ N}}{20 \text{ cm}^2}$$

$$= 40 \text{ N/cm}^2$$

At other times it is useful to *reduce* the pressure.

The weight of an earthmover is 1 000 000 N.

The total area of four normal tyres is 2 m².

$$Pressure = \frac{force}{area}$$

$$= \frac{1\ 000\ 000 \text{ N}}{2 \text{ m}^2}$$

$$= 500\ 000 \text{ N/m}^2$$

increasing the area reduces the pressure →

The total area of the earthmover tracks is 25 m².

$$Pressure = \frac{force}{area}$$

$$= \frac{1\ 000\ 000 \text{ N}}{25 \text{ m}^2}$$

$$= 40\ 000 \text{ N/m}^2$$

The same force over a bigger area produces a much *smaller* pressure.
40 000 N/m² is 4 N/cm², which is much less than the pressure of the studs.

Questions

1. Copy and complete this table. Don't forget the units!

Force	Area	Pressure
150 N	25 cm²	
60 N	15 m²	
5 N	0.1 cm²	

2. A pressure of 0.01 N/cm² is produced by a book with an area of 300 cm². What is the weight of the book?

3. A brick has a weight of 15 N. If the pressure it produces is 0.5 N/cm² what is the area in contact with the ground?

4. A hippopotamus produces a pressure of 250 000 Pa when it is standing on all four feet. If its weight is 40 000 N, what is the area of each foot?

Key points

- Pressure $= \dfrac{force}{area}$.
- Pressure is measured in N/m² (Pa, pascal) or N/cm².
- A small force over a big area produces a small pressure, and vice versa.

6.15 Using pressure

Objectives

- Describe how large pressure can be useful
- Describe how small pressure can be useful

Where is a lot of pressure useful?

 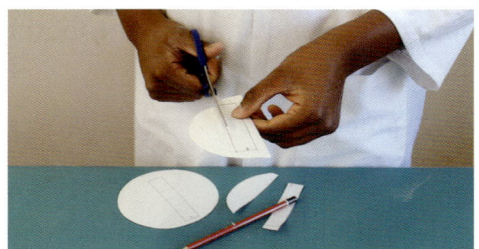

You can use ideas about pressure to explain how some tools work.

- The blades of knives and scissors have a very small area.
- The force that you apply produces a large pressure.
- If the blade becomes blunt then the area gets bigger.
- A bigger area means the pressure is smaller, and may *not* be high enough to cut things.
- The end of a nail is pointed, so it has a very small area.
- The pressure produced on a surface when you hit the nail with the hammer will be very large.

Lots of tools have sharp edges so that they can cut or make holes in things easily.

The force applied to a knife is 20 N

The area of the sharp knife blade is 0.5 cm².

$$\text{Pressure} = \frac{\text{force}}{\text{area}}$$
$$= \frac{20\,\text{N}}{0.5\,\text{cm}^2}$$
$$= 40\,\text{N/cm}^2$$

The area of the blunt knife blade is 1.25 cm².

$$\text{Pressure} = \frac{\text{force}}{\text{area}}$$
$$= \frac{20\,\text{N}}{1.25\,\text{cm}^2}$$
$$= 16\,\text{N/cm}^2$$

Where is small pressure useful?

It is possible to sink into surfaces like sand and mud if the pressure is too large.

Wading birds have wide flat feet to reduce the pressure on the ground. This means that they do *not* sink so much when they have to walk over a soft surface.

It is much easier for camels to walk across sand than it is for horses to walk across sand. The area of a camel's foot is much larger than the area of a horse's hoof.

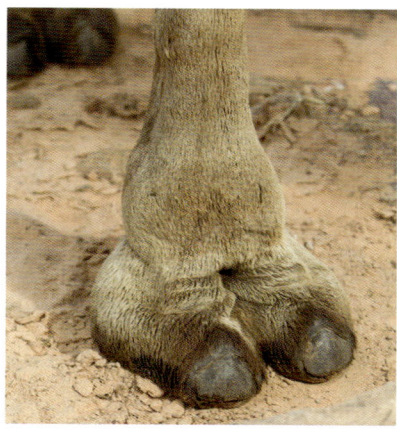

▲ Camels' feet have evolved not to sink in sand.

The weight of a camel is 5000 N.

The area of its four hooves is 2000 cm².

$$\text{Pressure} = \frac{\text{force}}{\text{area}}$$
$$= \frac{5000 \text{ N}}{2000 \text{ cm}^2}$$
$$= 2.5 \text{ N/cm}^2$$

The weight of a horse is 4000 N.

The area of its four hooves is 400 cm².

$$\text{Pressure} = \frac{\text{force}}{\text{area}}$$
$$= \frac{4000 \text{ N}}{400 \text{ cm}^2}$$
$$= 10 \text{ N/cm}^2$$

Even though a horse has a smaller weight than a camel, the pressure that it produces on the ground is four times bigger because the area of its hooves is smaller.

Handles and straps

Wide handles are more comfortable to carry than narrow ones. Their larger area reduces the pressure on your hands.

Straps on rucksacks are made wide to spread the force over a big area.

Bike tyres

Some bikes are designed to travel across mud and others are designed to be used on the road. The tyres of road bikes are much narrower than those of an off-road bike.

▲ *Wide straps reduce pressure.*

▲ *A mountain bike...*

▲ *... and a road bike.*

Questions

1. Explain in terms of force, pressure, and area why knives do not work when they are blunt.
2. Explain why the feet of camels and wading birds have the shapes that they have.
3. Two birds have the same mass but one has feet with twice the total area of the other. Describe the difference between the pressure that they produce on the ground. Explain your answer.
4. Describe what would happen if you tried using a road bike to ride over a soft surface. Explain your answer.

Key points

- Increasing the area reduces the pressure, which can be useful in some situations.
- Reducing the area can be useful if you need a large pressure.

6.16

Pressure in liquids

Objectives

- Explain liquid pressure in terms of particles
- Explain why liquid pressure increases with depth

Particles and liquid pressure

If you go swimming you can feel the pressure of the water. Your ears detect **liquid pressure**.

Pressure in liquids is different from the pressure produced when one solid is on top of another. Liquid pressure is due to the forces between the liquid particles and the surfaces of a container.

▲ *However you move your head under water, the pressure feels the same.*

- The particles in a liquid are very close together, so liquids cannot be **compressed**.
- We say that they are **incompressible**.

Isha fills a syringe full of water. He holds one finger over the hole so the water cannot get out. Then he presses down on the plunger. Nothing happens! It is *not* possible to force the liquid into a smaller space.

When Isha applies a force to a liquid the forces between the particles increase, and the forces act in all directions. Isha demonstrates that the pressure in a liquid acts in all directions.

- He fills a bag with water and makes holes in it.
- When he squeezes, the water comes out of the bag at 90° to the surface.
- The curved path of the water is due to gravity.

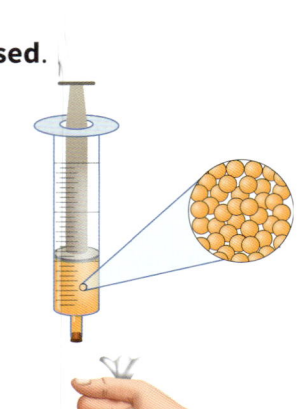

It is liquid pressure that produces **upthrust**, the force that keeps things afloat.

▲ *Liquid pressure forces the water out of the bag at right angles.*

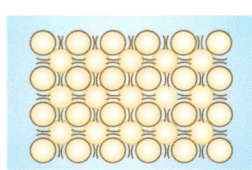

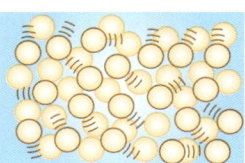

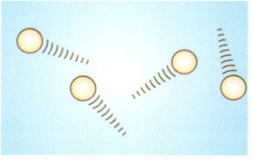

▲ *The particle model describes the different arrangements of particles in solids, liquids, and gases.*

Liquid pressure and depth

The pressure on a fish at the bottom of a lake is bigger than the pressure on a fish half-way down.

- Liquid pressure at a point depends on the weight of water above it
- There is twice as much water above a point at the bottom of a lake than there is half way down.
- The liquid pressure at the bottom of a lake is double the pressure half way down.

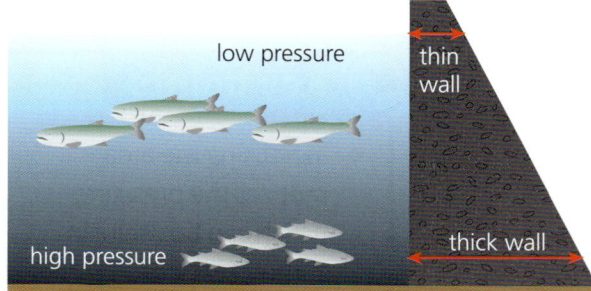

▲ *Dam walls are thicker at the bottom.*

This is why the bottom of a dam wall has to be many times thicker than the top. The pressure at the bottom is much, much bigger.

▲ *A diving vessel protects divers against the pressure in deep water.*

When people go deep underwater they use diving vessels or submarines with very strong walls to withstand the pressure.

Measuring pressure

The pressure in a liquid or gas is measured with a **pressure gauge**.
This contains a tube that is curled up. The tube straightens out as the pressure inside the tube increases. This moves a needle to show a reading on a scale.

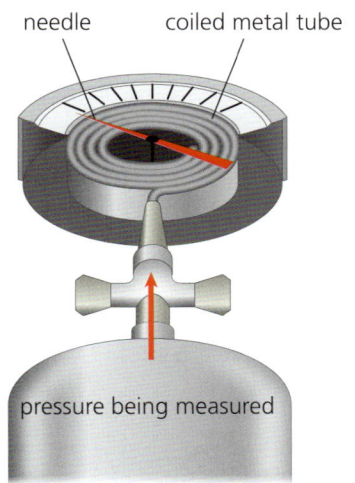

needle   coiled metal tube

pressure being measured

Questions

1. Describe what produces pressure in a liquid.
2. Explain why the pressure at the bottom of the Pacific Ocean is much bigger than the pressure near the surface.
3. A student fills a plastic bottle full of water. He makes a hole in the bottle half-way up, and another hole at the bottom.
   a. Copy and complete these sentences:
      The water will come out faster from hole
      _____. This is because _____.
   b. Explain why the water comes out at right angles to the surface of the bottle.
   c. Describe and explain what happens to the speed of the water leaving the bottle from the bottom hole as the water drains out.
4. Suggest why pressure gauges have a maximum pressure that they can measure.

hole A

hole B

Key points

- Pressure in liquids results from the forces between particles in the liquid acting over an area.
- Deeper in a liquid the pressure gets bigger because there is a greater weight of water above you.
- You can measure liquid pressure with a pressure gauge.

6.17 Pressure in gases

Objectives

- Explain gas pressure in terms of particles
- Explain the factors that affect gas pressure

David blows up a balloon. The balloon gets bigger and bigger. Eventually the balloon will burst. Why does this happen?

▲ *A balloon bursts.*

Gas pressure and particles

We can explain gas pressure using the particle model. In the particle model particles in a gas are far apart, and are moving fast.

- David pumps air particles into the balloon.
- Particles bump into each other and into the rubber of the balloon.
- The collisions with the walls of the container produce **gas pressure**.

All gases, including gases inside you, exert a pressure on the walls of their container.

What affects gas pressure?

Adding more gas

Air particles hit the outside of the balloon too.

- By blowing *more* air into the balloon, David *increases* the pressure inside it.
- If there are more collisions on the inside than on the outside the balloon will get bigger.
- When too much gas has been added, the balloon bursts.

If you put air into a container that has a fixed shape the pressure will increase.

Changing the volume or temperature

Gas pressure changes when the number of collisions between gas particles and the container walls changes. There will be more collisions if:

- the particles are moving faster
- the particles are in a smaller volume.

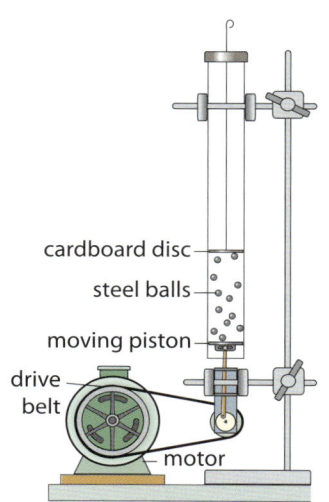

▲ *Model of gas pressure in which the cardboard disc is pushed up by the collisions of the steel balls.*

cardboard disc

steel balls

moving piston

drive belt

motor

Decreasing volume		Increasing temperature	
Particles in a big volume	Particles in a small volume	Particles in a cold gas	Particles in a hot gas
When the volume is smaller there are *more collisions*, and the pressure is higher		When the temperature is higher, particles move faster and there are *more collisions*, and the pressure is higher.	

Atmospheric pressure

The air around us exerts gas pressure on our bodies all the time. This is called **atmospheric pressure**. We do *not* feel the pressure because it is balanced by the pressure of the gases and liquids in our bodies pushing outwards.

Lots of air inside...

...Less air inside

◄ *Marshmallows contain pockets of air that expand when you pump out the air around them.*

▲ *You add air particles but the tyre doesn't get much bigger, so the pressure inside goes up.*

The atmospheric pressure at sea level is bigger than the atmospheric pressure high up a mountain.

- Gravity pulls the air particles towards the **Earth**.
- The air near the Earth's surface is compressed.
- If the particles are closer together, then the pressure is bigger.
- This means that there will be more collisions producing a bigger force and a bigger pressure near the surface.

People can use oxygen tanks in places where oxygen levels are low. The tanks contain oxygen gas compressed into a small volume. Some firefighters use oxygen tanks.

- The gas has to be compressed because otherwise they would not be able to carry the number of containers of oxygen that they need.
- The pressure inside the oxygen tanks is very high.
- The tanks are made of metal which can withstand the high pressures.

▲ *Oxygen compressed into tanks helps the firefighter to breathe.*

Questions

1. Describe in terms of particles what causes gas pressure.
2. Explain why you can compress a gas but you cannot compress a liquid.
3. Explain why a balloon might burst on a hot day.
4. Aircraft cabins are 'pressurised'. Explain why air is gradually pumped into the aircraft as it climbs, and then let out again when it lands.
5. Explain why the pressure inside a bicycle tyre increases as you pump it up.
6. Explain what would happen to the pressure in a gas if you compressed it into a volume that was one third of the size.

Key points

- Gas pressure is caused by particles colliding with the walls of the container.
- If you compress a gas into a smaller volume, or heat it, the pressure will increase.
- Atmospheric pressure decreases as you go up a mountain.

Diffusion in liquids and gases

Objectives

- Use the particle theory to explain diffusion in liquids and gases
- Describe evidence for diffusion

Rebekah is cooking dinner for her family.

Very soon, everyone in the house can smell the food. Why?

Food particles evaporate as Rebekah is cooking. They move around randomly in the air, and spread out. The food particles mix with air particles. Soon there are food particles all over the house. Some of the food particles enter your nose, which detects the smell.

air particle ———

——— food particle

▲ *Food particles diffuse into the air.*

Diffusion and particles

The random movement and mixing of particles is called **diffusion**. Particles in gases and liquids are moving. You do *not* need to move or stir a substance to make diffusion happen. The speed of mixing by diffusion depends on three factors:

- temperature
- size and mass of the particles
- the states of the substances that are diffusing.

Diffusion in gases

Diffusion happens quickly in a gas. This is because a particle can travel a long distance in a gas before it hits another particle.

▲ *Particles from an incense burner diffuse throughout a room.*

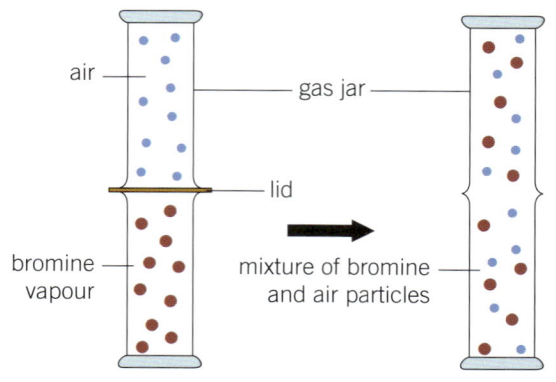

air — gas jar — bromine vapour — mixture of bromine and air particles — lid

▲ *When you remove the lid between the jars, the air and bromine diffuse into each other.*

Diffusion through liquids

Mo puts a crystal of potassium manganate(vii) in a beaker of water. He watches carefully. The purple colour starts to spread through the water. The next day Mo looks at the mixture again. The purple colour has spread evenly

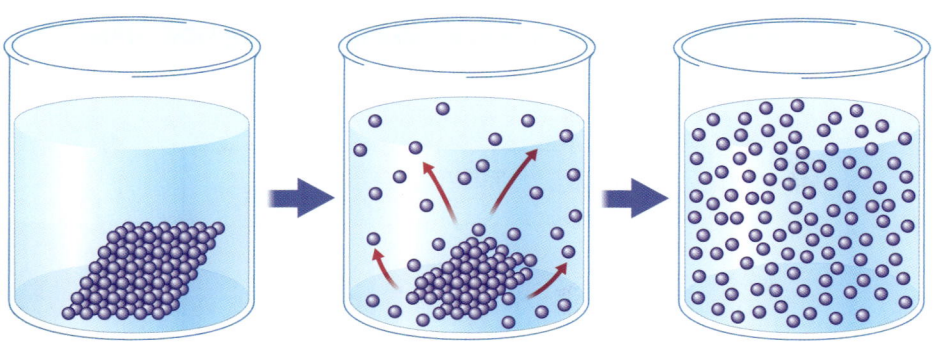

 The particles from the crystal diffuse into the water.

through the water. The particles of the purple crystal have moved away from the crystal and mixed with the water particles.

Diffusion happens more slowly in liquids than in gases. This is because particles are closer in liquids, they collide more often, and there are stronger forces between them.

Diffusion and temperature

Particles from warm food diffuse more quickly than those from cool food. The particles in a warmer gas or liquid move faster. Diffusion is faster in hotter liquids or gases.

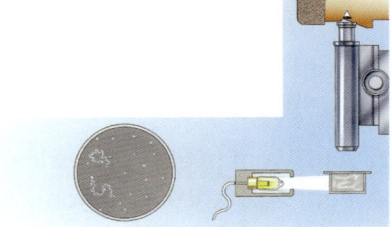

You can observe the 'Brownian motion' of smoke particles in air.

🏫 Science in context

Evidence for moving particles

In 1828, Robert Brown suspended pollen grains in water. He looked at them through a microscope. The pollen grains moved around quickly in a random motion now called Brownian motion. Why?

The pollen grains are pushed around by the random movements of the water particles around them. But how? Water particles are tiny compared to pollen grains. The answer lies in the speed of the water particles – on average, a water particle moves faster than 1600 km/h at 20 °C.

❓ Questions

1. Explain, in terms of particles, why diffusion happens.
2. Put the following situations in order of speed of diffusion from slowest to fastest:

   hot liquid   cold gas   hot gas   cold liquid

3. Explain how and why temperature affects the speed of diffusion.
4. Suggest why it is difficult for solids to diffuse.

📖 Key points

- Diffusion is the random movement and mixing of particles.
- Diffusion happens faster at higher temperatures.
- Diffusion is quicker in gases than in liquids.

1. Rafal went to visit his grandmother in a plane, which flew a distance of 1440 km.

   a. The journey took 2 hours. Calculate the average speed of the plane. **[2]**

   b. When the plane landed it had to slow down. Write down the scientific word for slowing down. **[1]**

   c. On the way home the plane has to fly into the wind. Describe and explain the effect on its average speed. **[2]**

2. Aadi went to visit Dev. The graph shows how Aadi's distance from home varies with time.

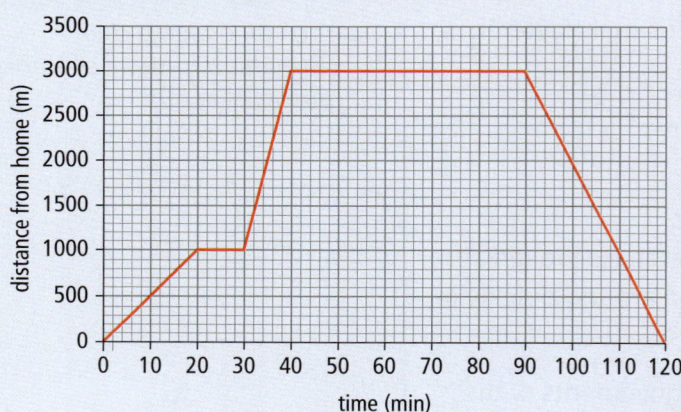

   Use the graph to deduce:

   a. the distance from Aadi's home to Dev's home **[1]**

   b. the time Aadi spent at Dev's home **[1]**

   c. on which part of the journey Aadi travelled the fastest **[1]**

   d. Aadi's speed walking back to his home in m/s. **[2]**

3. A student drops a stone off a cliff. Look at the graph of distance against time for the stone.

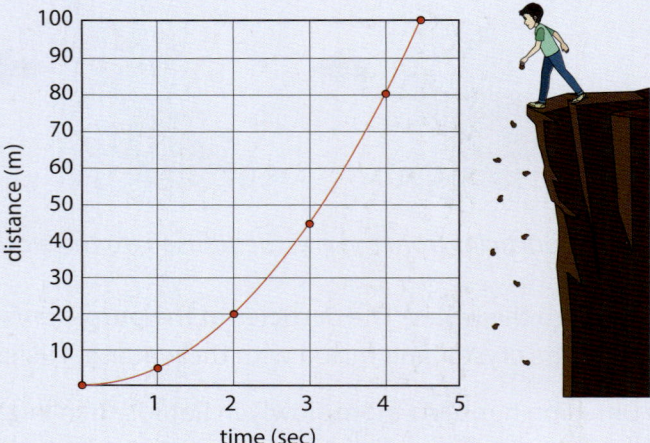

   a. Describe how you can tell from the graph that the stone was speeding up as it fell. **[1]**

   b. Calculate the average speed of the stone. **[2]**

   c. Is the average speed bigger or smaller than the speed of the stone just before it hit the ground? **[1]**

4. What is the difference between **balanced (B)** and **unbalanced (U)** forces?

   Copy the table below and put a tick in the correct column for each statement. **[7]**

	B	U
The object is not moving		
There is only one force acting on the object		
The object is accelerating		
The two forces are the same size but in opposite directions		
The object is slowing down		
The two forces are different sizes and in opposite directions		
The object is moving with a steady speed		

5. Arya is hanging some weights on an elastic band.

a. Give the letter of the correct table for collecting his results. **[1]**

A

Weight	Extension

B

Mass (kg)	Extension (m²)

C

Mass (kg)	Extension (cm)

D

Weight (N)	Extension (cm)

b. Define 'elastic'. **[1]**

c. Arya finds that if he hangs a weight of 2 N on the elastic band the extension is 1.5 cm, and when he hangs a weight of 4 N on it the extension is 4 cm. Is it a good idea to use an elastic band in a newtonmeter? Explain your answer. **[2]**

6. Two glasses are filled with liquids to the same height. The first glass is filled with water and the second glass is filled with fruit juice, which has a greater density than water.

Put these pressures in order from the smallest pressure to the biggest pressure.

A pressure at the bottom of the fruit juice

B pressure at the bottom of the water

C pressure halfway up the glass of water.

7. It is very important to make measurements that are both precise and accurate. Which of these measurements of time is the most precise? **[1]**

5.1 seconds   5.14 seconds   5 seconds

8. A car mechanic is using a hydraulic jack to lift a car. This is a simple diagram of a hydraulic jack.

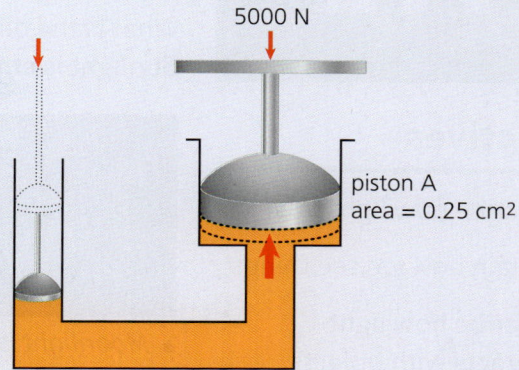

5000 N

piston A
area = 0.25 cm²

The weight of the car is 5000 N. The area of the piston that lifts the car is 0.25 m².

a. Calculate the pressure needed to lift the car. **[2]**

The area of the piston that the mechanic pushes down is 0.01 m².

b. Calculate the force the mechanic needs to use to lift the car. **[2]**

c. Explain why liquids, not gases, are used in machines that use pistons to lift cars. **[1]**

9. Calculate the moment of each of these forces:

a. a force of 10 N acting 2 m from a pivot **[2]**

b. a force of 2 N acting 0.4 m from a pivot **[2]**

c. a force of 0.1 N acting 0.2 m from a pivot. **[2]**

10. Jamal is playing with his friends on a see-saw. He sits at the end of the see-saw and his friends take turns to sit on the other end, an equal distance from the pivot. Jamal has a weight of 400 N.

a. Jamal's friend Maria, who has a weight of 350 N, sits on the other end. Describe what will happen to the see-saw. **[1]**

b. Jamal's brother has a mass of 200 N. Calculate where he should sit on the see-saw to make it balance. **[2]**

c. Ryan sits on the other end of the see-saw. He has a mass of 450 N. Can Jamal make the see-saw balance? **[1]**

Seeing things

What is the difference between light from the Sun and light from the Moon? Sunlight is **emitted** by the Sun. Moonlight is *not* emitted by the Moon.

Objectives

- Describe how light travels
- Explain how we see things
- Describe how light interacts with objects

▲ *Moonlight is light that came from the Sun.*

What is light?

Light is a way of transferring energy.

Light sources give out light. We say they are **luminous**.

Energy is transferred from the Sun to the Earth by light waves. Light can reach us from the Sun and other stars. It can travel through empty space because, unlike sound, light does *not* need a **medium**, a substance, to travel through. **Infrared radiation**, sometimes called heat or **thermal energy**, is very similar to light and is another method by which energy reaches us from the Sun.

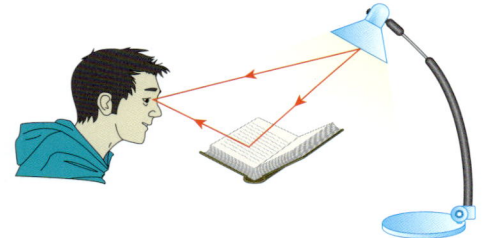

▲ *You see a book because it reflects light.*

How do we see things?

The Sun is luminous. It emits light. In the same way, you can see other stars at night because the light from them enters your eyes.

Objects that don't give out light are **non-luminous**. You can see them because light is reflected from them into your eyes. Sometimes people think that only mirrors reflect light, but all objects reflect at least some light. This is how we see them.

How does light travel?

You can demonstrate that light travels in straight lines using holes cut in card. You line up the holes so that light can travel in a straight line from the candle to your eye.

You can think of light travelling on a journey from a **source** to a **detector**.

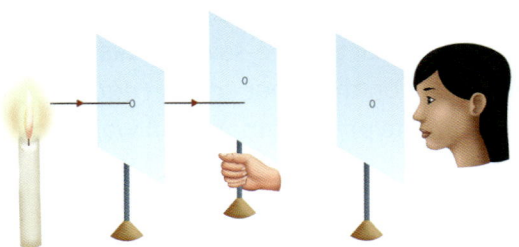

▲ *You cannot see the candle unless all three holes are exactly lined up.*

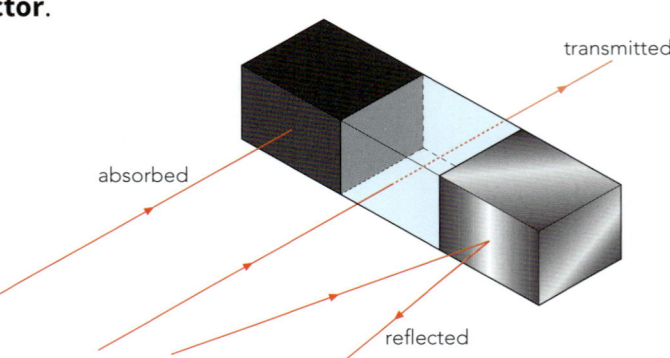

▶ *The behaviour of light depends on the type of material.*

Light can be **transmitted** by (pass through) objects it hits, or be **reflected** (turned back), or it can be **absorbed**. Light that is absorbed is neither transmitted nor reflected. It stays inside the object. Energy cannot disappear, so the energy of the light is transferred into thermal energy. Absorbing light makes the object heat up a bit.

Light spreads out as it moves away from the source. This means that less light will hit a certain area if it is further from the source. Some people think that light 'runs out', but that is *not* the case. The source appears dimmer as you move away from it because the light is spread out more.

What does light travel through?

You can see through some materials, but *not* others.

These words describe some ways in which different materials behave with light:

▲ *You can see through clear glass. It is described as **transparent**.*

▲ *Frosted glass lets light through, but you cannot see through it. It is described as **translucent**.*

▲ *Bricks do not let light through. They are described as **opaque**.*

📖 Key points

- Light travels in straight lines.
- We see things when light emitted by a luminous source or reflected by a non-luminous source enters our eyes.
- Light is emitted by sources and can be transmitted, reflected, or absorbed by transparent, translucent, or opaque materials.

The law of reflection

How is light reflected? How can you find out? We use a beam of light to work out the law of reflection.

Objectives

- State the law of reflection
- Use the law of reflection to describe how light is reflected

What is the law of reflection?

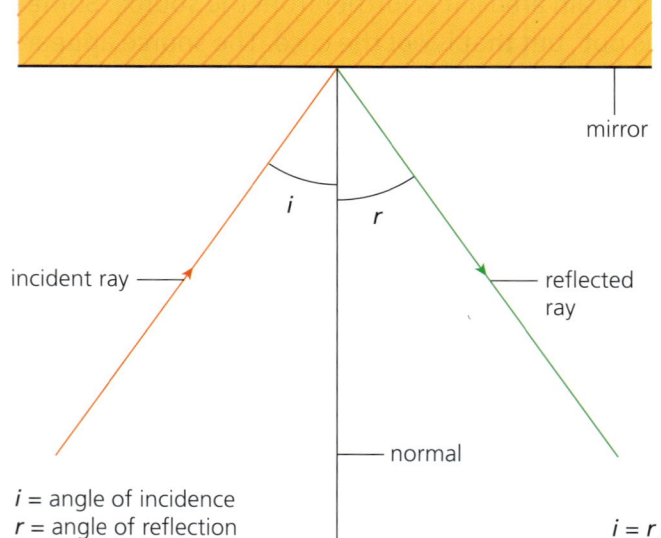

i = angle of incidence
r = angle of reflection
i = r

▲ *Beams of light reflect from a mirror.*

Light reflects off a smooth shiny surface, such as a mirror, in a regular way. A ray of light, called the **incident ray**, hits a mirror at a particular angle.

You can measure this angle.

- First you draw a line at right angles to the mirror at the point where the ray hits it.
- This line is called the **normal**.
- The **angle of incidence** is the angle between the normal and the incident ray, and is usually labelled *i*.
- The ray reflects off the mirror.
- The angle between the normal and the **reflected ray** is the **angle of reflection**, labelled *r*.

The angle of incidence is *not* the angle between the mirror and the ray.

The angle of incidence = the angle of reflection. This is the **law of reflection**.

Making measurements

Arjun and Diya are investigating the reflection of light in a mirror. They set up a ray box, a device that shines a ray of light. Arjun places a mirror to reflect the ray of light.

Diya draws a diagram of the normal and the mirror. Arjun puts the mirror on top of the diagram and shines the ray box at the mirror. He makes sure the ray hits the mirror where the normal is drawn.

How are we going to take measurements?

We could draw the mirror and the normal on some paper.

'How can we measure the angle of incidence and the angle of reflection?' wonders Diya. Arjun suggests drawing the rays on the diagram. He draws two dots on the paper under each ray.

He takes the ray box away and Diya joins up the dots to draw the rays. Now she can measure the angle of incidence and the angle of reflection.

They move the ray box to make different angles. Each time they draw the angle of incidence and the angle of reflection. Diya records their results in a table.

Angle of incidence (°)	Angle of reflection (°)
10	12
30	29
50	51
70	68

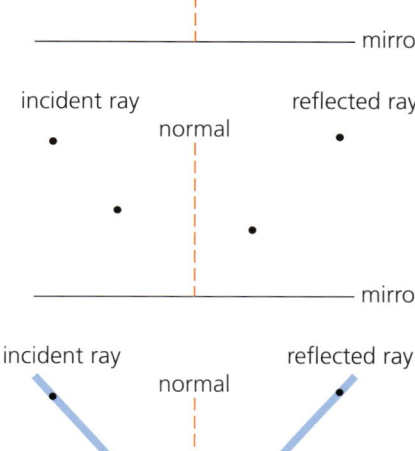

Mirrors in periscopes

Periscopes can be useful things. A periscope contains two mirrors. It enables you to look over the top of a wall, or over other people if you are at the back of a crowd.

▲ *A periscope is used in a submarine.*

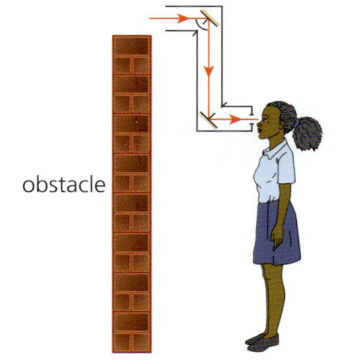

▲ *Parallel mirrors are used to make a periscope.*

Questions

1. **a.** Define the angle of incidence and the angle of reflection.
   b. State the law of reflection.
2. **a.** Name two things that Arjun and Diya did to get accurate measurements of the angles in their experiment.
   b. Suggest one improvement that they could make to their method that would make the results more accurate or precise.
3. Look at the drawing of the periscope. What are the angles of incidence and reflection at each mirror?
4. A student does an experiment with a mirror. She measures the angle between the *mirror* and an incident ray and finds that it is 30°.
   a. What is the angle of incidence?
   b. What is the angle of reflection?

Key points

- The angle of incidence equals the angle of reflection. This is the law of reflection.
- Mirrors are used in periscopes.
- To take accurate measurements about the law of reflection you need to mark carefully where the rays go.

Reflection and images

Objectives

- Use the law of reflection to explain how images are formed
- Describe the different types of image

Looking in a mirror

We often see reflections. Water can behave like a mirror and reflect animals or the sky.

◄ *You can see reflections in a variety of different places.*

▲ *A plane mirror produces an image that is the same size as the object.*

Adinda is looking in a flat mirror, or **plane mirror**. We call her reflection an **image**. It is the right way up, or **upright**, the same size, and the same colour as Adinda. If she moves closer to the mirror the image moves closer, and if she moves away the image moves away. The image always appears to be the *same* distance behind the mirror as Adinda is in front of it.

Types of image

Images can be:

- **virtual** – Adinda needs to look in the mirror to see her image, so it is virtual
- **real** – an image that you project onto a screen is real, because you don't need to look through anything to see it.

If Adinda holds up her left hand, the image she sees in the mirror appears to be holding up her right hand. We say that the image we see in a mirror is **laterally inverted**.

In these pictures the word AMBULANCE is written backwards on the front but not on the back. When a driver sees the ambulance in the rear-view mirror, the word appears the correct way round.

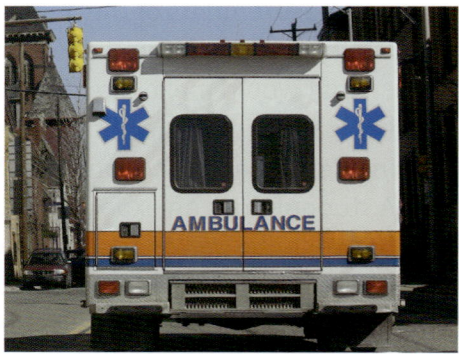

▲ *The word on the front can be read as a laterally inverted image.*

How is the image in a mirror formed?

You see an image because your brain assumes that light travels in straight lines, even when the light has in fact been reflected.

A **ray diagram** helps to show how the image is formed. A ray diagram is a model of what is happening. We choose two paths that the light takes to help us to work out where the image is.

- Light is emitted by a candle and spreads out.
- Some light from the candle is reflected from the mirror.
- The brain works out where the light appears to come from if it travels in straight lines.
- This is where it sees the image.

When you look in the mirror the light from the top of your face appears to come from the top of the image. The light from the bottom of your face appears to come from the bottom of the image. Your image is the right way up.

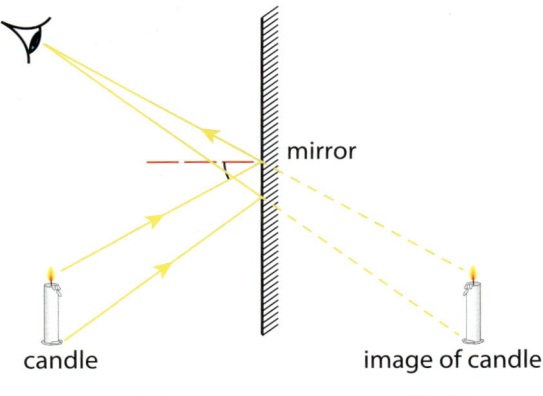

candle   image of candle

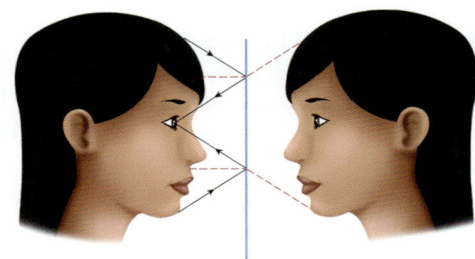

Adinda's face   virtual image

Types of reflection

All surfaces, rough or smooth, reflect some light. The image that you see in a mirror is very clear. An image on water will only be clear if the water is flat. If the water's surface is very rough you do *not* see an image at all. The light must be reflected in a regular way for you to see an image.

A piece of white paper reflects most of the light that hits it, but you cannot see an image in it. This is because the surface is rough and the reflection is **diffuse**. The light is scattered from the surface.

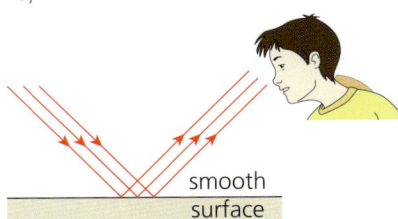

smooth surface

▲ *Reflection from a smooth surface (regular reflection).*

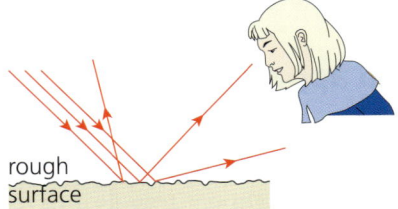

rough surface

▲ *Reflection from a rough surface (diffuse reflection).*

🥽 Science in context

Scientists have developed reflective materials that are used on clothing. This helps people to stay safe if they are working or cycling at night. Some roads have reflective studs to mark their edges.

📖 Key points

- You see an image when light reflects from a smooth surface such as a mirror, water, or glass.
- The image is virtual and laterally inverted.
- Reflections from rough surfaces are diffuse.

❓ Questions

1. **a.** Describe the difference between inverted and upright images.
   b. Describe the difference between virtual and real images.
2. Write down all the CAPITAL letters that look the same in a mirror.
3. Explain why you can see:
   a. your face in a shiny saucepan, but not in a painted white wall
   b. a faint reflection in the surface of a shiny plate or cup.
4. Explain why the image of the candle above is virtual.

Refraction

Refraction of light

Objectives

- Describe how light is refracted at a boundary between air and water

- Explain why light is refracted

▶ *Is this pencil really bent? Why does it look bent?*

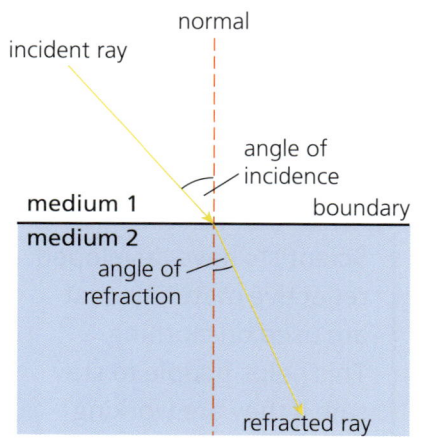

▲ *A ray diagram shows what happens when light is refracted.*

When light travels from one medium (substance) to another, such as from air to water, it can change direction.

- The 'bending' of light is called **refraction**.
- The change of direction happens at the boundary, or interface, between two media.

The angle between the normal and the ray reaching the boundary is the angle of incidence. The angle between the ray and the normal on the other side of the boundary is the **angle of refraction**.

- If light goes from a *less* dense medium to a *more* dense medium, like from air to water, the direction of the ray will move *towards* the normal.
- If light goes from a *more* dense medium to a *less* dense medium, like from water to air, then it will move *away from* the normal.

Real and apparent depth

Light is reflected off a rock at the bottom of the pool and travels to the surface. It changes direction when it travels into the air.

When light reaches your eye, your brain assumes that it has travelled in a straight line. So you see the rock as being in a different place. The depth that the rock appears to be is called the **apparent depth**, which is shallower than the **real depth**. Refraction explains why a pool looks shallower than it actually is.

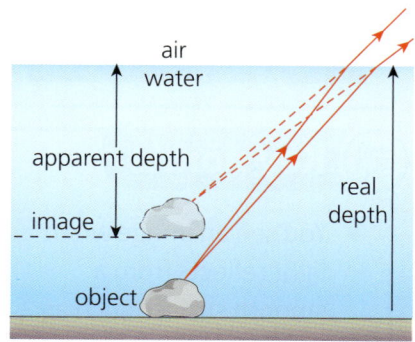

▲ *A pool looks shallower than it really is.*

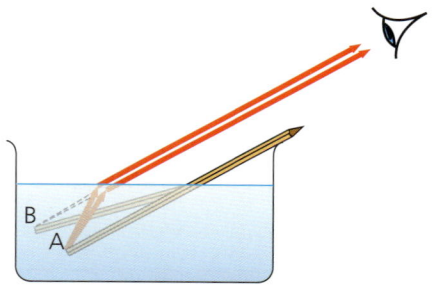

▲ *The bent pencil isn't really bent. The refraction of the light from the end of the pencil at A makes it look as if the light is coming from B.*

Why does refraction happen?

You can model what happens to light when it goes into water using people.

- People are marching in a carnival.
- They are moving from tarmac onto sand.
- As each person reaches the sand they slow down.
- This changes the direction of the row, because there is a change in speed.

The same thing happens to light when it is refracted.

- In air, light travels at about 300 million m/s, but in water it only travels at about 230 million m/s.
- Light going from air into water slows down.
- Light coming out of water into the air speeds up.
- In both cases the direction will change.

How much the light is refracted depends on how much the speed changes. The speed in different materials depends on the **density** of the material. A more dense material will slow light down more, so the light will be refracted more.

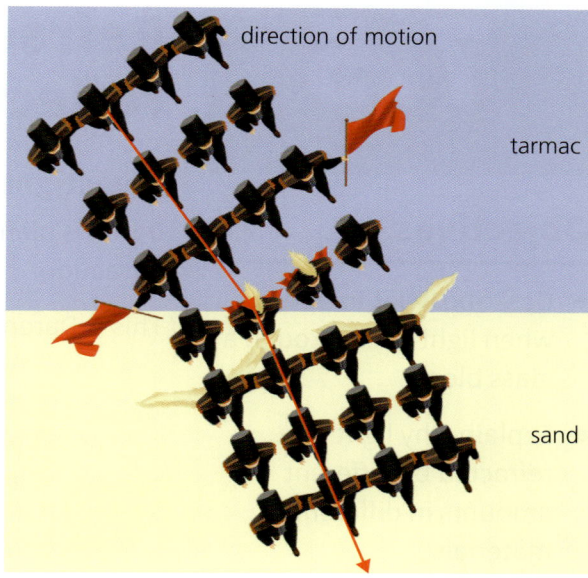

▲ *Rows of marching people make a model of refraction.*

The speed of light and refraction

When light goes from air into a denser medium such as water or glass, it slows down. Scientists work out a ratio for the two speeds called the **refractive index**.

$$\text{Refractive index} = \frac{\text{speed of light in a vacuum (empty space)}}{\text{speed of light in the medium}}$$

Refractive index does *not* have a unit because it is the ratio of two speeds. The speed of light in air is almost the same as the speed in a vacuum.

The refractive index is high if the light slows down a lot. The refractive index of diamond is 2.4, and the refractive index of glass is 1.5. This means light is refracted more when it goes into diamond than when it goes into glass.

▲ *Diamond is denser than glass, so light travels slower in diamond than in glass.*

Questions

1. Think about the carnival procession marching onto sand. If the rows marched straight towards the sand, and *not* at an angle, what would happen to:
   a. their speed
   b. their direction?
2. A hunter is standing on the edge of a river trying to spear a fish. Should he aim above or below where he sees a fish? Explain your answer.
3. Calculate the refractive index of water using the information on this page. What assumption are you making?

Key points

- Light changes direction when it goes from one medium to another.
- This is called refraction.
- Refraction is a result of light changing speed.

7.5 Refraction and total internal reflection

Objectives

- Describe what happens when light goes through a glass block
- Explain why light is refracted by different amounts in different materials

Investigating refraction

Daren is planning an investigation into refraction.

This glass block feels a lot heavier than this plastic one. I wonder if there is a difference in how they refract light.

This is Daren's **prediction**.

> I predict that the glass block will refract the light more than the plastic block. This is because the glass is heavier than the plastic.

Daren's friend Amadi looks at Daren's prediction. Daren then rewrites it:

> I predict that the glass block will refract the light more than the plastic block.
> This is because the glass is denser than the plastic. When light goes from air into a dense material, it slows down and is refracted. Denser materials will slow light down more, so light will be refracted more by the glass than the plastic.

You are right. I think that it is to do with the density. I have looked it up in a book and the glass is denser than the plastic.

But Daren, you haven't explained why you think that will happen using what we know about the speed of light.

When you make a prediction you are *not* just saying *what* you think will happen. You should say *why* you think that will happen, using scientific knowledge. The scientific knowledge that you use in this way is called a **hypothesis**. Daren can test his prediction by doing an experiment to collect data. Later, he can use the scientific knowledge that he has to explain his **conclusion**. A conclusion says what you have found out.

Refraction in a glass block

When a ray of light goes from air into glass, it is refracted.

- As the ray goes from the air into the glass it *slows down*. The ray bends *towards* the normal. The angle of refraction is *smaller* than the angle of incidence.

- As the ray goes from the glass into the air it *speeds up*. The ray bends *away from* the normal. The angle of refraction is *bigger* than the angle of incidence.

The ray going into the block is parallel to the ray coming out of the block. If the ray travels along the normal it does *not* change direction, but it will still slow down.

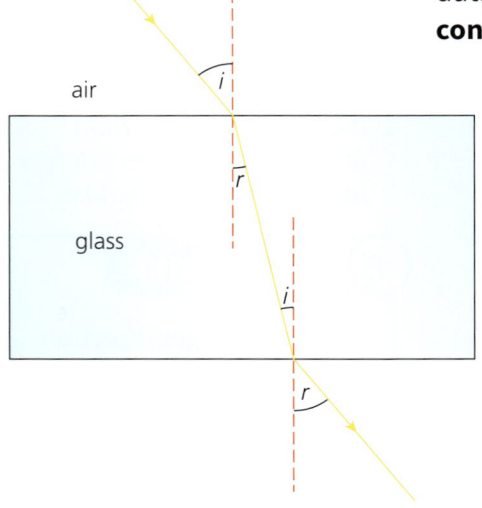

▲ *The path of a ray of light through a glass block.*

Total internal reflection

When light travels from a more dense substance into a less dense substance, such as from glass or water into air, the angle of refraction is *bigger* than the angle of incidence.

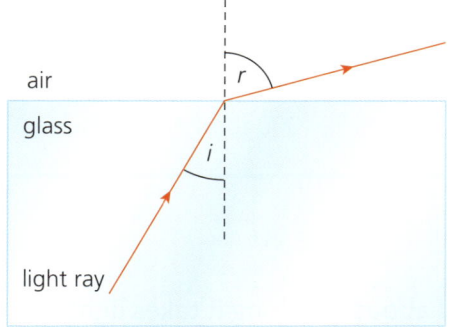

As the angle of incidence increases, so does the angle of refraction...

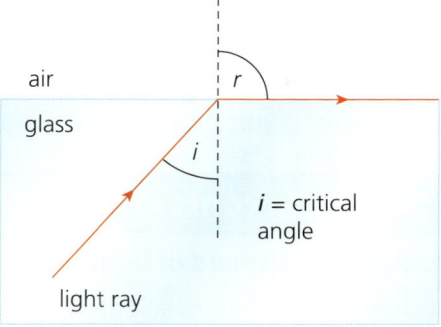

i = critical angle

... the angle of refraction reaches 90° when the angle of incidence is at the **critical angle**.

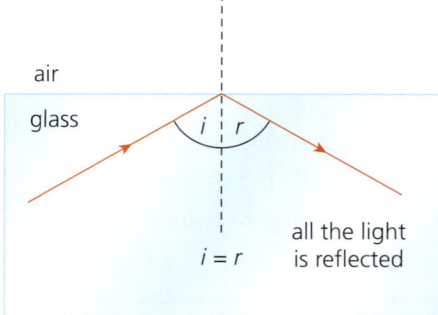

all the light is reflected

i = r

When the angle of incidence is bigger than the critical angle, there is **total internal reflection**.

Using total internal reflection

An **optical fibre** is a very thin fibre made of glass or plastic. It behaves like a 'light pipe'. If you shine light into one end, the light is totally internally reflected down the fibre, which means that all of the light stays inside the fibre.

An **endoscope** is used to see inside the human body. Light is sent down one bundle of optical fibres, and shines on a structure inside the body. Another bundle of optical fibres transmits the light reflected from the inside of the body to a camera, so the doctor can see an image and diagnose the problem.

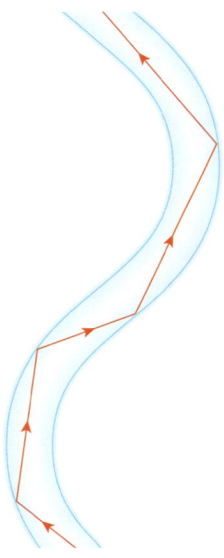

▲ Light is reflected inside an optical fibre.

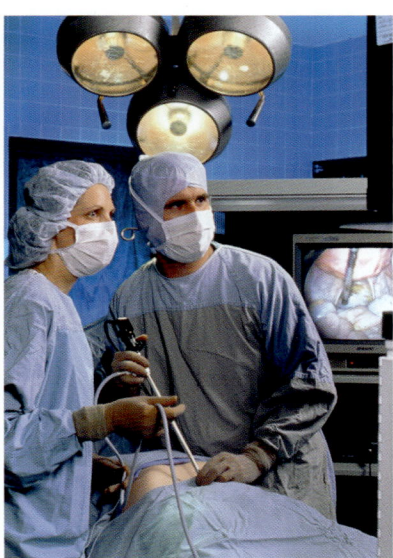

▲ Doctors can see inside a body using total internal reflection.

⑦ Questions

1. Explain why Daren's second prediction was better than his first.
2. Look at the diagram on page 138 showing the path of a light ray. Explain why the ray leaving the block is parallel to the ray going in.
3. Anyam drew a diagram to show how light travels down an optical fibre made of a material with a critical angle of 50°. Describe two things that are wrong with his diagram.

📖 Key points

- Light is refracted when it goes from air into glass, or glass into air.
- Denser materials refract light more than less dense materials.

Objectives

- Know how fast light travels

- Understand how astronomers use the speed of light to describe distances

The speed of light

How fast does light travel?

The **speed of light** is the fastest speed possible. Nothing travels faster than light.

There are lots of ways of writing down the speed of light.

300 million m/s   300 000 000 m/s   300 000 km/s   300 thousand km/s

Science in context

Measuring the speed of light

For thousands of years scientists argued about how long it takes light to travel. With sound you notice a time delay, but you cannot detect a delay between light being emitted and light being detected.

Over 2000 years ago the Greek philosopher Empedocles thought that light did take time to travel, but another Greek philosopher Aristotle disagreed – he thought that light travels instantaneously.

In 1021, the Arab philosopher Ibn Al-Haytham argued that, because light reflects off objects into your eye, it must take a certain amount of time to do so even if we cannot sense it.

It was not until there were methods for measuring very, very short periods of time that accurate measurements of the speed of light could be made.

The metre and the speed of light

For nearly 200 years a piece of metal was kept in Paris, France, that was exactly 1 m long. People would check the length of their rulers against this bar. In 1983 the definition changed. An international committee decided that one metre is 1/299792458 of the distance that light travels in 1 second. So the units of distance that we use are related to the speed of light.

Using speed to measure distance

Rajiv reads that light takes 1.3 seconds to travel from the Moon to the Earth. He wonders if he can work out how far away the Moon is.

$$Speed = \frac{distance}{time}$$

so distance = speed × time

= 300 000 000 m/s × 1.3 seconds

= 390 000 000 m or 390 000 km

Distances in astronomy

Suppose that you travelled at the speed of an aeroplane, 900 km/h. The table below shows how long it would take you to reach different objects in the Solar System.

Object	Distance in km	Time in an aeroplane	Distance in light-time
Moon	384 400	18 days	1.3 seconds
Sun	150 000 000	19 years	8 minutes
Mars	62 000 000–399 000 000	8–50 years	between 4 and 20 minutes
Neptune	4 500 000 000	600 years	4 hours 10 minutes

▲ *Saturn is over a billion (1 000 000 000) km from Earth.*

▲ *Andromeda is the nearest galaxy to the Milky Way.*

Light travels approximately 9 500 000 000 000 km in a year. We call this distance a **light year**.

It is much more convenient to write down huge distances in light minutes, light hours, or light years rather than metres or kilometres.

Object	Distance (km)	Distance (light years)
Voyager 1, furthest spacecraft	23 300 000 000	0.002 (21 light hours)
Proxima Centauri (our nearest star)	40 000 000 000 000	4.1
diameter of the Milky Way galaxy	946 000 000 000 000 000	100 000
Andromeda (our nearest galaxy)	23 700 000 000 000 000 000	2 500 000

Because light takes time to travel from faraway stars and planets, when astronomers study them they are looking back in time! They are seeing the Sun as it was 8 minutes ago. If you look up at the night sky, you are seeing stars and galaxies as they were when the light left them. For some objects this is thousands or millions of years ago!

Questions

1. Explain why looking up at the stars at night is like looking back in time.
2. Look at the table of light times. It takes light about 9.3 times as long to reach us from Saturn as it takes to reach us from the Sun.
   a. Calculate the distance to Saturn in light minutes.
   b. Approximately how long would it take to get to Saturn travelling at the speed of an aeroplane? (*Hint:* look at how long it would take to reach the Sun in an aeroplane.)
3. Mars and Earth both orbit the Sun. Use this information to explain why there are two distances from the Earth to Mars in the table above.
4. Suggest one of the problems with using a metal bar as the standard length for the metre.

Key points

- Light travels at 300 000 km/s. Nothing travels faster than light.
- It is convenient to measure the distance to very distant objects using light years rather than metres or kilometres.
- Measurements of the speed of light are difficult because the times are so short.

Dispersion

- Explain how a spectrum of light is produced
- Explain why we see rainbows

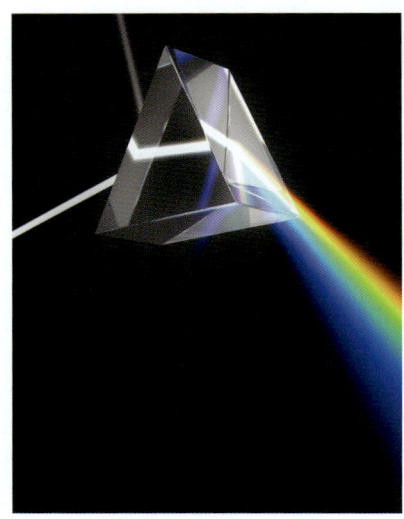

▲ *Some white light is reflected and some is dispersed by a triangular prism.*

▲ *You only see a rainbow when the Sun is behind you.*

When you look at a normal light bulb it looks white. We use rays from a light bulb to investigate how light is reflected and refracted. This 'white' light is really a mixture of colours. We see those colours when we look at rainbows.

Splitting light

When light goes into glass or water it slows down. If a ray of white light shines into a rectangular block of glass it will change direction or **refract**. We see a ray of white light entering and leaving the block.

If, instead of a block, we use a triangular piece of glass, called a triangular **prism**, at a particular angle of incidence the light emerging from the prism is split into colours.

- All the colours together are called a **spectrum**.
- The process of splitting white light into a spectrum is called **dispersion**.

We cannot see the light inside the prism because our eyes are outside. From the ray going in and the colours that emerge we can work out the paths that the different colours take inside the glass.

- Different colours of light are slowed down by different amounts.

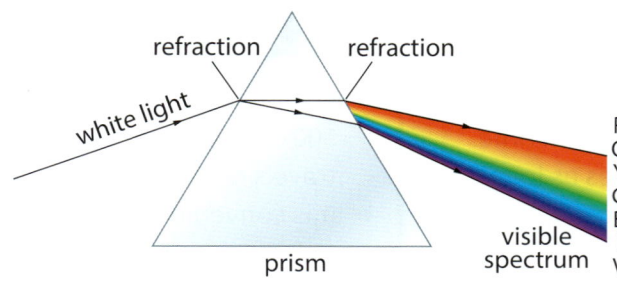

▲ *Dispersion is produced by refraction.*

- Red light is slowed down the least, so it is refracted the least.
- Yellow light is refracted more than red, and so on.
- Violet light is refracted the most, so it travels the slowest in glass.

The colours of the white light spectrum are:

| red | orange | yellow | green | blue | indigo | violet |

How rainbows form

You see a **rainbow** when you are facing falling rain with the Sun behind you.

Raindrops behave like small prisms. The sunlight from behind you is refracted, reflected, and dispersed so that you see a spectrum, which we call a rainbow.

About 700 years ago people didn't know what caused a rainbow to form. An Iranian scientist, Al-Farisi, correctly explained how the raindrops dispersed the white light. One of his students demonstrated his idea by directing white light through a large spherical glass container filled with water.

refraction and dispersion

light from the Sun

(total internal) reflection

rainbow seen in this direction

raindrop

refraction

▲ *You cannot see a rainbow if the Sun is in front of you.*

Recombining the spectrum

More than 350 years later, in 1666, Sir Isaac Newton demonstrated that light from the Sun could be split into colours. He also showed that he could recombine all the colours back into white light. He did this by using two prisms.

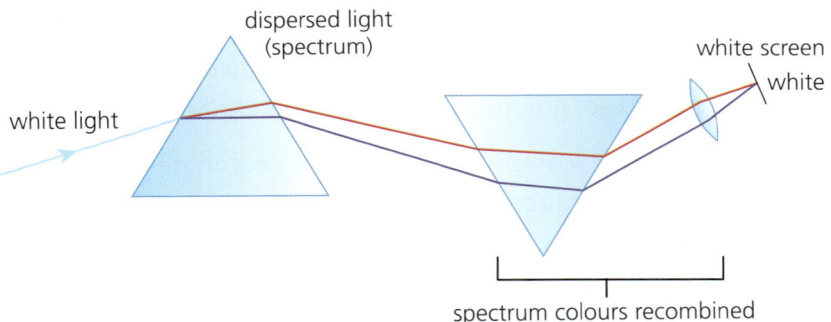

dispersed light (spectrum)

white light

white screen

white

spectrum colours recombined

▲ *Using a second prism shows that white light consists of a spectrum.*

? Questions

1. **a.** Write down the number of times that light is refracted when it passes through a prism.
    b. Write down the colours of the spectrum from the colour that is refracted most to the colour that is refracted least.
2. The spectrum of white light is described as continuous.
    a. Explain what 'continuous spectrum' means.
    b. Explain how there can be separate colours in a continuous spectrum.
3. Suggest which colour travels fastest in glass. Explain your answer.
4. You can have refraction without dispersion, but you cannot have dispersion without refraction. Explain why, and give an example.

📖 Key points

- A prism splits white light into the colours of the spectrum. This is called dispersion.
- A spectrum is formed when different colours are refracted by different amounts.
- The colours of the spectrum are red, orange, yellow, green, blue, indigo, and violet.
- Rainbows are formed when raindrops split sunlight into colours.

- Explain what happens when you mix light of different colours together
- Explain how filters work

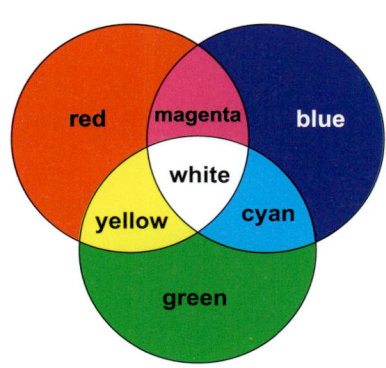

Colour

Colour addition – primary and secondary colours

In 1937 Ardeshir Irani made the first colour film in Hindi. It was not until the 1950s that colour film became popular. Now everything we watch on the Internet, television, or our phones is in colour.

You can make all the colours of light with just three **primary colours**. The primary colours are **red**, **green**, and **blue**. If you combine any two of the three colours you get a **secondary colour**. If you mix all three primary colours of light you get white light.

- **Red** and **green** light make **yellow** light.
- **Green** and **blue** light make **cyan** light.
- **Red** and **blue** light make **magenta** light.

It is *not* possible to make red, green, or blue light using any combination of other colours. That is why they are called primary colours. Your eye detects these three colours of light. You perceive different colours when different amounts of red, green, and blue are detected.

Colour displays on televisions, computers, and phones combine primary colours of light to produce the range of colours that you see.

▲ *Your phone emits three colours of light.*

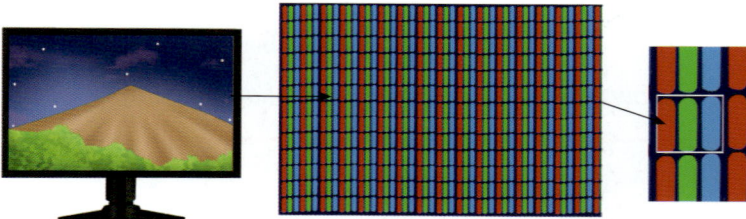

▲ *Screens emit three colours of light.*

Colour subtraction – filters

Coloured lights in a theatre can make a production very dramatic. But if a light bulb produces white light, how can we produce coloured lights for the stage?

Coloured filters are used to produce different colours from white light.

A **filter** absorbs some colours of light and transmits others.

- If you shine white light through a red filter, the filter will only let red light through.
- Red light is **transmitted**.
- All the other colours are **absorbed**.
- The filter has taken away all the colours *except* red.

This means that the light you see through a filter will be dimmer than the light without the filter because some light has been absorbed.

- If you shine red light through a green filter, then no light will get through.
- If you look at a white light through the red and green filter together you will see black.
- Your brain perceives no light as black.

▲ *Filters make beams of different coloured light.*

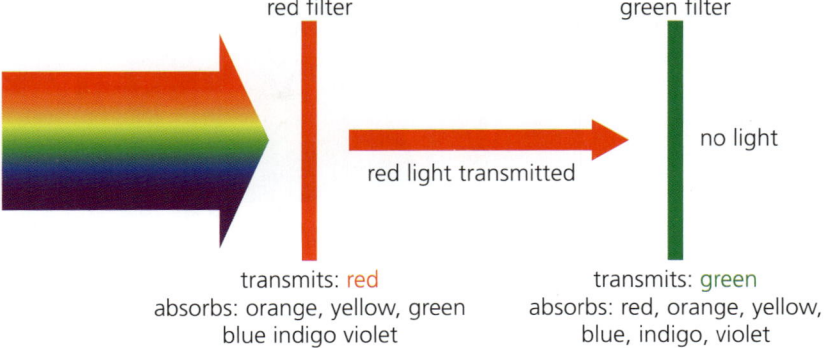

transmits: red
absorbs: orange, yellow, green blue indigo violet

transmits: green
absorbs: red, orange, yellow, blue, indigo, violet

▲ *If you combine filters of two primary colours, then no light will get through.*

Filters do *not* just come in primary colours – you can also have a secondary colour filter. A magenta filter will transmit red and blue light and absorbs green light.

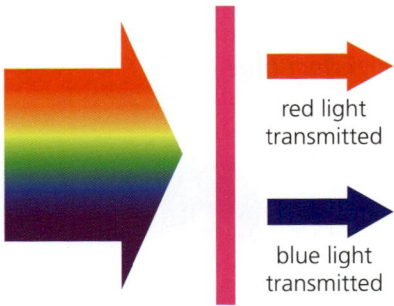
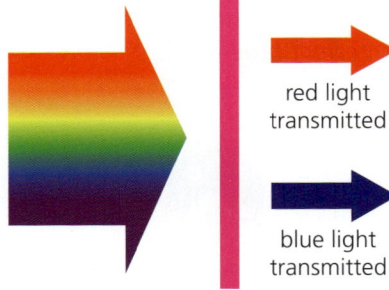

The primary colours are red, blue, and green because these are the colours that your eye detects. The cells in the back of your eye absorb different amounts of red, green, and blue light. This means that the human eye can distinguish between thousands or millions of different colours.

Questions

1. Deduce what colour light would be produced by mixing light in these situations. Copy and complete the table.

Combining...	Makes...
red + blue	
cyan + red	
blue + yellow	

2. **a.** Describe what a green filter does to white light.
   b. Some people think that the green filter has added colour to white light. Suggest how you could show that this is *not* true.
3. Describe and explain what you would see if you look at white light through a blue and a yellow filter at the same time.
4. Suggest a similarity between a phone screen and your retina.

Key points

- The primary colours of light are red, green, and blue. The secondary colours are yellow, cyan, and magenta.
- The primary colours add to give white light.
- Colour filters absorb all the colours of light except the colour that they transmit.

Looking at coloured objects

Objectives

- Explain why coloured objects look coloured in white light

- Explain why coloured objects look different colours in different colours of light

▲ The red shirt reflects red light and the blue shorts reflect blue light.

▲ White clothes reflect all colours of light.

▲ Mixing paint is not the same as mixing light.

▲ Most of the time we see coloured objects in white light.

Imagine that you are looking at a fabric that looks red. If the light from the Sun or lamps is a mixture of colours, why does only red light enter your eyes? Coloured objects *absorb* some colours of light and *reflect* other colours.

Coloured objects in white light

A red shirt absorbs all of the colours in white light except red. The red light is reflected into our eyes and we see a red shirt.

Paint vs light

- An object looks white because *all*, or nearly all, of the colours of light are reflected.

- An object looks black because *none,* or almost none, of the colours are reflected.

▲ Black clothes absorb nearly all of the light that hits them.

In science, the primary colours of light are red, green, and blue, because human eyes have cells that detect these colours best. These three colours of light are added together to give all the other colours.

In art, the primary colours of paint are red, yellow, and blue. The different colours of paint absorb different colours of light. Green paint absorbs all the colours except green, which it reflects. It acts like a filter and subtracts the other colours.

If you mix all the colours of light you get white light, but if you mix all the colours of paint, all the colours are absorbed and the paint looks black.

Coloured objects in coloured light

Filza is using filters to produce light beams of different colours. She shines coloured light on pieces of fabric of different colours. Here are her data.

	Red fabric	Blue fabric	Green fabric
Red light	red	black	black
Blue light	black	blue	black
Green light	black	black	green
Yellow light	red	black	green
Cyan light	black	blue	green
Magenta light	red	blue	black

Filza writes this conclusion down.

The 'colour' of an object is how it appears in white light. If the light shining on the object contains one or more of those colours, then that is the colour that you see. If the light shining on the object does not contain any of those colours, then it will look black. This is the case for both primary and secondary colours.

◀ The colour that an object appears depends on the colour of light reflected from it.

Questions

1. Write down which colours of light are reflected from an object that looks:
   a. white
   b. black.
2. You are looking at a yellow flower growing outside in the sunshine. Explain why it looks yellow.
3. Explain why a combination of red, yellow, and blue paint looks black.
4. A football team wears a yellow shirt and white shorts. Write down what the kit would look like:
   a. in red light
   b. in green light
   c. in blue light.

Key points

- A coloured object absorbs all the colours of light except the colour that it appears to be.
- Coloured objects appear different colours in different coloured light because they absorb certain colours and reflect others.

Changing ideas: Light

Objectives

- Recall that there can be different explanations for the same observations

- Explain why some explanations are accepted and others are not

Two ideas about light

Scientists make **observations** of the world around them. They describe what they see and try to make sense of their observations by asking questions and producing an **explanation**.

Sometimes two people who make the *same* observations can come up with *different* explanations. This was the case for two scientists who lived over 400 years ago: Isaac Newton and Christiaan Huygens.

▲ *Light is emitted by the Sun and by lamps – but what is light?*

Waves and particles

Isaac Newton showed that light could be split into colours using a prism, and that it can be recombined with a second prism. He also invented a telescope that used mirrors instead of lenses.

Newton thought that light was made up of **particles**. Lots of people had used the same idea over thousands of years.

- Nearly 2000 years ago, philosophers in India thought that light was made of a stream of high-speed 'atoms'.

- About 1000 years ago Ibn al-Haytham wrote a book about light that included the idea that light was made of particles.

- Newton explained many observations about light using the idea of particles.

Christiaan Huygens thought that light was a **wave**. The idea that light is a wave is much newer than the idea of light as particles. Huygens thought that light waves moved in some invisible medium, just as waves move through water. He too worked out some explanations for observations about light.

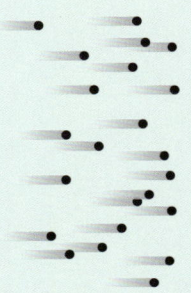
▲ *Newton thought of light as particles.*

▲ *Huygens thought of light as waves.*

Observation about light	Newton's explanation	Huygens's explanation
Light travels in straight lines.	Particles travel in straight lines.	Waves travel in straight lines.
Light is reflected at equal angles.	Particles of light bounce off the mirror like a ball off a wall.	Wave bounce off barriers in the way that sounds make echoes.
Light is refracted.	Particles near a boundary with a dense medium are pulled off course by the particles in the medium.	Waves slow down when they go into shallow water, and bend.

At the time the idea that light was a particle was very easy to understand because people could see balls bouncing off walls and imagine that light does the same thing.

But there was one observation that Newton found very difficult to explain with his particle model.

When light passes into a glass block, some of it is refracted but some of it is reflected. It is difficult to explain this observation using the idea of particles of light.

Which idea was accepted?

Even though the particle idea did *not* explain some observations very well, people believed Newton's explanation for over 100 years. This was because Newton had published a lot of explanations about gravity and was a very famous scientist. All his explanations had been very successful at explaining observations that astronomers made about stars and planets.

▲ *Newton struggled to explain why light does this.*

In 1801 another scientist, Thomas Young, did an experiment on light that could *not* be explained with the idea of particles. The wave theory became accepted as a better explanation of how light behaves.

That is *not* the end of the story. Another 100 years later scientists did a light experiment and made some observations that could *not* be explained with the idea of waves! Now it seems that you need both models to describe what light is.

▲ *Newton's ideas explained a lot of observations of the night sky.*

Questions

1. Explain why the idea of light as particles was popular at first.
2. Explain why Isaac Newton was famous.
3. You can use a football to model the particle idea about light.
   a. Describe how you could use a football to explain the reflection of light.
   b. Suggest how you can use a football to show how light is refracted. Explain your answer.
4. Newton tried to explain the fact that when light goes into a block of glass, some of it is reflected and some of it is refracted. He said that some particles of light bounce off and others go through the block. Is this a good explanation? Explain your answer.

Key points

- Scientists produce explanations for their observations about the world around them.
- Good explanations explain a wide range of observations.
- Sometimes explanations need to be changed if they do not explain new observations.

Objectives

- Describe how your eye works
- Describe how a camera works
- Compare the camera and the eye

The eye and the camera

How does your eye form an image? How is the eye like a camera? How is it *not* like a camera?

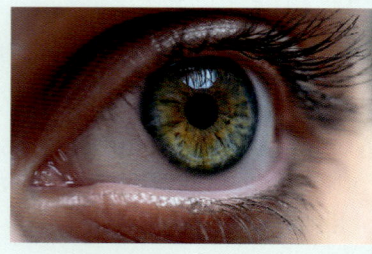

▲ *Nobody else has the same pattern in their iris as you.*

How do eyes work?

The **eye** is the human body's camera. If you look at a tree, the light from the Sun is reflected off the tree into your eye.

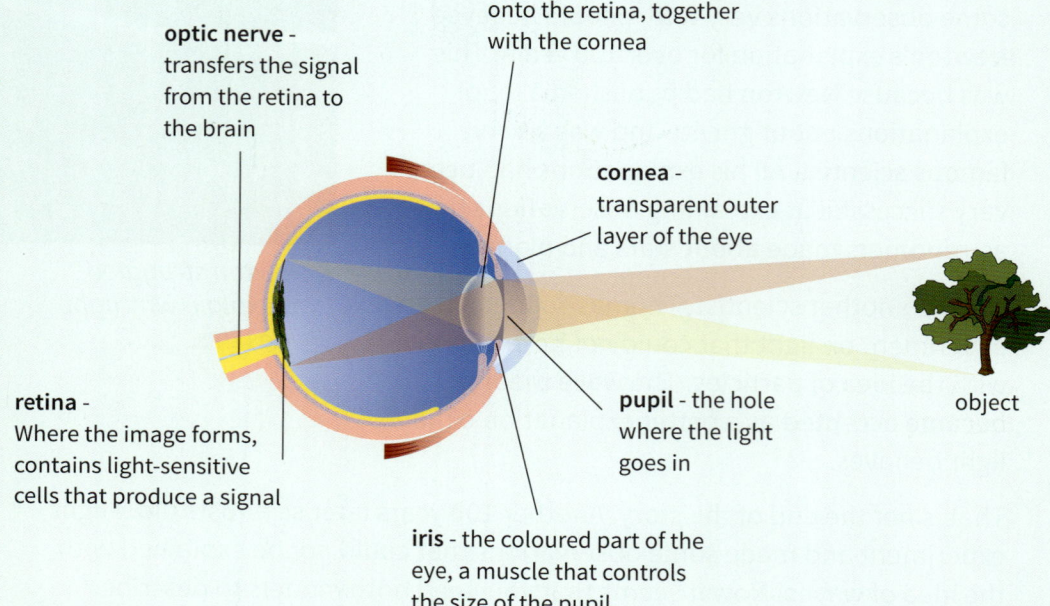

optic nerve - transfers the signal from the retina to the brain

lens - focuses the light onto the retina, together with the cornea

cornea - transparent outer layer of the eye

retina - Where the image forms, contains light-sensitive cells that produce a signal

pupil - the hole where the light goes in

iris - the coloured part of the eye, a muscle that controls the size of the pupil

object

- The **lens** is a small sack of jelly-like material.
- Your eyeball is filled with a liquid that keeps it nearly spherical.
- The image formed on your **retina** is upside down, or **inverted**.
- Your brain, your brain makes the image the right way up.
- The light-sensitive cells of the retina are called **rods** and **cones**.
- Rods are mainly around the edge of the retina and are sensitive to dim light.
- Cones are mainly found in the centre of the retina and they are sensitive to bright light and colour.

You can find out more about light-detecting cells in your *Cambridge Lower Secondary Complete Biology* Student Book.

The eye and colour

Your eyes can detect different colours of light because of three types of cones in your retina.

- Each cone is very sensitive to either red, green, or blue light.

- This is why red, green, and blue are the primary colours of light.

- The cones absorb the light, and send a signal to your brain.

- Your brain processes the signals and produces the colour image that you see.

- The human eye can distinguish between thousands or millions of different colours, but it is *not* possible to know if someone else perceives the same colour as you.

Some people have cones that do *not* work properly, and they don't see colours accurately. About 1 in 20 people (usually men) have the most common form of **colour blindness**. They find it hard to tell the difference between red and green.

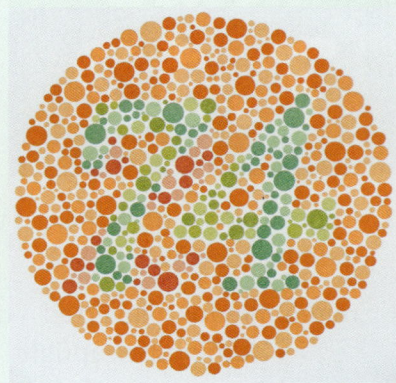

▲ *If you have normal colour vision, you will see the numbers 7 and 4 in this picture. If you have red–green colour blindness you may see 2 and 1.*

How does a camera work?

If you take a picture of a friend with a camera, light is reflected from your friend and goes into the camera. Your friend is the **object**, and an image is formed at the back of the camera. The image is focused by the lens and inverted, just as it is in your eyes.

In the past, cameras used photographic film to capture the image. Modern cameras, such as phone cameras, use electronic sensors to produce an image that can be stored or transferred easily.

▲ *Phones contain cameras.*

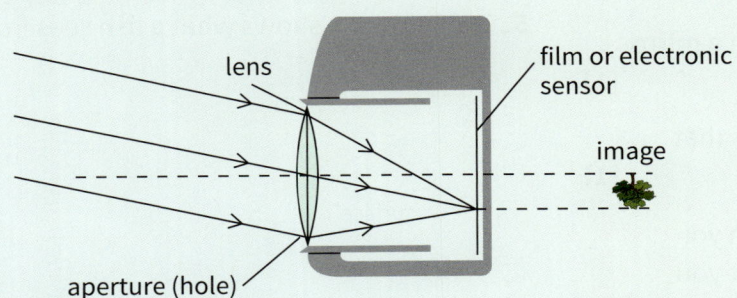

▲ *A lens in a camera focuses light into an image.*

Questions

1. Name the part of the eye that:
   a. controls the size of the pupil
   b. connects the eye to the brain.
2. The eye works in a similar way to a camera. Copy and complete the table below to compare the eye and the camera.

	Eye	Camera
Hole to let the light in		
Where the image is formed		

3. Describe one way in which the eye is *not* like a camera.

Key points

- Light from an object is focused by the lens and cornea on the retina of the eye, which contains light-sensitive cells.
- The retina sends a signal to the brain.
- The camera has an aperture like the pupil of the eye, and film or a sensor like the retina.

1. Choose words from this list to match each definition below.

   non-luminous inverted opaque transparent
   luminous upright translucent

   a. gives out light [1]

   b. does not allow light to travel through it [1]

   c. upside down (describing an image of an object) [1]

   d. you can see through it [1]

   e. does not give out light [1]

2. Copy and complete the sentences using the words from the list. You may need to use some words more than once.

   reflected transmitted absorbed transparent
   translucent opaque refracted

   a. When light is _____ from a mirror you see an image. [1]

   b. Light is _____ by objects that are opaque. [1]

   c. If you hold up a piece of white cloth you can see light coming through it, but you cannot see through it. It is described as _____, not _____. [2]

   d. You see non-luminous objects because light is _____ off them into your eyes. [1]

   e. Wood and concrete are examples of materials that are _____. [1]

   f. Water and glass are examples of materials that are _____. [1]

3. The primary colours of light are red, green, and blue. [1]

   Which of these statements is correct?

   A If you mix red and green light you get cyan light.

   B If you look at a white light through a green and blue filter you will see only black.

   C If you mix all three colours of light you get black. [1]

4. Here are some diagrams showing reflection. Which pictures correctly show how light is reflected from a mirror? [1]

   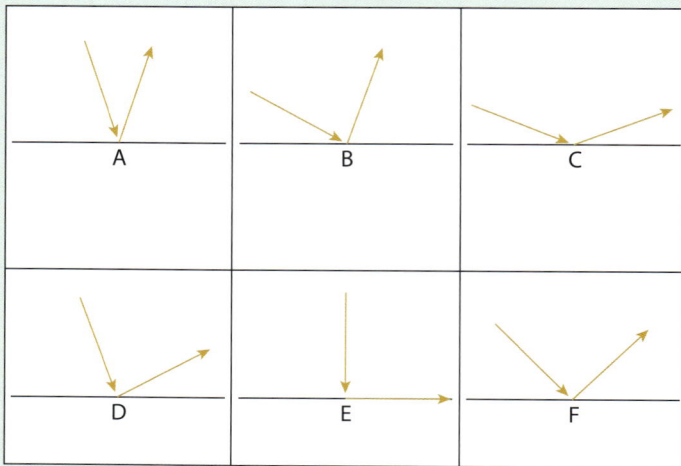

5. This diagram shows what a fish sees from below the water.

   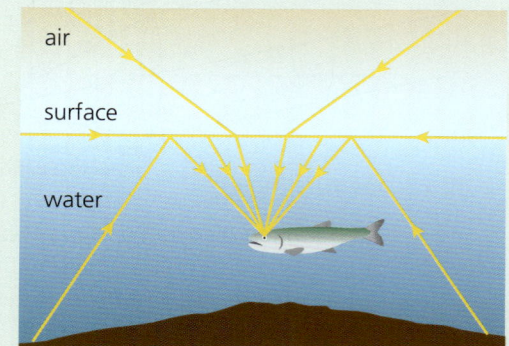

   a. Write down the scientific name for light changing direction as it goes from air into water. [1]

   b. Explain why the light changes direction. [2]

   c. Explain how it is possible for the fish to see the bottom of the lake as well as the water's surface and the bank above the water. [2]

6. Look at the diagram of a ray of light being reflected.

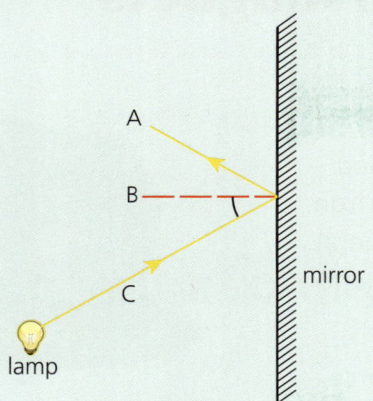

a. Which line, A, B, or C, is the incident ray? **[1]**

b. Which line, A, B, or C, is the reflected ray? **[1]**

c. Which line, A, B, or C, is the normal? **[1]**

d. What can you say about the angle between A and B, and the angle between B and C? **[1]**

7. Sometimes prisms are used instead of mirrors to reflect light, for example in binoculars.

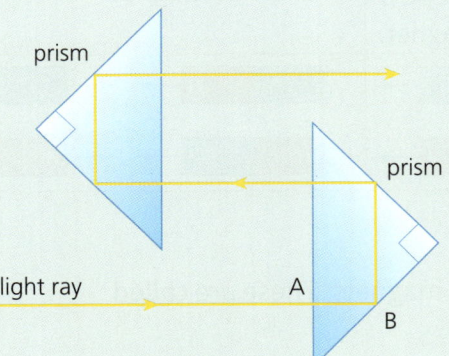

a. State the angle of incidence at A. **[1]**

b. Describe what happens to the ray at A. **[1]**

c. State the angle of incidence at B. **[1]**

d. Describe what happens to the ray at B. **[1]**

e. Write down what you can tell about the value of the critical angle for the material from which the prism is made. **[1]**

f. Write down what would happen if the critical angle were different from your answer to **e**. **[1]**

8. **TWS** Here are some statements about measuring the angle of refraction in a glass block.

A Put the glass block on a piece of white paper and draw around the glass block.

B Join the two rays with a straight line to show the ray inside the block.

C Remove the glass block, mark a point on one edge of the block, and draw a line at 90° to the edge at that point (the normal).

D Mark dots on the paper along the line where the ray goes on both sides of the block.

E Replace the block on the piece of paper and shine a ray of light at the block at the point where the normal meets the edge of the block.

F Remove the block and use a ruler to draw lines for both of the rays outside the block.

G Use a protractor to measure the angle of incidence and the angle of refraction.

a. Write the letters in order to show the correct order for the statements. The first and the last statements are in the correct place. **[1]**

b. Describe one way in which this method **TWS** ensures that the measurements of the angles are as accurate as possible. **[1]**

9. Jamila is in the front row watching an actor on a stage. She notices that the stage lights change the appearance of the actor's clothes. This is what she sees:

- In red light his trousers look black and his shirt looks red.

- In green light his trousers look green and his shirt looks black.

- In blue light his trousers look black and his shirt looks blue.

a. What colour are his trousers? **[1]**

b. What colour is his shirt? **[1]**

The properties of magnets

Objectives

- Describe the properties of magnets
- Know which materials are magnetic
- Use a model to explain the behaviour of magnetic materials

Magic rocks

A long time ago the Ancient Greeks discovered that a rock that they called **lodestone** attracted small pieces of iron. Lodestone is an iron-containing mineral that is naturally **magnetic**. The Ancient Greeks had discovered **magnetism**.

A magnetic force is a non-contact force – lodestone exerted a force on the pieces of iron even when they were *not* touching.

▲ *The first rocks that attract iron were found in Magnesia in Greece.*

Attraction and repulsion

Every **magnet** has a **north pole** and a **south pole**. The poles of magnets attract or repel each other.

Like poles repel	Unlike poles attract
• The north pole of a magnet repels the north pole of another magnet. • The south pole of a magnet repels the south pole of another magnet.	• The north pole of a magnet attracts the south pole of another magnet.

Magnetic materials

Some materials are attracted to magnets. These are called **magnetic materials**.

- **Iron**, **cobalt**, and **nickel** are all magnetic materials that are **elements**.
- Some of the **alloys** or **oxides** of these elements are also magnetic. You will learn more about alloys in your *Cambridge Lower Secondary Complete Chemistry* Student book.

Steel is magnetic because it contains mainly iron. Lodestone, or magnetite, contains iron oxide, which is also magnetic.

The strip on credit cards is made of a plastic that contains lots of very small iron pieces that behave like tiny magnets. These iron pieces are used to store information on the card.

A piece of magnetic material is attracted to a magnet, but it will *not* be repelled by it.

- Magnetic materials do *not* attract or repel each other.
- Only two magnets will repel each other.

▲ *Magnetic levitation happens when poles repel.*

▲ *The black strip on a credit card is magnetic.*

Why are some materials magnetic?

You can think of a magnetic material as being made up of lots of very small regions called **domains**. Each domain behaves like a tiny bar magnet.

When a magnetic material is *not* **magnetised**, the domains are pointing in lots of different directions. If you magnetise the material by putting it near a magnet or stroking it with a magnet, the domains line up. The material has been magnetised.

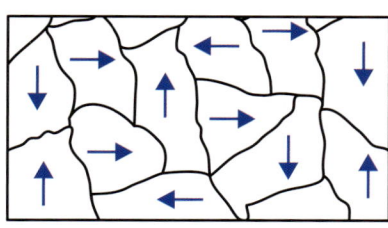

domains do not line up

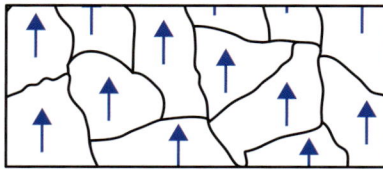

domains line up

Magnetising and demagnetising

Magnetising

Iron and steel are magnetised by being close to a magnet.

- Iron is easy to magnetise, but if you remove the magnet the iron is no longer magnetised.
- Steel is harder to magnetise than iron, but it stays magnetised when you remove the magnet.

A steel needle will be slightly magnetised in a magnetic field. To make a stronger magnet you can stroke the needle with a magnet repeatedly in the same direction. A piece of iron or steel magnetised in this way is a **permanent magnet**.

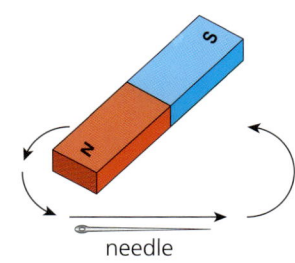

needle

▲ *You can magnetise a magnetic material by stroking it with a magnet.*

Demagnetising

You cannot turn a permanent magnet off. You can **demagnetise** it by:

- heating it up to a high temperature
- hitting it with a hammer.

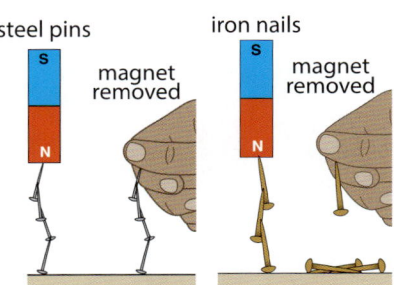

steel pins   iron nails

▲ *Steel stays magnetic, but iron does not.*

Questions

1. Sort the list below into materials that are magnetic and materials that are *not* magnetic.

   brass steel iron copper wool wood nickel cotton

2. You pick up some nails from a pile and find that they appear to stick together. Are they made out of iron or steel? Explain your answer.

3. The two magnets in the picture are attracted to each other.

   a. If A is a south pole, write down what poles B, C, and D are.
   b. Describe what will happen in the two experiments shown below.

   experiment 1:   experiment 2:

4. Use the idea of domains to explain why:
   a. if you break a magnet in two you get two magnets, each with a north and south pole
   b. you can demagnetise a permanent magnet by heating it.

Magnetic fields

Objectives

- Define 'magnetic field'
- Describe how you can find the shape of a magnetic field around a bar magnet

◄ *A dish of ferrofluid with two magnets underneath.*

Ferrofluid is a special magnetic fluid that makes beautiful patterns when it is near a magnet. But why does it move?

What is a magnetic field?

If you put a steel pin near a magnet, the pin will move. The magnet attracts the pin because it is exerting a force on it.

The region around the magnet where magnetic materials experience a force is called the **magnetic field**. The ferrofluid moves because it is in a magnetic field.

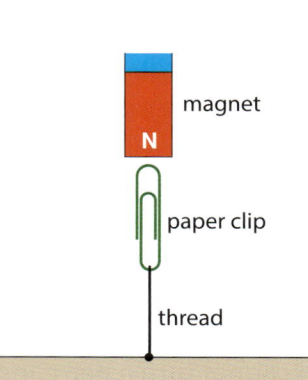

▲ *The paper clip is in the magnetic field of the magnet.*

'Seeing' magnetic field patterns with iron filings

How can we 'see' a magnetic field?

- Iron filings are very small flakes of iron.
- They become magnetised when they are in a magnetic field.
- They line up, just like paper clips, to show us the shape of the magnetic field.

We cannot see the magnetic field, only the effect of the field on magnets or magnetic materials.

You can investigate the magnetic field patterns around two magnets.

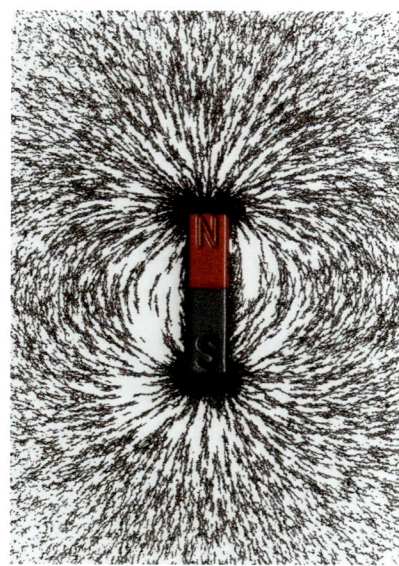

▲ *Iron filings show the shape of a magnetic field around a magnet.*

▲ *These two magnets are repelling each other.*

▲ *These two magnets are attracting each other.*

Point **X** is called the **neutral point**. At this point the magnetic fields of the two magnets cancel out.

'Seeing' magnetic field patterns with a plotting compass

A plotting compass is a very small compass. If you put it near a bar magnet, its needle will line up with the magnetic field of the magnet. You can use a plotting compass to see the shape of the field around a bar magnet.

We can show the pattern around a bar magnet by drawing **magnetic field lines** to show where the plotting compass lined up. The lines are *not* the field, but they show us the general shape of the field.

▲ *A plotting compass is a very small compass used to investigate magnetic fields.*

- The *spacing* of the lines indicates the *strength* of the field. The field is stronger where the lines are closer together.

- The *arrows* on the lines show the *direction* of the force. Magnetic field lines are drawn from the north pole to the south pole. When scientists had to decide how to show magnetic fields, they agreed that arrows would point towards the south pole.

Magnetic field lines are just a representation of the field. There is a field all around the magnet, including the region above and below it.

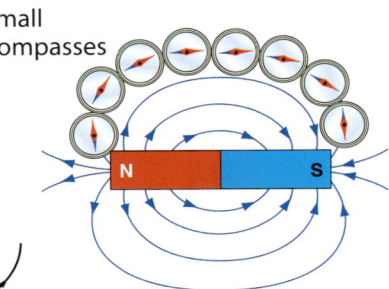

small compasses

Attracting and repelling

When you bring two magnets together, their magnetic fields interact. The field of one magnet exerts a force on the other. Magnets attract or repel.

Another way to think about the effect of magnets is that the magnets move so as to straighten the magnetic field lines.

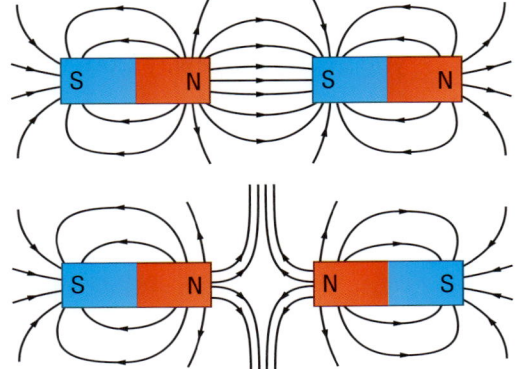

▲ *The field lines between attracting and repelling magnets.*

Questions

1. Define 'magnetic field'.
2. Write down where the magnetic field of a bar magnet is the strongest. Explain how you know.
3. Look at the pictures on page 156 of iron filings around two magnets repelling each other and two magnets attracting each other.
   a. Describe what would happen if you put a small steel ball at the neutral point.
   b. Explain your answer.
4. A student says that he has made a magnetic field with some iron filings. Write down what you would say to him.

Key points

- A region where magnetic materials experience a force is called a magnetic field.
- You can find the shape of a magnetic field using iron filings or plotting compasses.
- Magnetic fields exert forces on each other.

Electromagnets

Objectives

- Describe how to make an electromagnet
- Describe how to change the strength of an electromagnet

Bar magnets are fun to play with, but they have a fixed strength. A magnet with a strength that you can vary is far more useful.

You can make a magnet that you can switch on and off using a coil of wire and a battery. It is called an **electromagnet**.

A single wire makes a field...

When an electric current flows through a wire, a magnetic field is produced around the wire.

- Plotting compasses show the magnetic field around the wire.
- The magnetic field lines around a single piece of wire are circles.
- The magnetic field gets weaker as you get further away from the wire.

▲ *An electromagnet in the top of this toy keeps a metal Earth hovering below it.*

... and a loop makes it stronger

You can make a circular loop of wire and pass a current through it. The magnetic field lines at the centre are straight.

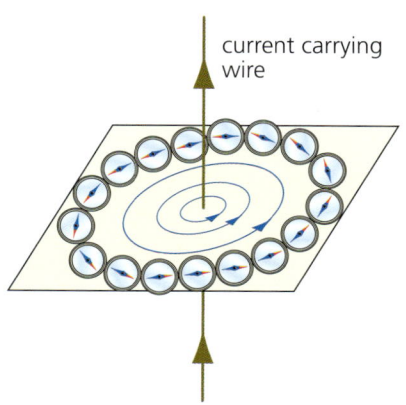

▲ *Plotting compasses show the shape of the field.*

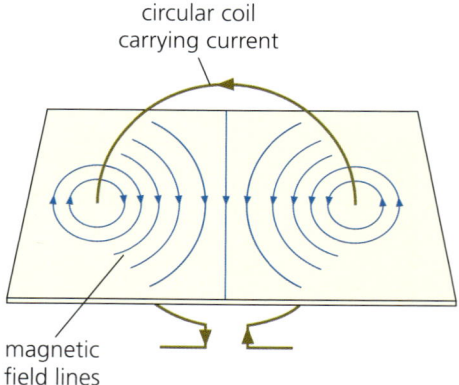

A coil makes it stronger still

You can make a coiled cylinder from *lots of loops* of wire.

- An electric current in the coil produces a magnetic field.
- The field has a similar shape to the field around a bar magnet.
- The switch is needed to turn the **solenoid** (coil) on and off because the magnetic field is only produced when a current is flowing in the wire.
- Coiling the wire concentrates the magnetic field inside the loops. The field is much stronger than for a single loop on its own.

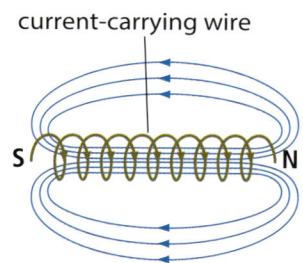

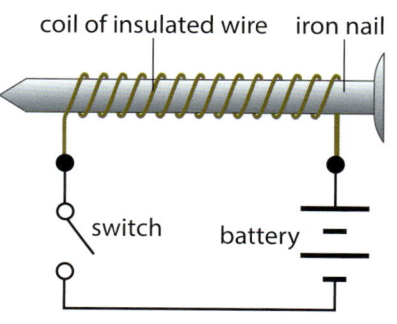

A coil of wire wound onto a **core** of magnetic material is an electromagnet.

- Most cores are made of iron.
- The core becomes magnetised by the magnetic field of the coil.
- This makes the electromagnet much stronger.
- The iron core does *not* remain magnetised when you turn the electromagnet off.

The strength of an electromagnet

The strength of an electromagnet depends on:

- the number of turns, or loops, on the coil – more turns will make a stronger electromagnet
- the current – more current will make a stronger electromagnet
- the type of core – using a magnetic material in the core will make a stronger electromagnet.

The current flowing in the wire depends on the power supply and the type of wire that is used in the electromagnet.

Permanent magnets and electromagnets

Electromagnets are more useful than permanent magnets.

- You can turn an electromagnet on and off.
- An electromagnet can be much stronger than a permanent magnet.

▲ *Electromagnets can move iron or steel in a scrapyard.*

▲ *Electromagnets can lift a train off its track to reduce friction.*

🏠 Science in context

Electromagnets are very useful. They can move large pieces of iron or steel in a factory, or cars in a scrapyard. When the current is turned off, the electromagnet drops the car.

Doctors can use a small electromagnet to remove iron or steel splinters from people's eyes.

❓ Questions

1. Describe what would happen to the magnetic field around a single piece of wire if the direction of the current was reversed.
2. Most electromagnets have a core.
   a. Explain why the core is usually made of iron.
   b. Describe what would happen if the core was made of steel instead.
3. Describe how you could use an electromagnet to sort a mixture of iron and copper pieces into two separate piles of iron and copper.
4. A student says that an electromagnet is stronger when it has an iron core because the electric current flows through the iron as well as the wire. Is this correct? Describe what you would say to the student.

📖 Key points

- A current flowing through a wire produces a magnetic field.
- An electromagnet is a coil of wire usually wrapped around an iron core.
- You can switch an electromagnet on and off by turning the current on and off.
- The magnetic field around an electromagnet is similar to that around a bar magnet.

Objectives

- Describe some uses of electromagnets
- Explain why electromagnets are used instead of permanent magnets

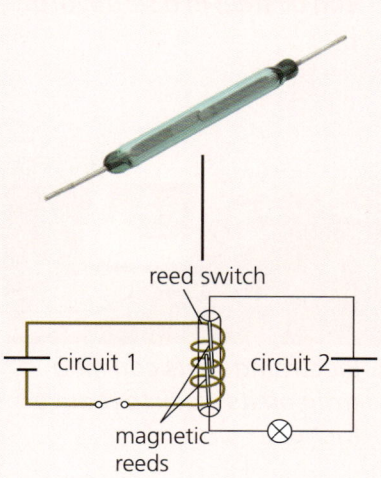

reed switch

circuit 1   circuit 2

magnetic reeds

▲ *If you close the switch in circuit 1 the lamp in circuit 2 will come on.*

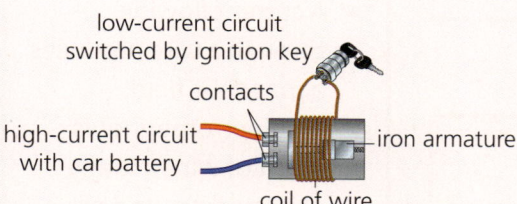

low-current circuit switched by ignition key

contacts

high-current circuit with car battery

iron armature

coil of wire

▲ *A starter motor.*

▲ *Electromagnetic catch on a fire door.*

Using electromagnets

Switching off

X-rays can damage your cells if you are exposed to them for too long. Radiographers, who operate X-ray machines, should *not* be near the machines when they are on.

The machine uses a high-voltage circuit. The radiographer turns it on and off using a **relay** instead of a normal on/off switch.

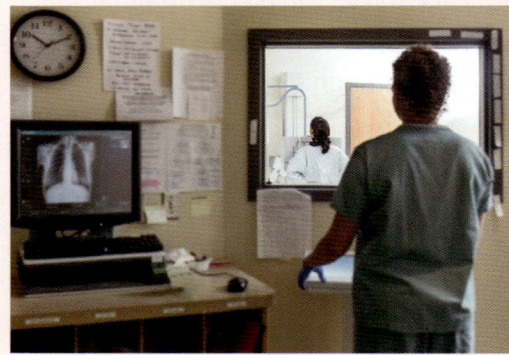

▲ *Some machines are switched on using a relay.*

Relays

A relay uses a small current in one circuit to switch on a circuit that could be dangerous.

You can make a simple relay with a **reed switch**. This is made of two pieces of magnetic material (called reeds) with a gap between them. If you wind a coil of wire around the outside and pass a current through it, the two reeds are magnetised. They attract each other and touch. If the reeds are part of another circuit, then current will flow in the second circuit when they touch.

A relay is often used to turn on the starter motor in a car. The starter motor works from the car battery, which produces a dangerously high current. The relay circuit that the driver switches on has only a small current.

When the driver turns the key:

- A current flows in the coil of wire.
- A magnetic field is produced.
- This magnetises the iron **armature**.
- The armature is attracted towards the contacts.
- When the contacts close, they complete the starter motor circuit.
- The car starts.

The current in the starter motor circuit could be dangerous if the driver had to switch it on directly. Using a relay is much safer.

Fire doors

In big buildings, fire doors are important to stop a fire spreading.

- If there is no fire, a current flows in the coil of an electromagnet on the wall.
- A piece of iron on the door is attracted to the electromagnet.
- This holds the door open so that people can pass easily.
- When the fire alarm is triggered, the circuit is broken.
- The current in the coil of the electromagnet stops and the door closes.

Electric bells

Some doorbells contain an electromagnet.

When a visitor presses the bell-switch, the circuit is complete.

- A current passes through the coil and the iron core is magnetised.
- The iron armature is attracted to the core and the hammer hits the gong, making a sound.
- This breaks the circuit at X so the iron core is no longer magnetised.
- The armature springs back and completes the circuit.
- The sequence starts again so the bell keeps ringing for as long as the visitor presses the bell.

Unlike in the relay, the armature moves backwards and forwards while the switch is pressed.

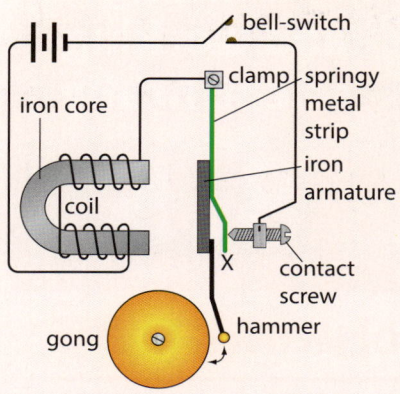

▲ An *electric bell circuit.*

Using magnets in medicine

A **magnetic resonance imaging (MRI) scanner** contains very strong magnets. These produce a very intense magnetic field that interacts with the hydrogen in the water molecules in your body.

MRI scans are safe and painless, and produce images that are much more detailed than X-rays. Doctors can use them to see structures inside the body that X-rays don't show, without having to cut the patient open, and to see how an organ works in normal conditions. Unfortunately, MRI scanners are very expensive, and the patient has to lie very still in a small space.

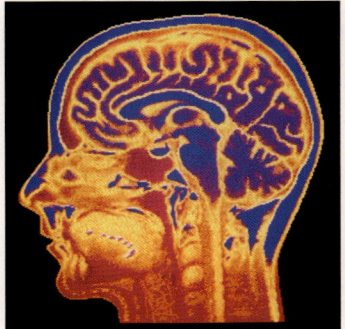

▲ *An MRI image.*

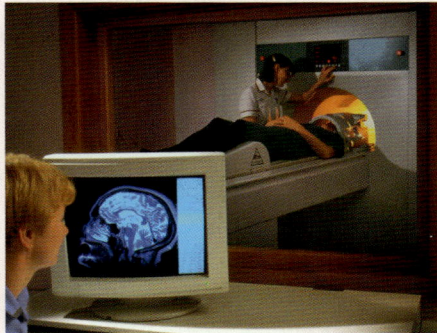

▲ *MRI scanners are noisy and cramped.*

Key points

- A relay is a type of switch that can be used in a low-current circuit to turn on a high-voltage or high-current circuit.
- Fire doors use electromagnets to close automatically.
- An electric bell works by continually making and breaking a circuit.
- MRI scanners use magnets to take pictures of organs in the body.

Questions

1. Describe a situation in which you might need to use a relay rather than a normal switch.
2. Look at the diagrams of the relay and the doorbell.
   a. Explain why the doorbell circuit is called a 'make-and-break' circuit.
   b. Describe how the doorbell circuit is different from the relay circuit.
   c. Write down what the two circuits have in common.
3. In magnetic levitation, trains are held above a track using electromagnets. Explain why permanent magnets are *not* used.
4. List two pros and two cons of MRI scanners.

Objectives

- Describe the difference between dependent and independent variables

- Describe how to show that you have controlled variables in an investigation

- Write an appropriate risk assessment for an investigation

Risk, variables, and tables: Investigating electromagnets

Electromagnet investigation

Aanjay and Citra made an electromagnet by winding insulated wire around an iron nail. When they turn the current on, the wire produces a magnetic field that magnetises the nail.

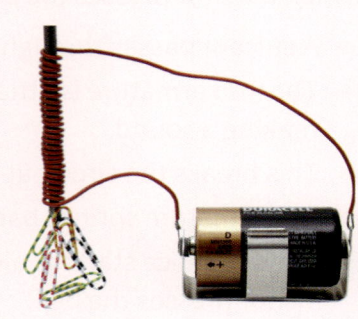

They are experimenting with metal paper clips.

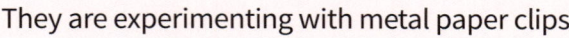

The number of paper clips tells us how strong the electromagnet is.

We could find out what affects how many paper clips the electromagnet can pick up.

Risk assessment

Citra tries the electromagnet and realises that the wire gets hot. She writes a **risk assessment**. A risk assessment says what can cause injury or damage, and how to reduce the chance of that happening.

The wire can get hot and burn our skin. We will only turn the current on for short periods of time.

We need to make sure that we only change one variable at a time.

What are the variables?

Aanjay and Citra think about the things that they could change **(independent variables)**. This is the list of variables that they think of first:

- the number of turns on the electromagnet
- the type of core
- the current in the wire.

They know that the current in the wire depends on the voltage and the type of wire. This is their final list of variables:

- the number of turns on the electromagnet
- the type of core
- the type of wire
- the voltage of the power supply or battery.

Yes, we need to say how we have controlled all the other variables

Controlling variables

Aanjay and Citra decide that they will change the number of turns on the electromagnet. Their **dependent variable** is the number of paper clips that the electromagnet picks up. They write down how they will control the other variables.

They could have listed the values of the control variables instead.

- type of core = iron
- voltage of the battery = 6 V
- type of wire = copper

In both plans it is clear which variable is the independent variable, which variable is the dependent variable, and how the other variables will be controlled during the investigation.

There are lots of other things that could change while the students are doing their investigation. However, they only need to control them if they will affect the strength of the electromagnet.

We are going to change the number of turns on the coil and measure the number of paper clips. We will keep the type of core, the voltage of the battery, and the type of wire the same. Here is the table for our results.

Number of turns	Type of core	Voltage of battery/V	Type of wire	Number of paper clips
5	iron	6	copper	
10	iron	6	copper	
15	iron	6	copper	
20	iron	6	copper	
25	iron	6	copper	
30	iron	6	copper	

Questions

1. Describe the two things that you should include in a risk assessment.
2. Eshe is investigating the link between the extension of an elastic band and the force that she applies. She changes the force and measures the extension. Suggest a risk assessment for this investigation.
3. A group of students plan to investigate how high a ball bounces on different surfaces. This is part of their plan.
   a. Which is the independent variable?
   b. Which is the dependent variable?
   c. Describe one way in which their plan could be improved.
   d. Draw the table of results that you would need for this investigation.

We are going to change the surface and measure the height of the bounce.

We are going to keep everything else the same.

We will use a tennis ball each time and drop it from the same height.

Key points

- A risk assessment lists what could go wrong and how you will reduce the chance that it does.
- You must clearly state how you have controlled other variables.

Review
8.6

1. For each of these statements about magnetic materials, write 'true' or 'false'.

   a. Magnetic materials will attract other magnetic materials.

   b. Magnetic materials will be attracted to magnets.

   c. Magnetic materials will be repelled by magnets.

   d. Magnetic materials will repel other magnetic materials. [4]

2. You have a small piece of each of the materials listed below. Which of them would be picked up by a magnet?

   A nickel   **E** zinc

   B steel   **F** cobalt

   C iron   **G** magnesium

   D copper   [1]

3. A student is playing with two magnets on the desk. She puts a blue magnet on the desk and brings a green magnet near to it. When she does this the blue magnet moves.

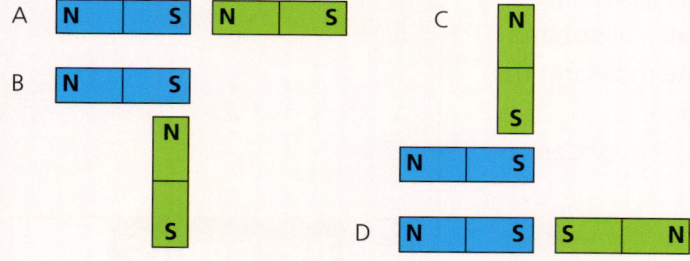

   a. For each diagram, write down the direction in which the *blue* magnet will move, using the words up, down, left, or right.

   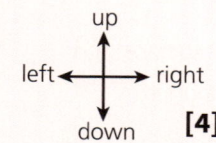

   [4]

   b. There is one direction that you have *not* used to answer part **a**. Draw a diagram of two magnets to show how you could make the blue magnet move in that direction. [1]

4. Copy and complete these sentences. You may need to use the same word or phrase more than once.

   a. We can draw _____ _____ lines to show the pattern of the _____ _____ around a bar magnet. [2]

   b. If there are lots of lines close together, that means that the _____ _____ is very _____. This happens near the _____ of the magnet. [3]

   c. All magnets have a north and a south _____. [1]

5. Icha wants to investigate the magnetic field around two magnets.

   a. Name two methods that she could use to find out about the shapes of the magnetic fields. [2]

   She puts two magnets on the table so that they are repelling.

   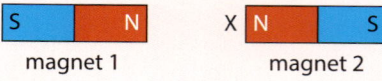

   b. X is the neutral point. Describe what is meant by 'neutral point'. [1]

   c. Deduce which magnet is stronger. Explain your answer. [2]

   d. Apart from the methods in part **a**, describe a method that you could use to find the position of the neutral point. [1]

6. Here is a diagram of a compass when there is no magnet near it and when it is near the end of a bar magnet.

   A compass with no magnet near it

   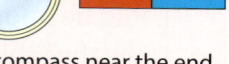

   B compass near the end of a bar magnet

   Explain what is happening to the needle in each picture. [2]

7. Explain what is meant by:

  a. a permanent magnet [1]

  b. a magnetic material [1]

  c. a magnetic field [1]

  d. magnetic field lines [1]

  e. an electromagnet. [1]

8. A student has watched a teacher use an electric bell.

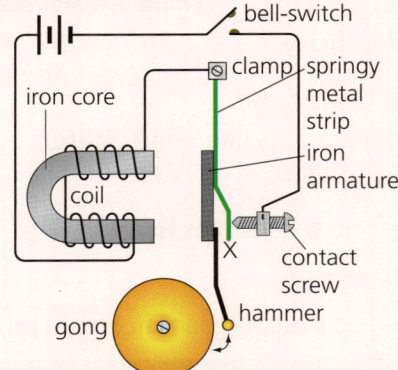

  a. Describe what happens to the coil of wire when the switch is closed. [1]

  b. Describe what happens to the coil when the circuit is broken. [1]

  c. Describe and explain what would happen if the core was made of steel, not iron. [1]

9. Abasi wants to make an electromagnet.

  a. Describe how he could make a simple electromagnet using a piece of wire, a nail, a battery, and a switch. [2]

  b. He wants the electromagnet to be strong. Describe two things he could do to increase the strength of the electromagnet. [2]

  c. He and his friends make four different electromagnets.

    Put electromagnets A, B, C and D in order of strength, starting with the weakest. Write the letters in the correct order.

    A 50 turns, low current, copper core

    B 100 turns, high current, iron core

    C 50 turns, low current, iron core

    D 100 turns, low current, iron core [3]

10. Which of these statements are true? There may be more than one. [1]

  A You can make a magnetic material into a magnet by stroking it with another magnet.

  B You cannot change the strength of an electromagnet.

  C The magnetic field of a current-carrying wire gets stronger as you move away from it.

  D You can demagnetise a magnet by hitting it with a hammer.

  E You can pick up cars with an electromagnet but not with a permanent magnet.

11. A student wants make the core of an electromagnet from a metal that will be magnetised while there is a current in the wire of the electromagnet, but will *not* stay magnetised when the current is switched off.

  a. Choose the correct metal or metals that could be used from the list below. [1]

    iron or steel   copper   iron

        copper or iron   steel

  b. Explain why the other metals would *not* be suitable. [2]

12. Ikenna is going to investigate how the type of core affects the strength of an electromagnet.

TWS

  a. Name the independent variable. [1]

  b. Name the dependent variable. [1]

  c. Name a variable that must be controlled. [1]

  d. He has a choice of small paper clips, large paper clips, or iron filings to measure the strength of the electromagnet. Which should he choose? Explain your answer. [1]

TWS **e.** Draw a table for Ikenna's results. [1]

The world's energy needs

Objectives

- Describe the difference between renewable and non-renewable energy sources

- Describe the ways humans use energy resources

- Describe how the world's energy needs have changed and are likely to change in the future

Will there be enough energy for humans in the future? To answer this we need to know what energy resources we have and how we use them.

▲ *The Earth has limited energy resources.*

Primary and secondary energy sources

It is useful to divide energy resources into **primary energy sources** (or resources) like coal and **secondary energy sources** like electricity.

Some primary sources can be found in the ground, such as coal, **oil**, **gas**, or **uranium**. Other sources like **wind**, **water**, and **biomass** are in the environment around us.

Primary sources are converted into secondary sources by:

- **power stations**, which generate electricity
- **refineries**, which produce fuels such as **petrol** (gasoline) from oil.

Primary energy sources	Primary energy converters	Secondary sources/carriers
coal	power stations (e.g. coal)	electricity
oil and natural gas		fuels such as petrol/gasoline
sunlight	generators (e.g. wind)	hydrogen gas
uranium	solar cells	
wind	refineries (to produce petrol/gasoline)	
water (tides, downhill flow)		
biofuel (wood, waste)		
geothermal (using thermal energy from under the ground)	hydrogen gas production	

▲ *We use a lot of energy to cool ourselves down...*

Renewable and non-renewable energy sources

Some primary energy sources such as coal, oil, gas, and uranium will run out. They are **non-renewable**. We cannot get more during our lifetime.

Other sources, such as wood, are **renewable**. We can grow more trees if we need more wood. Some people think that renewable resources can be used again, but that is *not* correct. 'Renewable' means you can replace it soon.

Uses of energy resources

The main uses of primary and secondary energy resources are:

- transportation (e.g. using petrol in cars)
- industry (e.g. using electricity to run machinery)
- heating and cooling, using electrical devices in the home (e.g. using electricity to run air conditioning units).

▲ *... or to heat ourselves up.*

Changes in the world's energy needs

150 years ago most people in the world were using primary energy sources that they could cut down or dig up, like wood or coal.

The chart on the right shows how the world's energy needs have been met in the last 150 years. It shows the energy in exajoules (EJ). One exajoule is one million, million, million joules.

There has been a huge increase in the use of non-renewable energy sources like coal, oil, gas, and uranium (nuclear), but the use of biomass such as wood has stayed roughly the same. The use of renewable sources like wind and water is still very small (the bands at the top).

This chart shows predictions for the demand for electricity in 2030. A kWh is another unit of energy.

1 kWh = 3 600 000 J

Both charts show **secondary data**. The data were *not* collected by you in a laboratory, but collected and displayed by other people. Data you collect yourself are **primary data**.

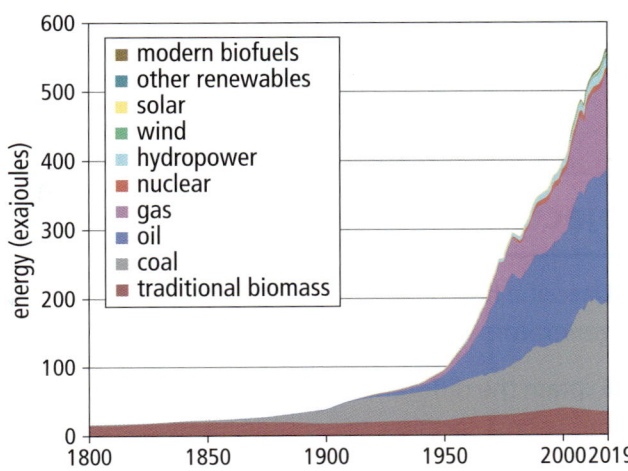

▲ *Primary energy sources used throughout the world.*

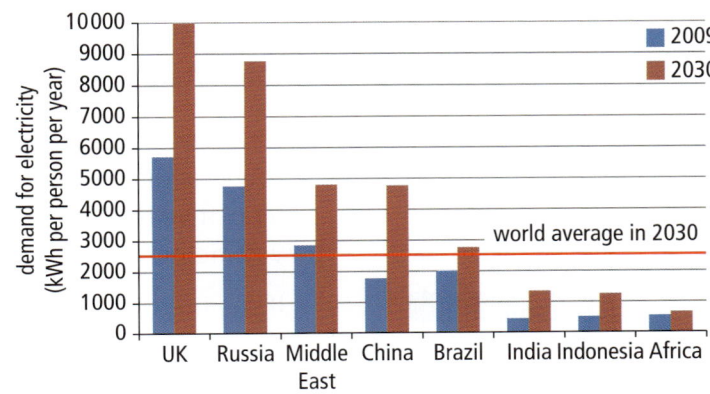

▲ *Predicted demand for electricity by some regions and countries of the world.*

Questions

1. There are primary and secondary energy sources.
   a. Explain the difference.
   b. Write down whether petrol (gas or gasoline) is a primary or a secondary source. Explain why.
   c. Write down whether oil is a primary or a secondary source. Explain why.
2. Describe the difference between a renewable and a non-renewable source of energy.
3. Look at the graph of global energy supply.
   a. Write down which energy source was used the most in 1900.
   b. Write down which energy source was used the most in 2000.
   c. By how much has energy supply increased between 1900 and 2000? Choose from these options: it doubled, it trebled, it is ten times bigger.
4. Look at the electricity demand chart. Write down which country or countries will double or more than double their demand for electricity between 2009 and 2030.

Key points

- Primary energy sources are coal, oil, gas, and uranium.
- Secondary energy sources are electricity and fuels such as petrol/gasoline and hydrogen.
- Non-renewable resources will run out, but renewables will *not*.
- The demand for primary energy sources needed to generate electricity is increasing rapidly.

Generating electricity

What do we mean when we say that 'electricity is generated'? We know that to make a component work in a circuit we need a voltage. Generating electricity means producing a voltage so that a current is possible.

Objectives

- Describe how generators work

- Explain the difference between a simple generator and a generator in a power station

Inside a generator

You saw in Chapter 8: Magnetism on page 158 that when a current flows in a wire it produces a magnetic field around the wire. There is a link between electricity and magnetism.

I wonder how electricity is produced in a power station.

Bulan takes a coil of wire and connects it to a voltmeter.

If she moves the magnet towards the coil there is a positive reading on the voltmeter.

If she holds the magnet still the reading on the voltmeter is zero.

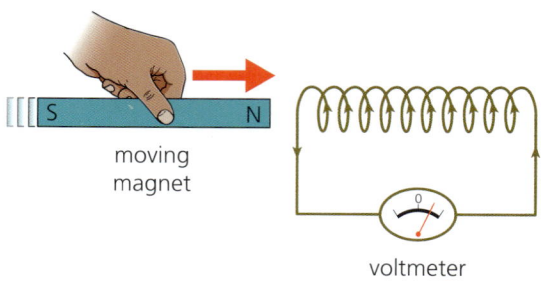

moving magnet

voltmeter

magnet not moving

voltmeter

▲ *Either the magnet or the coil has to move to induce a voltage.*

If she moves the magnet away from the coil there is a negative reading on the voltmeter.

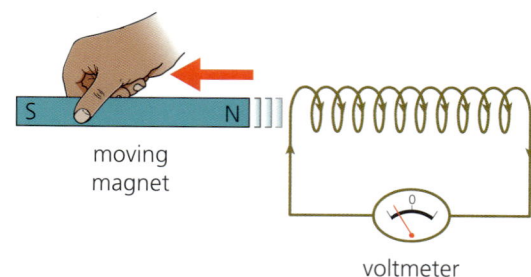

moving magnet

voltmeter

We say that a voltage has been produced or **induced** in the coil of wire. If you replace the voltmeter with a lamp or other circuit component, a current will flow in the circuit.

Bulan tries the experiment again but this time she holds the magnet still and moves the coil of wire towards and away from the magnet. She sees the same thing happening on the voltmeter.

- If she moves the magnet towards the coil the reading is positive.
- If she moves it away from the coil the reading is negative.

She realises that to induce a voltage you need *movement*, but it does *not* matter whether the magnet is moving or the coil is moving.

Inducing a bigger voltage

Bulan's friend Utari says that she has worked out how to make the reading on the voltmeter bigger. Utari is correct. To make the induced voltage bigger, she can:

- move the magnet (or the coil) faster
- use a coil with an iron core
- use more turns on the coil
- use a stronger magnet.

I think that the reading will be bigger if you move the magnet faster.

A simple generator

Simple **generators,** like **dynamos,** usually contain a coil of wire and a magnet that spins when you turn the handle. The moving magnet induces a voltage.

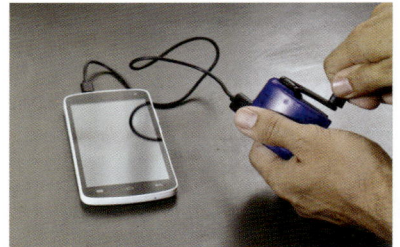 

 You can generate electricity by winding a handle.

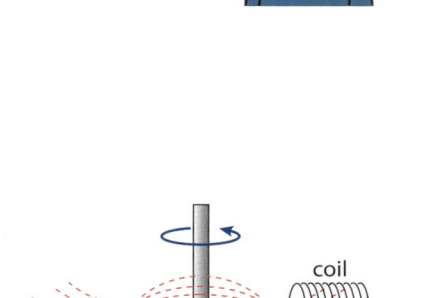

▲ Spinning a magnet near a coil lights a lamp.

Power station generators

The generators inside a power station are very, very big. They use moving electromagnets instead of permanent magnets. This is because electromagnets can produce a much stronger magnetic field than a permanent magnet, so the induced voltage will be bigger.

Science in context

The ability to generate electricity has had a huge impact on human society.

▲ Generators in a power station.

Questions

1. Describe what is meant by an 'induced voltage'.
2. Utari moves a magnet into her coil of wire with the north pole first. This induces a voltage, and the voltmeter reading is positive. Then she moves the magnet into the coil with the south pole first. Write down the type of voltage produced.
3. Many people use battery-powered lights on their bicycles. You can use a dynamo (a little generator) instead. As you pedal, the wheel moves the dynamo, which powers the lights.
   a. What is one disadvantage of using a bicycle dynamo instead of a battery-powered light?
   b. What is one advantage of using a bicycle dynamo instead of a battery-powered light?

Key points

- When a magnet moves near a coil of wire, or a coil of wire moves near a magnet, a voltage is induced.
- Large-scale generators are made using very large coils and powerful electromagnets.

Non-renewable resources: Fossil fuels

Objectives

- Describe how fossil fuels were formed
- Explain how a fossil-fuel power station works
- Describe uses of fossil fuels

Why are **fossil fuels** called 'fossils'? How does that make them non-renewable?

We depend on fossil fuels but they will run out, and this may be a problem in the future.

▲ Oil is a fossil fuel.

Uses of fossil fuels

The main uses of fossil fuels are:

- generating electricity (all three are used, but mainly coal and gas)
- transportation (oil is used to produce petrol and other fuels)
- heating (all three can be burned to heat houses).

Fossil-fuel power stations

The stored energy in fossil fuels can be used to generate electricity.

The diagram below shows how a **coal-fired power station** works. Oil- and gas-fired power stations work in the same way.

▲ Gas is a fossil fuel.

▲ Coal is a fossil fuel.

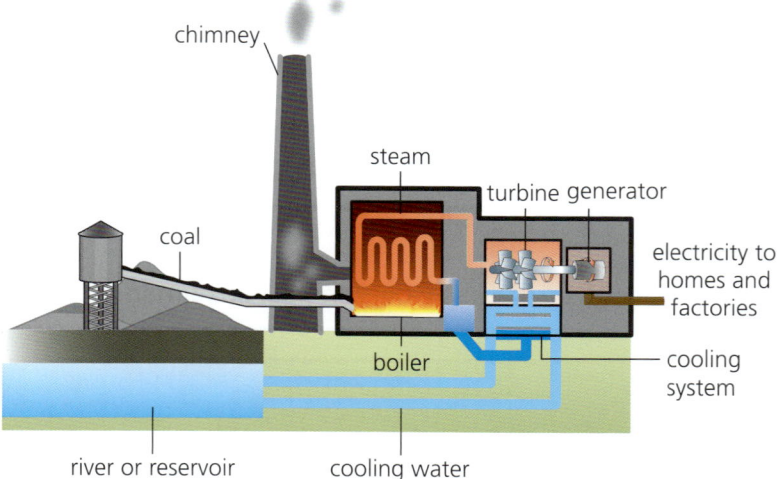

- In the power station the coal, oil, or gas is burned to turn water into steam.
- The steam drives a **turbine**, which is like a giant fan spun by the steam.
- The turbine is connected to the generator to produce a voltage.
- The steam is turned back to water in the condenser.

Fossil fuel power stations have advantages and disadvantages.

Advantages	Disadvantages
electricity produced constantly	produces carbon dioxide
very reliable	contributes to climate change
fossil fuels readily available	produces pollution

You can find out more about pollution in your *Cambridge Lower Secondary Complete Chemistry* Student Book.

Most of the energy wasted in generating electricity heats the surroundings. You could say that a power station is better at heating than generating electricity. Some power stations transfer hot water to nearby homes and factories.

How were fossil fuels formed?

Fossil fuels formed from plants and animals that lived millions of years ago.

Coal is made from fossilised trees, and oil is made from fossilised sea creatures.

Scientists are developing new ways to find fossil fuels underground. They can only make an estimate of the amount of coal, oil, and gas there is in the world, and the effects of burning it.

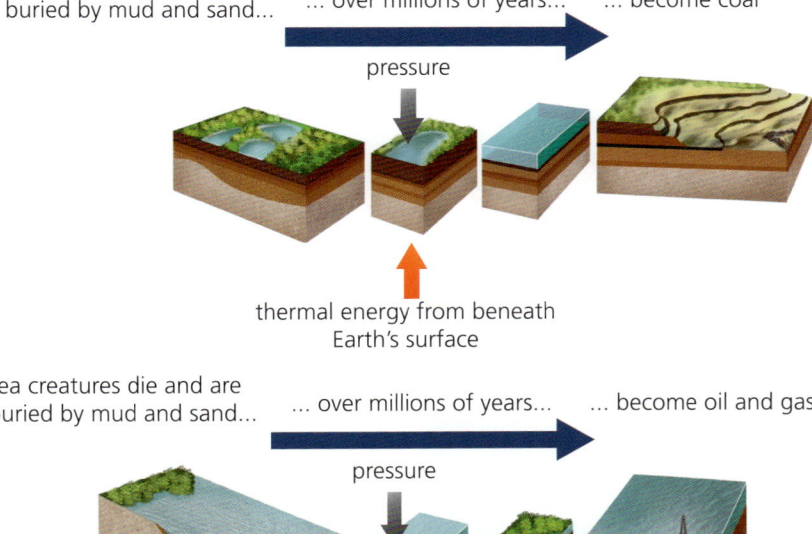

trees die and are buried by mud and sand... ... over millions of years... ... become coal

pressure

thermal energy from beneath Earth's surface

sea creatures die and are buried by mud and sand... ... over millions of years... ... become oil and gas

pressure

thermal energy from beneath Earth's surface

▲ Fossil fuels take millions of years to form.

Science in context

The impact of electricity generation on the environment is one of the biggest challenges that we face. Societies rely heavily on a ready supply of electricity to industries and homes. However, the serious effects of climate change are becoming obvious.

Key points

- In a fossil-fuel power station the fuel is burnt to turn water into steam.
- Steam turns the turbine, which turns a generator to generate electricity.
- Burning fossil fuels produces carbon dioxide and pollution.
- Coal, oil, and gas are the remains of plants and animals that lived millions of years ago.

Questions

1. Copy and complete these sentences.

   When fuel is burned in a power station the energy is transferred to turn _____ into _____. This drives a _____, which drives a generator to produce electricity.

2. Explain why coal, oil, and gas are called fossil fuels.
3. Explain why fossil-fuel power stations are built near rivers or the sea.
4. Explain why you might want to build a coal power station even though it produces pollution and carbon dioxide.
5. Describe one similarity in the formation of coal and of oil. Describe one difference.

Renewable resources: Solar and geothermal

Objectives

- Describe how the energy from the Sun can be used to generate electricity
- Explain how energy from the Earth can be used to generate electricity

This is the Mithapur solar power station. There are over 100 000 solar cells covering 0.4 square kilometres. India has about 300 sunny days a year and could generate all the electricity it needs using solar cells.

Using energy from the Sun

We mainly use energy from the **Sun**, or **solar energy**, for:

- heating homes and water heating
- generating electricity using solar cells.

▲ *Solar cells generating electricity from solar energy.*

Solar cells

A **solar cell**, or photovoltaic cell, converts the energy of light directly into a voltage. Unlike a generator, a solar cell has no moving parts.

- Each solar cell on its own produces a small voltage.
- Lots of solar cells are connected to produce a big voltage.
- The efficiency of solar cells has increased from 5% in the 1970s to more than 30% today.
- This means that now fewer solar cells can produce the same voltage.

The table lists some advantages and disadvantages of solar cells.

▲ *For some objects like this satellite, solar is the only source of energy.*

Advantages	Disadvantages
can be used in individual houses/ small devices with low power needs	don't work at night
can be used in remote locations	expensive to produce
can work on cloudy days	greenhouse gases made during production
renewable – will *not* run out	

Solar water heating

Solar water heating uses **solar panels** instead of solar cells.

- The heater consists of pipes inside a glass-topped panel.
- The pipes are painted black and are usually made of copper.
- Cold water from a tank flows through the pipes.
- The water is heated by the Sun.
- Hot water flows back to the tank.

hot water out

storage tank

cold water in

pipe inside the panel is painted dull black

insulation

glass-covered box

dull black surface

▲ *Energy from the Sun can heat water.*

Using energy from the Earth

Geothermal energy, from the Earth, is mainly used for:

- heating homes and water using heat pumps
- generating electricity by pumping water underground to heat it.

Geothermal power

The rocks beneath Earth's surface are very hot. You can find out more about the structure of the Earth in your *Cambridge Lower Secondary Complete Chemistry* Student Book. If you drill down into the Earth the temperature rises by about 25 °C for every kilometre down. In a geothermal power station:

- you pump water into pipes deep underground
- the water is converted into steam
- the steam drives a turbine, which drives a generator.

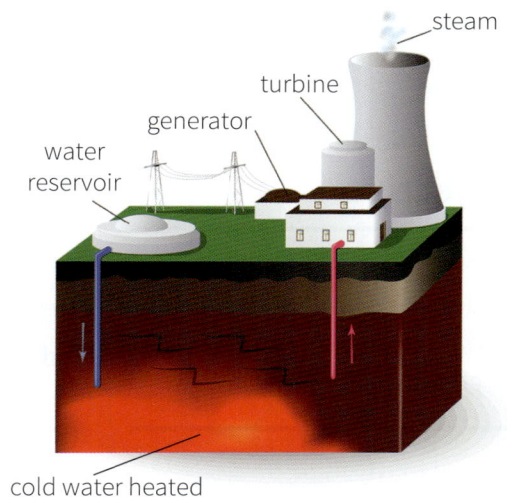

steam
turbine
generator
water reservoir
cold water heated to produce steam

Advantages	Disadvantages
renewable – will *not* run out	expensive to build
does *not* produce greenhouse gases while running	greenhouse gases released during building

Heat pumps

Some people use the thermal energy from the Earth to heat their houses in the winter, or cool them down in the summer, by using a **heat pump**.

- It pumps water through pipes in the ground.
- In the winter the water gains a little thermal energy, which can be transferred to heat the house.
- In the summer the ground is cooler than the house, so the same system can be used to cool the house.

▲ *Pipes are buried deep in the rock below this geothermal power station in Olkaria, Kenya.*

Questions

1. Describe the difference between a solar (photovoltaic) cell and a solar hot water panel.
2. Explain why you need to connect lots of solar cells together.
3. Estimate how many times more efficient a modern solar cell is than an early version from the 1970s.
4. Look at the coal-fired power station on page 170. Compare it with the geothermal power station above. Describe one similarity and one difference between the two power stations.
5. The power output of the Mithapur solar power station is about 25 MW. The power output of the Olkaria II power station is about 100 MW. Calculate the approximate number of solar power stations that would produce the same power as the Olkaria II geothermal power station.

Key points

- Solar energy can be used to produce electricity directly using solar cells.
- Energy from the Sun or Earth can also be used to heat water.
- Energy from deep inside the Earth can be used to turn water into steam. The steam can turn a turbine, which can drive a generator.

Renewable resources: Water and wind

Objective

- Describe how wind, waves, tides and water behind dams can be used to generate electricity

People have harnessed water and the wind to power machines for thousands of years. Windmills and water mills were used to pump water and grind flour. Now we mainly use water and wind to generate electricity.

Hydroelectricity

Hydroelectricity is another method of generating electricity that does not use fossil fuels. The photo shows Indonesia's largest hydroelectric power station, Cirata, 100 km south-east of Jakarta.

- A river fills up a big **reservoir** behind a dam.
- Gates are opened to allow water from the reservoir to flow down pipes inside the dam.
- The water in the pipes turns the turbines.
- The turbines drive generators to produce electricity.

▲ A hydroelectric power station: the dam is 125 m tall.

Advantages	Disadvantages
produces electricity on demand	expensive to build
renewable - will *not* run out	greenhouse gases made during building
does *not* produce greenhouse gases while running	needs a suitable place, usually with mountains and lakes
	expensive to connect to houses
	flooding valleys can displace people and affect the environment

Wave power

Wave power is generated with a chamber on the shoreline. The waves force water up and down inside it. The motion of the water forces air backwards and forwards through a turbine, which makes a generator spin to generate electricity.

- Wave power does *not* produce a constant supply of electricity.
- The chamber needs to be very strong to resist storms.
- You need a suitable site to build the chamber.

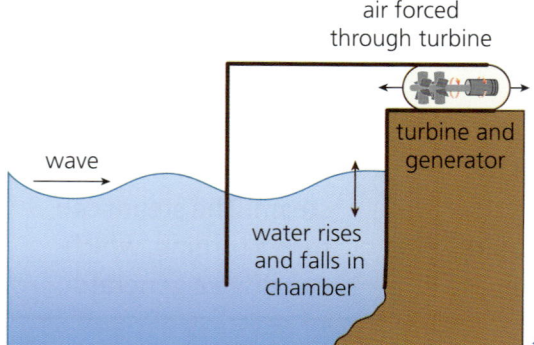

◀ A wave power station.

Tidal power

The largest **tidal power** station in the world is the Sihwa Lake tidal power station in South Korea. A sea wall contains ten large underwater turbines. This is sometimes called a 'barrage'. As the tide comes in the turbines spin and they drive generators to generate electricity. They spin again when the tide goes out. Tidal power is a bit like hydroelectric power.

▲ *A tidal power station.*

- There are tides every day so a tidal power station can produce electricity for about 10 hours each day.
- They can affect the environment. For example fish cannot swim upriver to breed.
- Tidal barrages are very expensive to build.
- As with hydroelectricity and wave power, greenhouse gases are produced when barrages are constructed but *not* when they run.

Wind

A **wind turbine** is a large windmill that is used to generate electricity. Although it is called a wind turbine it contains a turbine *and* a generator. Groups of wind turbines make **wind farms**. A wind turbine can produce a power of about 2 MW.

The Tamil Nadu wind farm in India produces just under 6000 MW from its collections of wind turbines.

- The output of a wind turbine is *not* always predictable because the wind does *not* always blow.
- Some people like the way that wind turbines look, but others do *not*.
- They can be noisy and can kill birds.
- They can be expensive to build if they are offshore (in the sea).

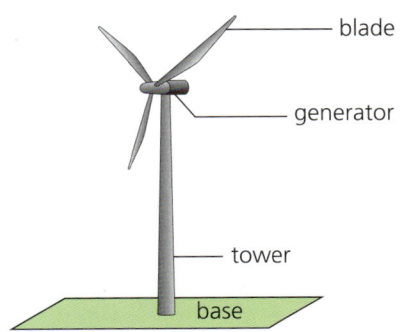

blade

generator

tower

base

▲ *A wind turbine.*

 **Key points**

- Water in lakes, wind, waves, and tides can generate electricity.
- These are all renewable.
- Moving air or water is used to turn a turbine, which drives a generator.
- There are advantages and disadvantages to all of these methods.

Renewable resources: Biofuels and bioplastics

Objectives

- Describe how we use plants to make plastics
- Describe how we use plants to make fuels

▲ *Soybeans are used to make biodiesel.*

▲ *Sugar beet is used to make bioethanol.*

Plastics made of oil are thrown away in huge amounts, and plants could be the solution. Imagine a plastic container that will decompose and *not* pollute the environment!

Fuels made from plants are also a possible solution to the problem of running out of fossil fuels.

What is a biofuel?

Plant and animal material are called **biomass**. Some biomass, such as wood, can be used as a **biofuel**. A biofuel is a fuel from living things, such as trees, or animal waste. These fuels have been used for thousands of years.

▲ *A plastic fork will not decompose for a thousand years.*

The Sun is the source of the energy in biofuels.

- Plants use **sunlight** to make glucose by **photosynthesis**. They turn some of the glucose into starch and cellulose.
- Starch is a complex carbohydrate that plants use as a store of energy.
- They often store starch in their roots and stems to help them survive difficult conditions. They also store some starch in seeds and other cells.
- All plants have cellulose in their cell walls to help support them.

There are different types of biofuel.

- Wood is one of the most common biofuels used throughout the world for thousands of years.
- **Biodiesel** is a biofuel made from plant oils.
- **Bioethanol** is a biofuel made from sugar.
- **Biogas** is a biofuel that can be piped from landfill (rubbish) sites, or other sites where biomass is decomposing.

▲ *Gas produced from waste is biogas.*

All sources of energy have a cost. The energy from biofuels is *not* free. You need to buy a furnace to burn the fuel and a generator to generate electricity. Some people build biogas digesters, which use waste material from plants and animals to produce gas that can be burned. Some of the advantages and disadvantages of biofuels are listed in the table.

Advantages	Disadvantages
can be used to generate electricity in remote locations	produce carbon dioxide when they burn, which can contribute to climate change
fuels can be obtained from waste products	more expensive than fossil fuels
produce less pollution when they burn	need a lot of land to grow crops for some fuels
fossil fuel power stations can be converted to use biofuels	most cars are *not* designed to run only on biofuels

 Science in context

All methods of generating electricity have risks or consequences for humans and for the environment. This includes renewable energy sources.

What is a bioplastic?

Plastic waste, such as cups and bags, does *not* break down easily. **Bioplastics** are made from starch. Starch can be broken down by decomposers, so bioplastics are **biodegradable**. You learned about decomposers in your *Cambridge Lower Secondary Complete Biology* Student Book.

▲ *These drinking glasses are made from maize starch.*

▲ *This packaging and these utensils are made from starch.*

▲ *These cups are made from polylactic acid, obtained from plant carbohyhdrates.*

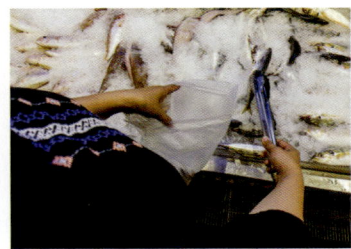

▲ *This bag is made from starch.*

What are the problems with biofuels and bioplastics?

At first glance biofuels and bioplastics look like the solution to a lot of problems with resources and pollution. However, producing biofuels and bioplastic needs land to grow plants, where food could be grown instead.

You might think that using biofuels and bioplastics does *not* contribute to climate change. However, carbon dioxide is still released when machinery is used to farm the crops, when biofuels are burnt, and when bioplastics are manufactured.

 Key points

- Biofuels are fuels made from biomass, which is plant and animal material or waste.
- Biofuels include wood, bioethanol, biodiesel, and biogas.
- Bioplastics are made from starch instead of oil.

② Questions

1. Define 'biofuel' and 'bioplastic'.
2. Explain why biodiesel is a renewable fuel.
3. Write down two similarities and two differences between biofuels and fossil fuels.
4. Suggest why most plastic cups are *not* made of bioplastic.

1. Sort the following energy resources into the correct columns in the table. **[3]**

   electricity coal oil power station solar cells
   wood petrol water refineries

Primary energy sources	Primary energy converters	Secondary sources/carriers

2. Explain what is meant by:

   a. photosynthesis **[1]**

   b. fossil fuel **[1]**

   c. hydroelectricity **[1]**

   d. solar cell. **[1]**

3. Here is the graph showing the changes in the world's energy consumption over two centuries.

   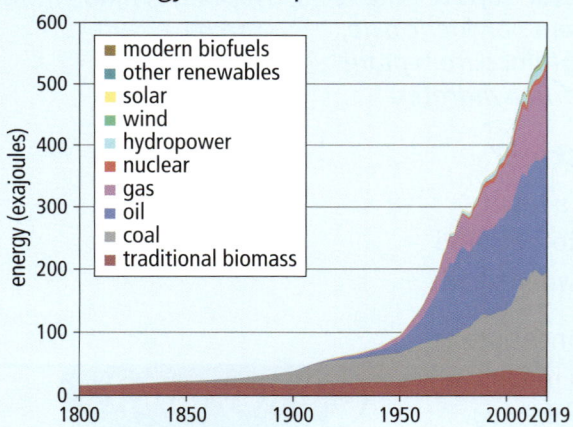

   a. Name the main fuel that is used as 'traditional biomass'. **[1]**

   b. Use the graph to predict the total world energy consumption in 2030 in exajoules. **[1]**

   c. Name two 'modern biofuels'. **[2]**

   d. Material used for biofuels can also be used for bioplastics. Give one use of a bioplastic. **[1]**

   e. Give one advantage of using bioplastics over plastics from oil. **[1]**

   f. Suggest and explain whether bioplastics contribute to climate change. **[2]**

4. A student is investigating solar cells. She has seen that people use solar cells on the roofs of their houses to generate electricity. She wonders how clouds affect the output of the cell. **TWS**

   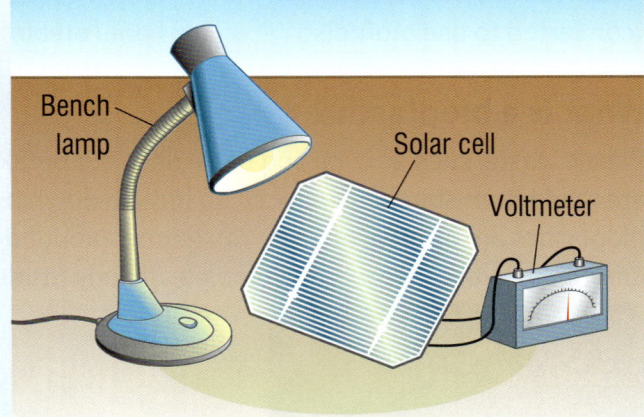

   She puts a solar cell on the desk below a lamp and connects it to a voltmeter. She puts thin layers of clear plastic film on top of it to see whether the voltage output of the solar cell is affected by them.

   a. State the question that she is investigating. **[1]**

   b. Draw a table that she could use for her results. **[3]**

   c. State and explain the type of graph that she could draw. **[2]** **TWS**

5. Here is a list of energy sources.

A	coal	**G**	gas
B	sunlight	**H**	tidal
C	oil	**I**	biomass
D	hydroelectricity	**J**	waves
E	uranium	**K**	electricity
F	wind	**L**	geothermal

Give the letter or letters of the sources that are:

a. renewable resources [1]

b. non-renewable resources [1]

c. things that you burn to produce steam [1]

d. things from underground [1]

e. things that involve water. [1]

One of them is the odd one out.

f. Which one, and why? [2]

6. Are these statements about generating electricity true or false?

a. If you hold a magnet near a coil of wire a voltage will be induced. [1]

b. You move a magnet out of a coil and the induced voltage is positive, so when you move it in it will also be positive. [1]

c. It doesn't matter whether you move the coil or move the magnet to induce a voltage. [1]

7. Draw a diagram of a fossil fuel power station. Label each of the following.

a. the turbine [1]

b. the generator [1]

c. the steam [1]

d. the burning fuel [1]

e. the water. [1]

Write down which of the items from your diagram:

f. you would also find in a geothermal power station [1]

g. are also needed to generate electricity using the wind. [1]

8. Look at the list of energy sources in question 5. Here is a list of definitions, advantages, and disadvantages. For each one write down which energy source(s) each statement applies to. You may need to use the same source twice.

a. It uses water stored behind dams. [1]

b. It contributes to climate change by producing carbon dioxide. [1]

c. Building it can cause damage to the environment from flooding. [1]

d. It is unreliable because it depends on the weather. [1]

e. It produces electricity on demand or all the time. [1]

9. Here are some secondary data about solar cells, showing the area of the cells, the power produced, the energy produced per year, and the mass of carbon dioxide that would be saved because you are *not* burning fossil fuels when you use the cells.

Area (m^2)	Power (kW)	Energy produced per year (kWh)	Mass of carbon dioxide saved per year (kg)
10	1	2000	500
20	2	4000	1000
30	3	6000	1500
40	4	8000	2000

a. Calculate the area of solar cells needed to power a 2 kW kettle. [1]

b. A family wants to save 7000 kWh per year. Calculate the area of solar cells needed. [1]

c. Calculate how much carbon dioxide you would save if you used 25 m^2 of solar cells on your roof. [1]

d. The data are provided by a company which sells solar cells. Do you think that the energy produced as shown in the table is the maximum or the minimum energy that you would get? [2]

Explain your answer.

10.1

Galaxies

Objectives

- Describe what is in a galaxy
- Describe what is in the Universe

The *Voyager 1* and *Voyager 2* space probes were launched from Earth in 1977. In 1979, *Voyager 1* passed Jupiter. Since then, they have both passed through the **Kuiper belt** – the rocks and dust beyond Neptune – and beyond the edge of the Solar System. *Voyager 1* has now travelled further than anything else made by humans, and is about 23 billion kilometres away from us. Both *Voyagers* will spend over 10 000 years travelling through the **Oort cloud** of objects weakly held by the gravity of the distant Sun. When they exit that, they will still be inside the **Milky Way galaxy**.

▲ *Several images taken by* Voyager 1 *were combined to give this picture of the surface of Jupiter and its moon Europa casting a shadow.*

▲ *The Sombrero galaxy is one of thousands of millions of galaxies in the Universe.*

What is a galaxy?

A **galaxy** consists of

- stars – usually thousands of millions of them
- planetary systems – astronomers think that most stars have planetary systems
- stellar dust and gas – from stars that have exploded, or left over from the formation of the Universe.

Our galaxy is just one of thousands of millions of galaxies in the Universe.

Our galaxy

The word 'galaxy' comes from the Greek word for 'milky'. If you look at the night sky from a dark place, you will see a band of stars. These are stars in our galaxy, the Milky Way.

- The nearest star to our Sun is Proxima Centauri. It has a diameter seven times smaller than our Sun's diameter.
- There is at least one exoplanet orbiting Proxima Centauri, making it our nearest solar system.

Our Solar System and the Proxima Centauri solar system are part of the Milky Way galaxy.

- Most of the stars that you can see are stars in the Milky Way.
- Astronomers think our Sun and the other stars in our galaxy are all orbiting a huge **black hole** at the centre.
- The Milky Way is a spiral galaxy. It is shaped with curved arms making a disc with a bulge in the middle.

▲ *The Milky Way is the hazy white band through the centre of the sky.*

The images that we have of the Milky Way disc are models – they are *not* photographs that anyone could take from inside it using a camera. Scientists have used observations with telescopes to work out the shape of our galaxy.

The hazy white band in the sky is the concentration of stars that we see when we look along the disc. When we look in other directions there aren't as many stars.

Other galaxies

Not everything that you see in the night sky is a star in the Milky Way. Some of the bright, fuzzy objects are other galaxies beyond the Milky Way.

- One of the most spectacular galaxies is the Andromeda galaxy.
- Andromeda is the nearest spiral galaxy to our own.
- You can also see two of our other neighbouring galaxies, the Large and Small Magellanic Clouds.
- Not all galaxies are spiral.

You need a telescope to see the galaxies in any detail. With the naked eye they look like fuzzy discs.

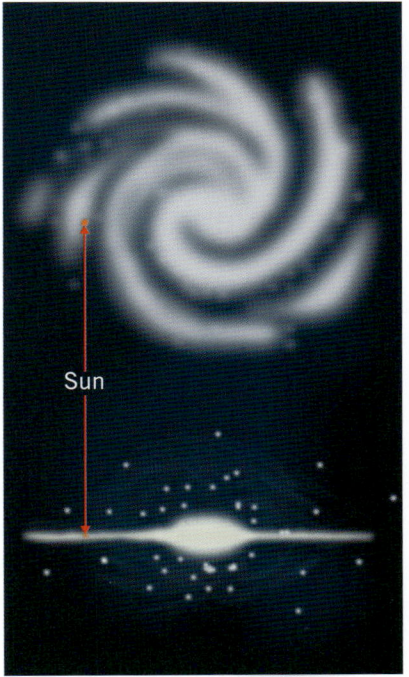

▲ *We are on a planet in a solar system located on a spiral arm of the Milky Way galaxy.*

The Universe

The Universe is everything that exists. All the stars, planets, moons, comets, asteroids, meteors, and everything else make up the Universe, including you! Some people ask what is outside the Universe. This is *not* a question that science can answer because there may be no 'outside'. We don't know how large the Universe is or whether it is even **infinite**, but we do know that we can see for 13.7 billion light years in all directions.

▲ *The Triangulum galaxy is a spiral galaxy 2.7 million light years from Earth.*

Key points

- Our Solar System is surrounded by gas and dust in the Kuiper belt and Oort cloud.
- A galaxy consists of stars, planetary systems, stellar dust, and gas.
- There are thousands of millions of stars in a galaxy.
- There are thousands of millions of galaxies in the Universe.

Questions

1. Imagine you are in a spacecraft moving away from the Earth. You reach the orbit of Neptune. Copy and complete these sentences.

   At the edge of the Solar System I enter the _____ _____. Beyond that I will be in the _____ _____, where we think some comets come from. If I carried on, I would travel out of our _____, which is called the Milky Way.

2. Describe the difference between a solar system and a galaxy.
3. Explain why it is *not* possible to count all the stars in the Milky Way.

Objectives

- Describe asteroids
- Describe how asteroids form
- Describe evidence for asteroid impacts on celestial bodies

Asteroids

In 2017, for the first time, an object from **interstellar** space was detected in our Solar System. At first astronomers thought it was a **comet**. Then they classified it as an **asteroid**. What are asteroids, and where do they come from?

▲ 'Oumuamua is an unusually shaped asteroid that passed through our Solar System.

What is an asteroid?

An asteroid is rock that orbits the Sun. It is smaller than a planet, but some are large enough to be called a 'minor planet'.

- Asteroids can range in size from less than 10 m to over 500 km in diameter.
- Most asteroids are irregularly shaped, although some are nearly spherical.
- The total mass of all the asteroids in our Solar System is less than the mass of the Moon.
- More than 100 asteroids also have moons!

The word 'asteroid' means 'star-like shape'. The person who named them over 200 years ago did not know what they were, but could see them using a telescope.

How do asteroids form?

Asteroids are left-over debris from the formation of the Solar System. As you learned in Chapter 5:

- The Solar System formed when gravity pulled dust and gas together about 4.6 billion years ago.
- The pieces of rock that were left over are what we call asteroids.
- Jupiter's gravity may have broken up some smaller planets into pieces of rock that became asteroids.

We now know that most stars have a planetary system, just like the Sun. Those systems probably have asteroids for the same reason that there are asteroids in our Solar System.

Where are the asteroids?

In our Solar System there are over a million asteroids. Most of them are located in the **asteroid belt** between Mars and Jupiter, but some are in the same orbit as larger planets, such as Jupiter.

Some are in orbits that bring them close to Earth. Astronomers watch these carefully in order to give a warning if one of them might hit the Earth.

▲ Asteroids have many shapes and sizes.

▲ The position of the asteroid belt (not to scale).

Collisions between asteroids can break off pieces of rock that become meteors when they hit the Earth.

Asteroid, meteor, comet: what's the difference?

It can be hard to remember the differences between objects that can be seen in the sky.

Object	Definition	Approximate size
Asteroid	A large piece of rock orbiting the Sun	Smaller than a planet, larger than a meteoroid
Meteoroid	A smaller piece of rock that has broken off from an asteroid or comet	1 m in diameter or smaller
Meteor	A smaller piece of rock that has entered the Earth's atmosphere	1 m in diameter or smaller
Meteorite	A meteor that reaches the ground (is *not* burnt up in the atmosphere)	1 m in diameter or smaller
Comet	An object made of ice and dust orbiting the Sun that has a tail when it gets close to the Sun	about 10 km in diameter

Meteorites are sometimes found on the ground. These are fragments of meteoroids that have mostly vaporised in the atmosphere.

▲ *Part of a 500 kg meteorite made of rock that landed in Chelyabinsk, Russia, in 2013…*

▲ *… and a 37 g meteorite made of iron and nickel found in Arizona, USA.*

Questions

1. Copy and complete these questions:
   An asteroid is a piece of _____ that orbits the _____.
   Asteroids usually have _____ shapes. Most asteroids are in an orbit between _____ and _____.
2. Explain how asteroids formed.
3. Explain why the Moon is *not* an asteroid, even though it is made of rock.
4. Explain why asteroids are *not* stars.

Key points

- Asteroids are irregularly shaped pieces of rock, mostly in orbit between Mars and Jupiter.
- Asteroids are pieces of rock left over from the formation of the Solar System.

Magnetic Earth

A compass is useful if you want to know which way is north. It works because there is a magnetic field around the Earth. Why is there a magnetic field, and what would happen if there was no magnetic field?

Objectives

- Explain what a compass does
- Describe the Earth's magnetic field
- Explain why the Earth has a magnetic field

▲ *The red end of the compass needle points north.*

What is the shape of the Earth's magnetic field?

If you suspend a magnet on a piece of cotton, the magnet will line up in the same direction each time. This direction is north–south. A **compass** needle points north because the needle is a small magnet. It swings and lines up with the Earth's magnetic field. The Earth's magnetic field exerts a force on it.

- Compass needles actually point to the Earth's *magnetic* pole.
- This is *not* the same place as the Earth's *geographic* North Pole, which is at the 'top' of the Earth's axis of rotation.
- The magnetic north pole is in Canada, about 400 km away from the geographic North Pole.
- The magnetic north pole is moving at about 30 km a year from Canada towards Russia.
- The compass needle moves in two directions, because the Earth's magnetic field can have a 'dip' as well.

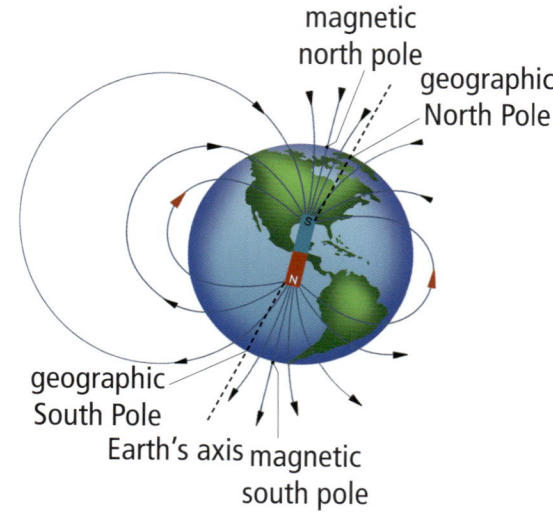

▲ *The Earth's magnetic field.*

Science in context

People have used compasses to navigate since ancient times. Without compasses a lot of the exploration of the world would have been a lot more difficult, or taken a lot longer.

Scientists have also learned that some birds and insects rely on the Earth's magnetic field to navigate.

Why is there a magnetic field around the Earth?

The Earth behaves as if there is a huge bar magnet inside it. The magnetic field pattern is like that of a bar magnet. If a compass needle points north, then the *south* pole of the Earth's magnet must be in the north! This means that you draw the arrows on the Earth's magnetic field lines pointing from the magnetic south pole to the magnetic north pole.

There is no bar magnet inside the Earth, and scientists think that the outer core of the Earth is made of iron that can move. This produces a magnetic field like a current flowing in a wire. Scientists are not sure why the magnetic north pole moves as it does.

There is evidence that the Earth's magnetic field has 'flipped' in the past so that the magnetic north and south poles changed places. This happened 780 000 years ago, and may happen again in the future.

crust
mantle
outer core
inner core

▲ *The outer core is like an electromagnet.*

Why is the Earth's magnetic field important?

If the Earth did *not* have a magnetic field then there would be no life on Earth. The magnetic field protects the Earth. The Sun emits radiation in the direction of the Earth. This is called the **solar wind**. It is *not* the sort of wind that you feel when air moves around you. It is made of electrically charged particles that can damage or kill cells, so living organisms would die.

The Earth's magnetic field deflects the solar wind so most of it does *not* reach the surface of the Earth.

Not all of the solar wind is deflected. Some of it can reach the atmosphere at the north and south poles. This produces an **aurora** – dramatic lights in the sky called the northern and southern lights.

▲ *Most of the solar wind is deflected by the Earth's magnetic field.*

▲ *The southern lights seen from Antarctica.*

 Questions

1. Explain why a compass needle moves.
2. Describe the shape of the Earth's magnetic field.
3. A student uses a magnet that can rotate in three dimensions. She notices that it points diagonally downwards.
   a. Suggest why.
   b. Suggest where she would have to go to see the needle point directly downwards. Explain your answer.

Key points

- The Earth behaves as if it has a bar magnet inside it.
- Compasses point north because the needle is a magnet that lines up with the field.
- Scientists think that the Earth's magnetic field is produced by moving currents of iron in the Earth's core.

1. Copy and complete these sentences using the words below. You may need to use the words once, more than once or not at all.

   millions billions gas rock planetary
   galactic

   a. A galaxy contains stellar dust and _____, stars, and _____ systems. **[2]**

   b. A galaxy contains _____ of stars. **[1]**

   c. The Universe contains _____ of galaxies. **[1]**

2. Give the letters of the true statements. **[1]**

   A Galaxies contain stars.

   B Stars contain galaxies.

   C Our galaxy is called Andromeda.

   D Our galaxy is a spiral galaxy.

3. Here is a diagram of the solar system.

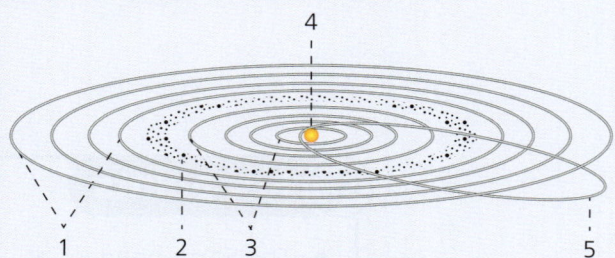

   a. Give the number of the asteroid belt. **[1]**

   b. Give one similarity between an asteroid and a planet. **[1]**

   c. Give one difference between an asteroid and a planet. **[1]**

   d. Give one reason to expect that other solar systems in our galaxy have asteroids. **[1]**

4. **a.** Give the name of the force that pulled together stellar dust and gas to produce asteroids. **[1]**

   b. Match each object in the table below with the correct definition by giving the letters and the numbers. **[4]**

A	Asteroid	1	A smaller piece of rock that has entered the Earth's atmosphere
B	Meteoroid	2	A large piece of rock orbiting the Sun
C	Meteor	3	A piece of rock that reaches the ground (is not burnt up in the atmosphere)
D	Meteorite	4	A smaller piece of rock that has broken off from an asteroid or comet

5. The Earth has a magnetic field. Give an explanation for the origin of this magnetic field. **[2]**

6. The Earth has a magnetic field shaped as if it had a bar magnet through its centre.

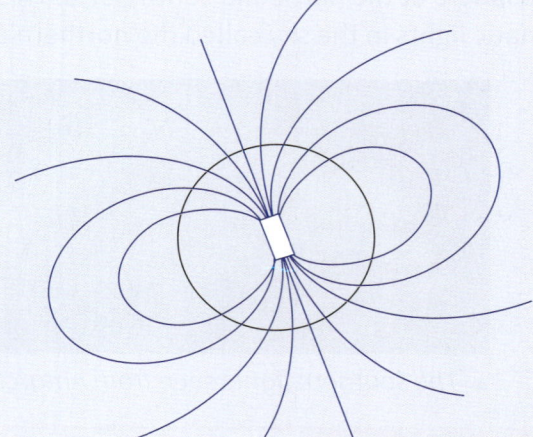

   a. Copy the diagram of the Earth with a bar magnet at the centre and label the poles of the imaginary magnet. **[1]**

   b. Explain your choice of labels. **[1]**

   c. Add arrows to the magnetic field lines. **[1]**

d. A compass needle shows north by lining up with the Earth's magnetic field. What would happen if you were trying to use a compass near a large piece of lodestone? [1]

7. An astronomer uses a telescope to look at some astronomical objects.

a. Suggest which of the objects is an asteroid. [1]

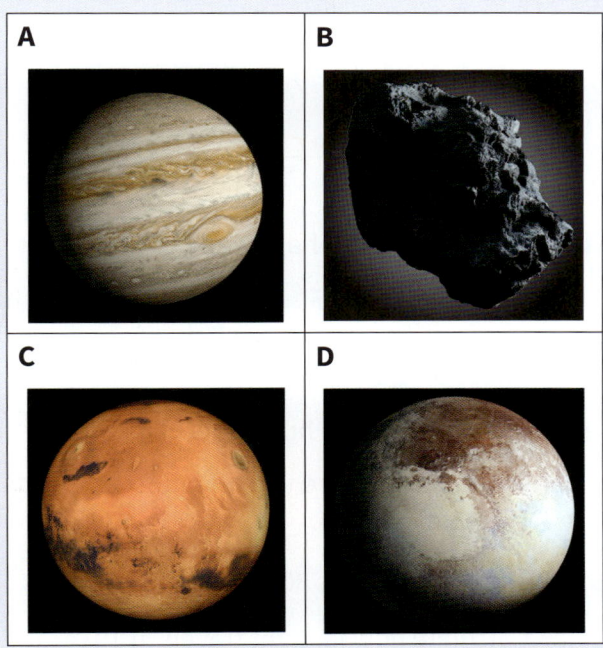

b. Explain your answer to part **a**. [1]

c. Give the names of the two planets that are likely to be either side of the location of the asteroid. [2]

8. Here is a list of astronomical objects.

A an asteroid in our Solar System

B the Andromeda galaxy

C a planet around a star called Proxima Centauri

D a star in our galaxy called Pegasi

E Jupiter.

a. Give the letters of two things that were made from the same gas and dust. [1]

b. Give the letters of objects that are roughly spherical. [1]

c. Give the letters of objects that were made from gas and dust. [1]

9. An astronomer uses a telescope to look at some astronomical objects.

a. Space probes have been sent to investigate the planets of the Solar System. They have gathered information about the surface, the atmosphere, the gravitational field strength, and the magnetic field strength.

This table shows the strength of the planets' gravitational fields and magnetic fields compared with those of the Earth.

Planet	Made of...	Gravitational field strength	Relative magnetic field strength
Mercury	Rock	0.37	0.006
Venus	Rock	0.89	0
Earth	Rock	1.00	1.0
Mars	Rock	0.37	0
Jupiter	Gas	2.31	19 519
Saturn	Gas	0.90	578
Uranus	Gas	0.87	48
Neptune	Gas	1.10	27

i. Suggest a link between what the planet is made of and the magnetic field strength. [1]

ii. Suggest a possible reason for the link in part **i** using what you know about the probable cause of the Earth's magnetic field and about particle theory. (*Hint:* remember what you learned about electromagnets.) [2]

b. A student wishes to investigate the link
TWS between gravitational field strength and magnetic field strength.

i. Suggest and explain which type of graph they should plot. [2]

ii. Explain why it would be difficult to plot the points for magnetic field strength. [1]

iii. Do you expect to see a link between
TWS the magnetic and gravitational field strength? Explain your answer. [2]

c. Space probes have observed auroras on other planets. Suggest and explain which ones. [2]

1. Look at this distance–time graph for a cyclist.

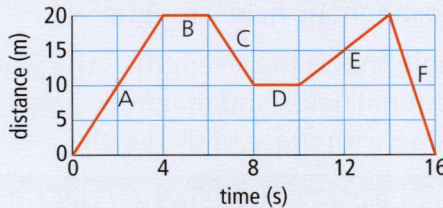

    a. Calculate the speed for each of the sections of the graph A–F. **[6]**

    b. Explain why the graph is not very realistic. **[1]**

2. A cyclist is cycling at a steady speed.

    a. Name two drag forces acting on the cyclist. **[2]**

    b. Compare the drag forces acting on her with the forward force that she applies through the pedals. Explain your answer. **[2]**

She crouches forward and makes herself more streamlined, but continues to pedal with the *same* force.

    c. Describe what will happen to her speed. Explain your answer. **[4]**

3. A fisherman uses a nylon fishing line to catch a fish and pull it out of the water.

    a. The original length of the line is 60 cm. When the fish is on the line and out of the water it is 62 cm. Calculate the extension. **[1]**

    b. He notices that the fish feels a lot heavier when he has taken the fish out of the water than it did when it was in the water. Explain why. **[1]**

    c. The fisherman uses the line to catch a second fish with half the mass of the first one. What will happen to the extension? Choose from these options: **[1]**

        It will halve.   It will stay the same.
        It will double.

    d. Write down one assumption that you have made in working out the answer to part **c**. **[1]**

4. The first person to stand on the Moon took 3 days to get to the Moon from the Earth. The distance to the Moon is 384 400 km.

    a. Calculate the average speed of the spacecraft. **[2]**

    b. Explain why the speed that you have calculated is the *average* time. **[1]**

Here is the distance–time graph for the first 10 seconds of the launch.

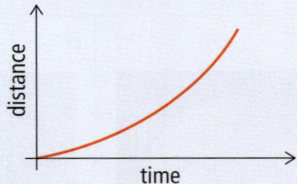

    c. Describe what is happening to the rocket during this time. **[1]**

    d. Describe how you can tell that from the distance–time graph. **[1]**

The Moon has been hit by lots of asteroids.

    e. Write down where you find most asteroids in the Solar System. **[1]**

5. The speed of light is 300 000 km/s.

    a. Calculate how far light travels in 10 seconds. **[2]**

    b. Write down how far 10 light seconds is. **[1]**

    c. Explain why you do not need to do a calculation for part **b** of this question. **[1]**

    d. Put these speeds in order from fastest to slowest: **[1]**

        A the speed of light in glass

        B the speed of light in air

        C the speed of sound.

6. Here is a diagram showing a ray of light that is hitting a piece of glass in a window.

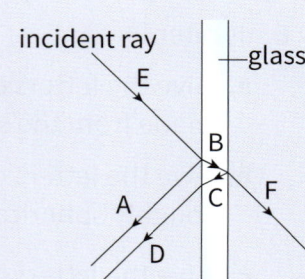

a. Give the letters for *two* reflected rays **[1]**

b. Give the letters for *two* refracted rays. **[1]**

c. The incident ray E is brighter than ray A. Why? **[1]**

d. Explain why you can see the colour of the clothes that you are wearing when you look in the mirror. **[2]**

e. Explain why it is harder to see colour in the reflection of a window at night. **[2]**

7. Rani has three painted rods. One is a magnet, one is copper, and one is iron.

The first rod (rod A) was attracted to the both poles of the magnet. The second rod (rod B) was attracted to the north pole but repelled from the south pole. If she turned the rod around the opposite thing happened. The third rod (rod C) did not move at all.

a. Identify the rods. **[1]**

b. Explain how Rani can make an electromagnet with some wire and a battery. **[1]**

c. Which rod, if any, could she use for the electromagnet? Explain your answer. **[2]**

d. Describe what Rani will notice if she moves a compass around the magnetic rod and her electromagnet. Explain your answer. **[2]**

e. Describe three differences between the electromagnet used in a scrapyard and an electromagnet you use in class. **[3]**

8. A car weighs 20 000 N and the area of each tyre in contact with the ground is 300 cm².

a. Calculate the pressure that the car exerts on the ground. **[2]**

b. If you pumped up the tyres so that there is *less* area in contact with the ground, how would the pressure change? Explain your answer. **[2]**

A mechanic needs to lift the car up. Here is a diagram of a hydraulic pump. You can put a car on the ram and use the handle to lift the car.

The mechanic presses the handle and the oil pushes the large piston up. The valves let more oil in so that he can pump the handle again.

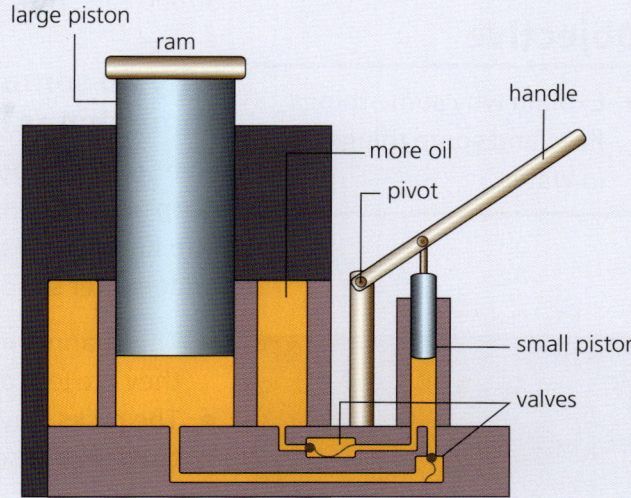

c. Describe what would happen if you used air instead of oil. **[1]**

d. The area of the large cylinder is 100 cm², and the area of the small piston is 5 cm². Calculate the pressure you will need in the oil to raise the car. **[2]**

e. Calculate the force the mechanic needs to apply to the small piston to produce that pressure. **[2]**

To make the job easier the handle is part of a lever.

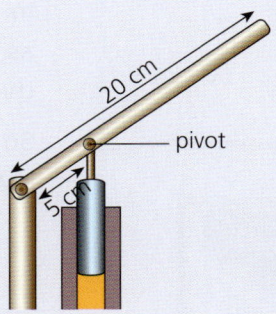

f. Describe what is meant by a lever. **[1]**

g. You worked out the force that needs to be applied to the piston in part **f**. What force does the mechanic need to apply to the handle? **[2]**

Floating and sinking

Icebergs can have a mass of more than 100 000 tonnes, which is the same as over 50 000 cars. How can they float?

Objective

- Explain why some things float and some things sink in water

▲ *A massive iceberg floats.*

Why do some things sink and some things float?

Floating metal buoys mark out the deep-water channel for boats coming into harbour, so that they don't hit underwater sand or rocks.

- The channel markers float because they are less dense than the water.
- The rocks stay on the bottom because they are denser than the water.

An object will float on a liquid if it has a **density** that is *less* than that of the liquid. We calculate density using this equation:

$$\text{density} = \frac{\text{mass}}{\text{volume}}$$

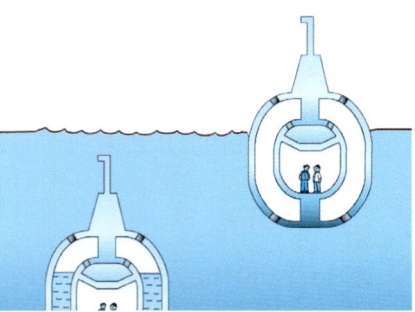
▲ *Channel markers made of iron can float.*

▲ *A coconut floats.*

A coconut has a mass of 400 g and a volume of 500 cm³.

$$\text{density of coconut} = \frac{400\,\text{g}}{500\,\text{cm}^3}$$
$$= 0.8\,\text{g/cm}^3$$

The density of water is 1.0 g/cm³. The density of the coconut is less than that, so it floats.

Imagine what would happen if you had a ball with exactly the same volume as a coconut but with twice the mass. The density of the ball would be twice that of the coconut, and it would sink.

Some people think that rocks always sink and wood always floats. It is not always easy to predict whether something will float or sink.

▲ *Pumice is a rock that floats, but sandstone sinks...*

▲ *... and ironwood sinks.*

- Pumice is a volcanic rock that floats – its density is about 0.3 g/cm³.
- Ironwood is a type of wood that sinks – its density is 1.3 g/cm³.

If you are swimming, it is much easier to float in salt water than in a swimming pool. This is because salt water is denser than the water in the swimming pool. Liquids have different densities.

Liquid	Density of liquid (g/cm³)
Pure water	1.0
Salt water	1.2
Petrol/gasoline	0.7
Bromine	3.12
Mercury	13.6

▲ Humans float very well in salt water like the Dead Sea.

Floating and forces

Another way of thinking about why something floats or sinks is to compare the **upthrust** with the weight. You learned about upthrust on page 109. The object floats if the upthrust is equal to the weight. We can explain upthrust with the particle model.

- Particles in liquids and gases collide with solid objects.
- Particles in the air collide with a floating object from above.
- Particles in the water collide with a floating object from below.
- The force exerted on the object depends on the *difference* between the forces due to the particles.
- The particles in a dense liquid will exert a bigger force than those in a less dense liquid, because they are more massive.

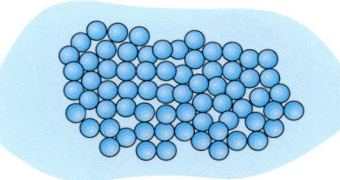

▲ Molecules in liquid water.

Ice and water

Ice is made of water, so how can it float? When water freezes, the water molecules form a solid structure that takes up more space than the water. The ice is less dense than the water. This is unusual – most solids are denser than their liquids.

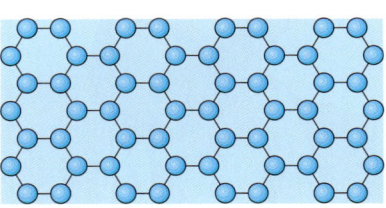

▲ Molecules in ice.

Questions

1. Explain why a person floats in a swimming pool in terms of density.
2. An object has a mass of 100 g and a volume of 90 cm³.
   a. Calculate its density.
   b. Write down whether it would float or sink in water. Explain your answer.
3. The diagram shows what happens in a submarine when it dives below the surface of the water.
   a. Explain why it is able to dive.
   b. Suggest and explain what happens when the submarine needs to return to the surface.

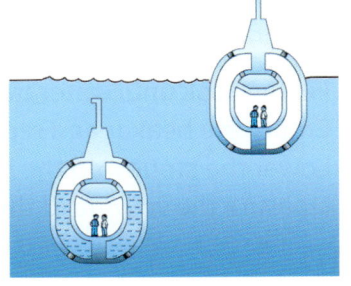

Key points

- An object will float on water if it is less dense than water.
- Objects float because liquids exert a force on them, called upthrust.
- Collisions of liquid and gas particles produce upthrust.

11.2

Objectives

- Describe how ideas about density have been used

- Describe how scientists worked in the past and how they work now

▲ Al-Biruni investigated many topics.

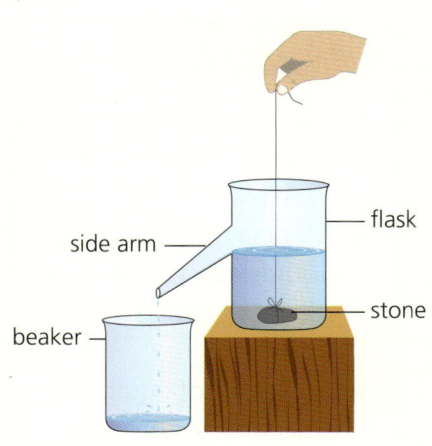

▲ Al Biruni's equipment.

Using ideas about density

For thousands of years people have prized jewellery. They made necklaces and rings from metals such as gold that they found in streams, and gemstones such as red rubies and green emeralds. But quartz and glass could be mistaken for gemstones. How could people tell whether they were real?

Identifying gemstones

Gemstones are **minerals** found in the Earth that have distinctive colours.

▲ Are these gemstones or glass?

Over 1000 years ago Al-Biruni was born in a country that is now a region of Uzbekistan. He was a Muslim scientist and mathematician who wrote books on a wide range of topics from astronomy to geography.

He was very interested in geology and minerals, and he knew that people needed to be able to identify gemstones correctly.

▲ Quartz and rough diamond can be hard to tell apart.

- Quartz is the most common mineral on Earth.
- People could make a lot of money by pretending that a lump of quartz was diamond.
- It isn't easy to tell the difference between quartz and diamond just by looking at them.

Al-Biruni wondered if he could identify stones by their density. He knew how to measure their masses, but the stones did *not* have regular shapes. Finding their volume was more difficult. He designed a flask with a side arm to measure the volume accurately, as shown in the diagram. This was his method.

- Fill the flask until water overflows through the side arm into a beaker
- Empty the beaker and replace it under the side arm.
- Lower the stone into the flask.
- Measure the volume of the water that overflows into the beaker.

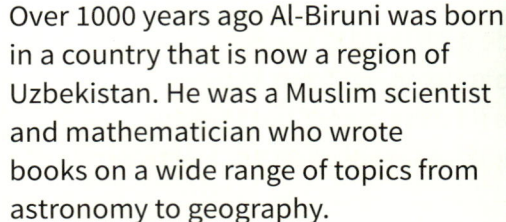

Al-Biruni used the equation for calculating density to work out the density of the gemstones.

$$\text{Density} = \frac{\text{mass}}{\text{volume}}$$

The table shows the densities that he worked out for some minerals.

Mineral	Al-Biruni's measured density (g/cm³)	Modern value for density (g/cm³)
quartz	2.58	2.58
ruby	4.01	4.4
pearl	2.7	2.7

As well as measuring density, Al-Biruni built on the work of earlier scientists by classifying gemstones according to:

- their colour
- whether white light splits into a spectrum when it goes through the gem
- their hardness
- their crystal shape.

One of the most important things about Al-Biruni's density measurements was their **precision** (see pages 94–95). It was not until over 700 years later that scientists in Europe made measurements of density with the same level of precision.

Modern science

Scientists and jewellers can still use Al-Biruni's technique to work out the density of an object to help to identify it.

Today there are also other techniques for identifying gemstones. Scientists can find the **refractive index** of a gemstone, or measure how well it conducts electricity. You learned about refractive index in Chapter 7: Light on page 137.

▲ *Today most research is published by teams of scientists.*

Al-Biruni had many research areas, building on the ideas of other scientists. He designed new techniques to make more accurate measurements. He worked alone. Today scientists usually specialise in one area of science, often with a team of people. They share their ideas in publications that are read all over the world and build on each other's ideas.

Questions

1. Explain why Al-Biruni wanted to find a method to measure the density of gemstones.
2. Scientists work in lots of different ways.
   a. Describe one way in which Al-Biruni's method of working was similar to the way scientists work today.
   b. Describe one way in which it was different.
3. Suggest why scientists did *not* improve on the precision of Al-Biruni's measurements for hundreds of years.

Key points

- Density can be used to identify materials.
- In the past most scientists worked alone.
- Today most science is done by teams of people.
- Today, as in the past, scientists build on the work of other scientists.

1. A student wants to find the density of a brick. He measures the sides of the brick.

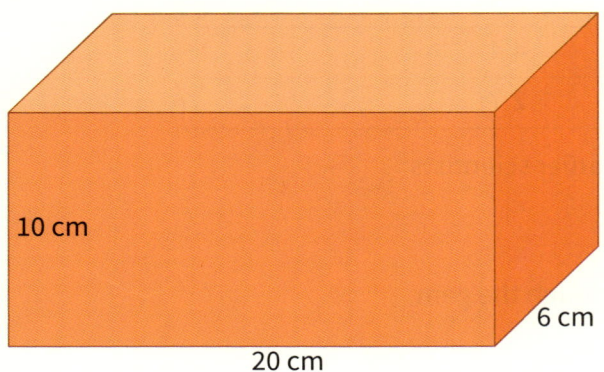

10 cm

6 cm

20 cm

a. Calculate the volume of the brick. **[2]**

b. He measures the mass of the brick and finds that it is 2.4 kg. Calculate the density of the brick. **[2]**

c. Water has a density of 1 g/cm³. Explain why the brick does *not* float on water. **[1]**

2. The table below lists the densities of some materials.

Material	Density in g/cm³
Air	0.001
Ice	0.9
Iron	7.9
Plastic	1.4
Wood	0.7
Water	1.0

a. Write down which of these statements is true.

i. 10 cm³ of iron has a smaller mass than 10 cm³ of wood. **[1]**

ii. 2 cm³ of ice has a bigger mass than 1 cm³ of water. **[1]**

iii. 1 cm³ of plastic has a bigger mass than 100 cm³ of air. **[1]**

iv. 5 cm³ of plastic has a smaller mass than 1 cm³ of iron. **[1]**

b. Explain which solid materials would float in:

i. water **[1]**

ii. mercury, which has a density of 13.6 g/cm³. **[1]**

3. A fish can move up and down in the water. It has a sac inside it called a swim bladder. If the fish fills the sac with oxygen from its gills it will float upwards. Explain why. **[2]**

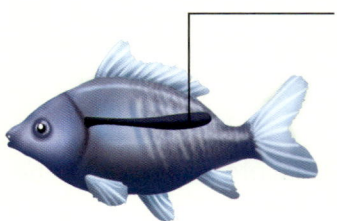

swim bladder allows fish to float or sink

4. Which of the statements about the way that scientists work today is *not* correct? **[1]**
TWS

A Scientists share their ideas and scientific evidence in publications that others read.

B Scientists use creative ideas to interpret the evidence that they collect in experiments.

C It is unusual for scientists to read about the work of other scientists.

D It is possible to make measurements that are more precise and accurate using new technology.
TWS

5. Mercury is a liquid metal with a density of 13.6 g/cm³. Water has a density of 1.0 g/cm³.

a. Gold has a density of 19.3 g/cm³.

i. Will it float or sink in mercury?

ii. In water? **[1]**

b. Silver has a density of 10.5 g/cm³.

i. Will it float or sink in mercury?

ii. In water? **[1]**

c. Wood has a density of 0.8 g/cm³

i. Will it float or sink in mercury?

ii. In water? **[1]**

6. You are planning an investigation. Which of these things do you *not* do as part of the planning? **[1]**

    A Decide on a question to investigate.

    B Work out how to take precise and accurate measurements.

    C Write down what you found out.

    D Decide which measurements to make.

    E Describe how to work safely.

7. A thermometer designed in 1666 and now known as a Galileo thermometer uses changes in density with temperature. As the temperature drops, small glass bulbs containing different amounts of air and liquid move upwards in another liquid. Suggest why. **[2]**

▲ *A Galileo thermometer when it is warm (left), and when it is cold.*

Hot and cold

A scorpion is a cold-blooded animal, and the person holding it is warm-blooded. You cannot tell the temperature of the scorpion or the person by looking at them. This image of the scorpion is taken with a special camera. It produces a **thermal image** that shows the differences in temperature.

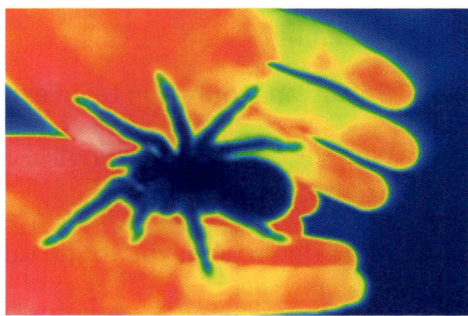

▲ *A thermal image shows that the scorpion is colder than the person holding it.*

What is the difference between energy and temperature?

The **temperature** tells us how hot or cold something is. We use a **thermometer** to measure temperature. A liquid inside a very narrow glass tube expands when it is heated.

- Thermometers used to be made with liquid mercury.
- Mercury is poisonous, so now the liquid in thermometers is alcohol.
- Digital thermometers do *not* use a liquid – instead they use a sensor to measure the temperature directly.

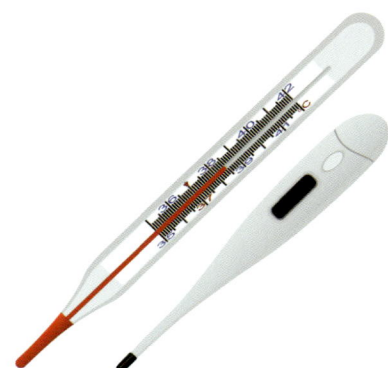

▲ *A liquid and a digital thermometer.*

We measure temperature in **degrees Celsius** (**°C**).

- Human body temperature is 37 °C.
- The coldest temperature ever recorded on Earth was –89.2 °C, in Antarctica in 1983.
- The hottest temperature ever recorded was 56.7 °C in Death Valley, USA in 1913.

Heating a substance such as water causes its temperature to rise. We transfer energy produced from burning fuel to the water. Energy, and temperature are *not* the same.

- A cup of warm water and a swimming pool can have exactly the same temperature.
- There is much more energy in the thermal store of the swimming pool than there is in the thermal store of the cup of water.
- Energy depends on the amount of material, but temperature does *not*.

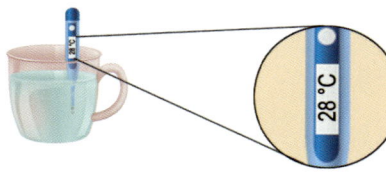

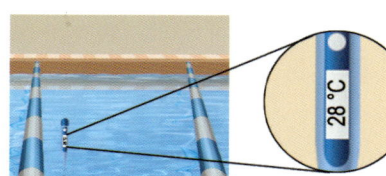

▲ *The water is at the same temperature.*

What happens when you heat materials?

The process of heating changes the motion of particles. If you heat a solid, the particles in the solid vibrate more. In liquids and gases the particles move faster. The temperature has increased. Temperature is related to the average speed of the particles.

You cannot say that the individual particles in a solid, liquid, or gas get hotter. Each particle can only move or vibrate faster. The temperature of a solid is a property of the solid, *not* of the individual atoms or particles that make it.

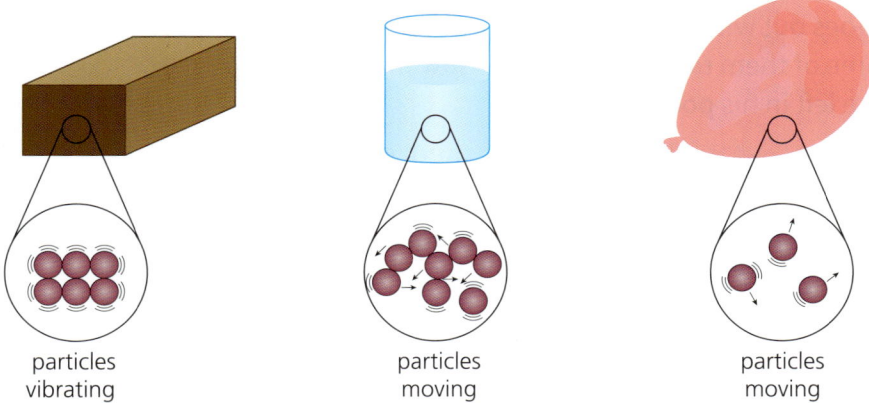

particles vibrating   particles moving   particles moving

▲ *Particle arrangements in solids, liquids, and gases.*

The energy needed to raise the temperature of an object depends on:

- *the mass of the object* – greater mass means more particles, so more energy is needed
- *the type of material of the object* – if the particles are more massive, more energy is needed
- *the temperature rise needed* – a bigger temperature rise means more energy is needed.

How do things cool down?

Energy is always transferred from a hot object to a cold object. If you put a cold object in contact with a hot object the temperature of the cold object will increase, and the temperature of the hot object will decrease. Energy has been transferred. This is called **dissipation**.

If you put a hot pan in cold water the water gets hotter. Energy is transferred from the pan to the water. Eventually the pan and the water have the same temperature. The water and the pan are in **thermal equilibrium**.

▲ *Energy is transferred from the water to the ice, so the ice melts and the water cools.*

Questions

1. Describe the difference between temperature and energy.
2. Copy and complete these sentences.

   When you heat a liquid, the particles in it move _____. The same thing happens when you heat a _____. When you heat a _____ the particles just vibrate more. Dissipation is when energy moves from a _____ to a _____ region.
3. Explain why it takes more energy to heat 1 kg of cold water than 0.5 kg of cold water to the same temperature.
4. Explain why it takes longer to boil a kettle of water than to warm the same kettle of water to a lower temperature.

Key points

- Temperature is a measure of how hot something is, measured in degrees Celsius.
- Thermometers are used to measure temperature.
- When you heat a solid, liquid, or gas, the particles move faster or vibrate more.
- Energy is dissipated when it moves from a hot to a cold region.

Conservation of energy

Objectives

- Write down the law of conservation of energy
- Apply ideas about energy conservation and dissipation to different situations

▲ *Energy is conserved if you consider the surroundings and the buildings together.*

Energy is a bit like money

Jamal goes out with four coins in his pocket, and uses three of them to buy some food. How many coins are left in his pocket?

- He does *not* need to look in his pocket to know that he has one coin left.
- If he gets home and no longer has a coin, he knows that he must have lost it.
- He will *not* find more than one coin in his pocket.

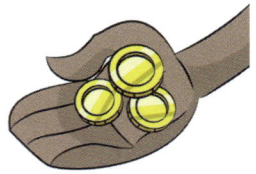

Coins cannot appear or disappear. Energy is like the coins.

This is called the **law of conservation of energy**. This law says that energy cannot be created or destroyed. It can be transferred in a process. If you know how much energy you have at the start of a process, then you must have this same amount of energy at the end. This law is one of the most important laws in science.

This law allowed scientists to work out how energy is stored and transferred. If they found a process in which there seemed to be less energy at the end than there was at the beginning, then they knew there must be another type of energy involved.

Useful and wasted energy

What happens when you turn on a light? The light bulb emits light, and it also gets warmer.

- The **useful energy** is transferred by the light.
- We can think of the energy heating the surroundings as **wasted energy**.
- Energy that is **dissipated** is usually wasted energy.

| 100 J transferred electrically | → | 20 J transferred by light |
| | ↘ | 80 J heating the surroundings |

▲ *This diagram show what happens every second in one type of light bulb.*

Suppose only 20 J of the 100 J of energy transferred to the light bulb is then transferred as light. Most of the energy is wasted.

In *all* processes some energy is wasted.

- A car engine has moving parts.
- There is friction between moving parts.
- The friction makes those parts get warm.
- Energy is dissipated – it is transferred to the surroundings, which heat up. All the energy heating the surroundings is wasted.

'Wasted' doesn't mean that the energy is 'lost', just that it is *not* the energy that we want.

Energy that is transferred by heating is not always wasted energy. When we cook food we want it to get hot, but the air around the pan heats up as well, which is *not* useful.

Energy conservation and efficiency

Some light bulbs transfer most of the energy into light. We say that they are **efficient**. This is the energy transfer diagram for an energy-efficient light bulb. Less of the energy transferred to the light bulb is wasted.

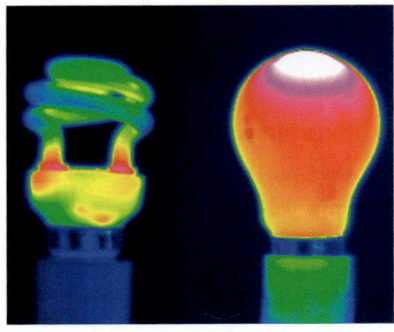

▲ *The energy-efficient light bulb on the left does not get as hot as the normal light bulb.*

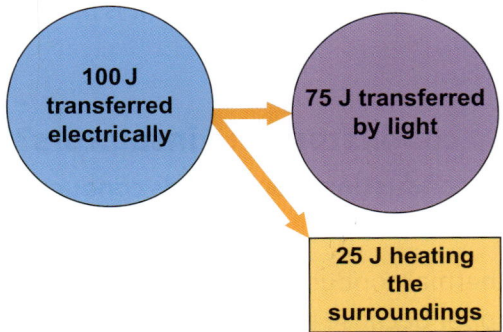

100 J transferred electrically → 75 J transferred by light

25 J heating the surroundings

▲ *The energy transfer diagram for an energy efficient light bulb.*

The efficiency of a machine or device tells you how much of the energy that you put in is transferred as useful energy. If a lot of the energy that you put in is wasted, then the machine is *not* very efficient. Most energy that is wasted is dissipated.

Questions

1. **a.** State the law of conservation of energy.
   b. Compare energy conservation and energy dissipation.
2. Copy and complete these sentences using the words below.

   wasted useful thermal

   A machine is more efficient if it transfers more _____ energy than _____ energy. A lot of the wasted energy in a machine with moving parts is transferred as _____ energy due to friction.
3. Write down the useful energy and the wasted energy in these items:
   a. hairdryer
   b. television
   c. kettle.
4. Describe what you would notice if you stood next to an energy-efficient light bulb compared with a normal light bulb.
5. Explain why more efficient electrical devices save you money.

Key points

- The law of conservation of energy says that energy cannot be created or destroyed, only transferred or stored.
- In all processes you have the same amount of energy at the end as you had at the beginning.
- In all processes some energy is dissipated, usually heating the surroundings.

Energy transfer: Conduction

Objectives

- Know the names of some conductors and insulators
- Explain why some materials feel warmer than others

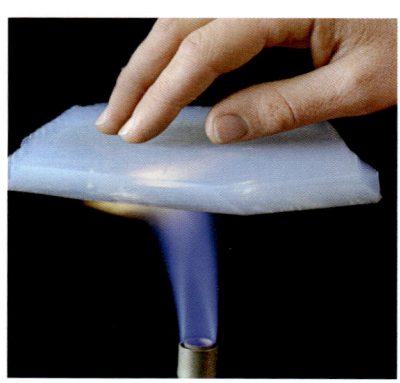

▲ Energy transfers very slowly through a special material called aerogel.

▲ Plastic conducts heat less well than metal.

The metal is cold but the plastic is warm.

Neela wants to light a Bunsen burner. She takes a piece of wood, called a splint, and lights it from the teacher's Bunsen burner. She carefully carries the splint back to her Bunsen burner and lights it. The splint is on fire at one end, but she can hold it without burning her hand. This is because the energy from the flame is *not* transferred easily down the splint to her hand.

▲ Only one end of the splint is hot.

What is the difference between conductors and insulators?

Energy is transferred through a solid by **conduction**. Some solids conduct energy better than others.

- Metals, like copper, are very good thermal conductors.
- Cooking pans are often made of copper because it quickly conducts the energy to the food in the pan.

Many non-metals are poor conductors of energy. They are **insulators**. This does *not* mean that they do *not* conduct at all, but that energy is transferred very slowly through them.

- Materials such as paper, cloth, wood, and plastic are all insulators.
- You stir hot food with wooden spoons so you do not burn yourself.

Why do some objects feel hot and others feel cold?

Bicycle handlebars are made of metal but the ends are often covered in plastic. Neela touches the metal of her handlebars, and then the plastic.

- Neela thinks the metal is colder than the plastic.
- Our skin detects the *transfer* of energy rather than the temperature.
- When Neela puts her hand on the metal, energy is quickly transferred away from her hand to the metal. She says it feels cold.
- When Neela puts her hand on the plastic, energy is *not* transferred quickly away from her hand because the plastic is an insulator. She says it feels warm.
- The plastic is at the same temperature as the metal.

If you touch something hotter than your skin, it feels warm because energy moves from the object to your hand.

How do metals conduct?

The diagram shows a piece of metal being heated at one end. The metal atoms on the left are vibrating a lot and the atoms on the right are *not* vibrating very much. The left end of the bar is hot and the right end of the bar is cold, so energy is transferred from the left to the right. Electrons also transfer energy in metals.

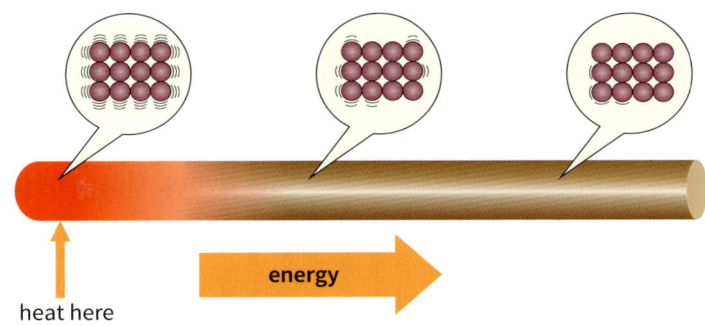

heat here

energy

Do liquids and gases conduct energy?

Liquids like water are poor conductors of energy compared with metals. You can see this if you heat one end of a test tube of water. An ice cube held down by a weight will *not* melt, even if the water at the top is boiling.

A diver wears a wetsuit to dive in cold water.

- Inside the wetsuit a small amount of water forms a thin layer between the diver's skin and the thick elastic suit.
- The water is *not* free to flow away.
- Energy is *not* easily transferred from the diver to the sea water so the diver does *not* feel cold.

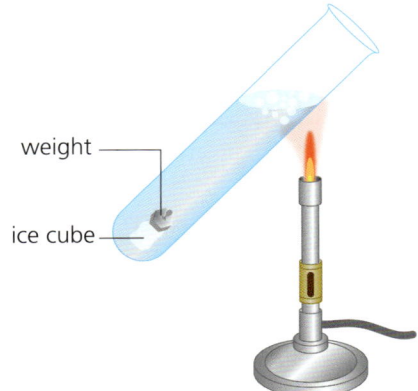

weight

ice cube

▲ *The ice cube will not melt.*

🏊 Science in context

Air is a good insulator if it is trapped and cannot move.

- Clothing, for example padded jackets, and bedding can be made from materials that trap air.
- Gases, like air, do *not* transfer energy easily because their particles are far apart.

A vacuum is an even better insulator. Scientists designed windows with two panes of glass and a near vacuum between the panes. This 'double glazing' helps insulate houses, and reduce fuel bills.

❓ Questions

1. **a.** Describe the difference between a conductor and an insulator.
    b. Give an example of a conductor and an insulator.
2. Describe what would happen if you heated water in a saucepan made of a material that is *not* a good conductor of energy.
3. Some divers use drysuits rather than wetsuits. They wear clothing underneath that traps a layer of air between their skin and the suit.
    a. Explain how a drysuit keeps a diver warm.
    b. Which keeps a diver warmer, a wetsuit or drysuit? Explain your answer.
4. A student says that her blanket keeps her warm because it traps heat. What would you say to her?

📖 Key points

- Energy is transferred through a solid by conduction.
- Metals are good conductors of energy, but plastic and wood are poor conductors; they are insulators.
- Liquids and gases that are not free to move around are poor conductors of energy.
- Objects feel cold when they conduct energy away from your skin.

Energy transfer: Convection

If there is a fire in a house, firefighters may crawl along the floor to look for survivors. This is because there is less smoke near the floor than higher up. Why is that?

Objectives

- Describe how energy is transferred by convection
- Describe how convection currents form in everyday situations

▲ *Convection currents form in the water as it is heated.*

▲ *In a fire the cleanest air is closer to the floor due to convection.*

What is convection?

When you heat the bottom of a saucepan of water, all of the water gets warm, *not* just the bottom part. The liquid in the pan is free to move so energy is transferred by **convection**.

The bottom of the pan is heated by the flame. Energy is transferred through the pan to the water at the bottom of the pan by conduction. This is because the bottom of the pan is solid.

- The water in contact with the bottom of the pan gets warmer.
- The molecules in the warmer water are moving faster than the molecules in the cooler water above.
- The molecules in the warmer water move further apart.
- The warmer water becomes less dense.
- The warmer water rises (floats up).
- Cooler, denser water moves down to take its place.

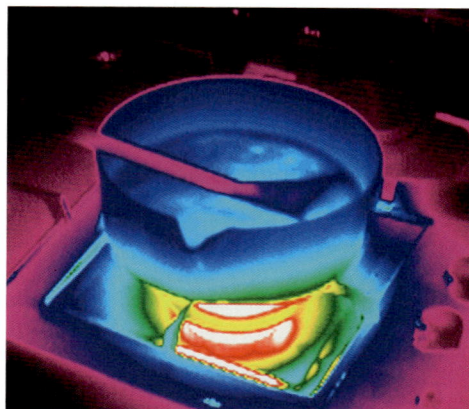

▲ *Energy transferred through a pan of liquid.*

Eventually all the water in the pan is circulating up and down. The circulation of water that is set up in this way is called a **convection current**. Convection currents can happen in all liquids and gases that are free to move.

In a fire the hot smoky air expands and rises. That is why firefighters crawl on the floor.

Where do convection currents form?

The atmosphere is heated by the Sun, and also by the warm land beneath it. The land heats up more quickly than the air or the oceans.

Sea breezes

If you live near the coast you might notice that a breeze comes off the sea during the day, but changes direction at night. We can explain this using convection.

During the day:

- The air just above the ground gets hot.
- This hotter air expands and rises just like the hot water in a pan.
- The rising air is replaced with cooler air from above the sea.
- The breeze comes off the sea onto the land.

During the night:

- The ground cools much more quickly than the sea.
- Now the air above the sea is warmer and rises.
- This pulls cooler air from above the ground to replace it, so the breeze comes off the land.

Atmospheric currents

Convection also happens on a bigger scale. The Sun heats the Earth more at the equator than it does nearer the poles. The air near the equator rises and is replaced by cooler air. This causes the air in the atmosphere to move. Weather patterns across the world are affected.

Thermals

You might see birds using the convection currents produced when the Sun heats the ground. Convection currents in the air are called **thermals**.

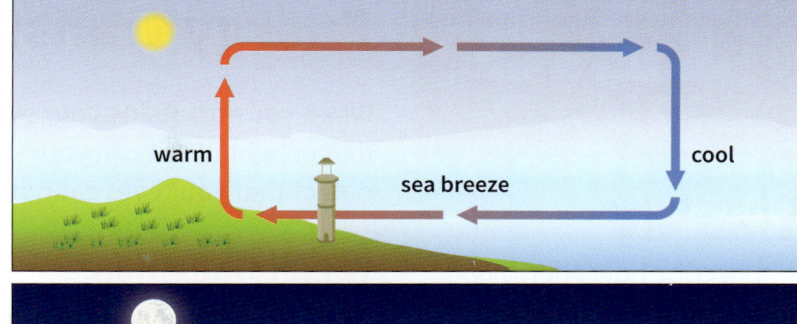
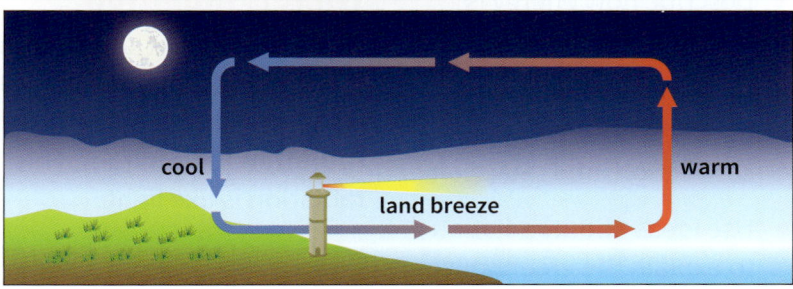

▲ *The breeze by the sea or ocean changes direction at night.*

▲ *Birds of prey use thermals caused by convection to climb high and look for prey.*

Questions

1. Describe the difference between conduction and convection.
2. Explain what 'hot air rises' means, in terms of the particles in a gas.
3. This diagram of the Earth shows a convection current in the atmosphere (not to scale).
   a. Draw a clockwise or anticlockwise arrow to show the direction of the convection current.
   b. Explain your answer.

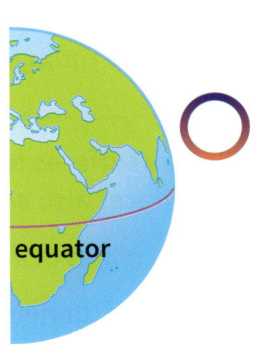

Key points

- When a gas or liquid gets hot it expands and becomes less dense.
- The less dense gas or liquid rises and is replaced by cooler gas or liquid.
- This sets up a convection current.
- A convection current in the atmosphere is called a thermal.

Energy transfer: Radiation

Objectives

- Describe how energy is transferred by radiation
- Describe the differences between conduction, convection, and radiation
- Use ideas about radiation to explain the greenhouse effect

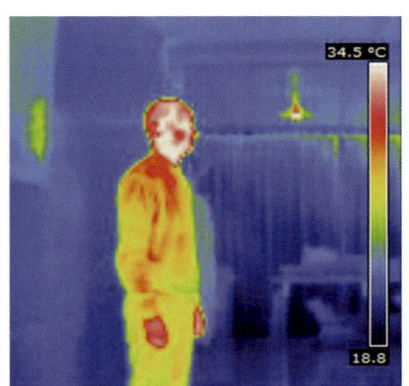

▲ *Humans emit infrared radiation.*

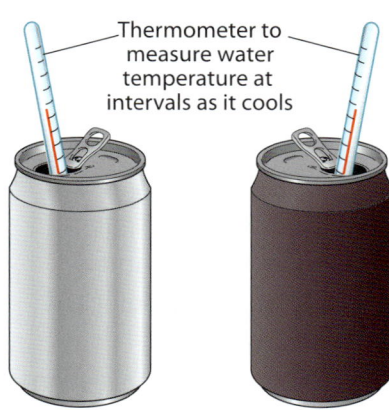

Thermometer to measure water temperature at intervals as it cools

▲ *The dark surface emits more infrared radiation than the silver surface.*

When you go outside your eyes detect light. What can your skin detect that your eyes cannot?

▲ *Your skin can feel warmed by the Sun.*

What is radiation?

There are several sorts of radiation, but the sort that your skin can feel is thermal radiation, or **infrared radiation** (or infrared). This radiation is like light:

- It travels from the Sun to the Earth.
- It travels as a wave.
- It can be reflected, transmitted, and absorbed.

All objects emit radiation. The type of radiation that they emit depends on their temperature. The wire inside a torch bulb can get hot enough to give out infrared *and* light. You will only ever emit infrared radiation.

How is radiation different from conduction and convection?

Energy can be transferred by conduction and convection. Both methods need a material for the energy to travel through.

Infrared radiation does *not* need a material to transfer energy.

Light and infrared reach the Earth from the Sun by travelling through space. There is no material in space. We say it is a **vacuum**.

Light can be...	Infrared can be...
emitted by hot objects	emitted by hot objects
reflected well by shiny surfaces	reflected well by shiny surfaces
absorbed by objects and your retina	absorbed by objects and your skin
detected by a camera	detected by a **thermal imaging camera**
transmitted through a vacuum as a wave	transmitted through a vacuum as a wave

Dull, dark surfaces tend to absorb more infrared than light, shiny surfaces. Dark surfaces also *emit* more infrared than light surfaces.

Some examples of infrared radiation in everyday life are:

- Marathon runners use foil wraps to keep warm after they stop running.
- Pale-coloured houses stay cooler in the summer.
- Dark or black clothes will dry first on a washing line.

What is the greenhouse effect?

When radiation from the Sun reaches the Earth:

- About 50% of the radiation is reflected by the Earth's atmosphere.
- The rest of the radiation gets through and is absorbed by the land and oceans.
- The Earth itself emits infrared outwards, and some of that is reflected back in by the atmosphere.
- The rest is transmitted out into space.

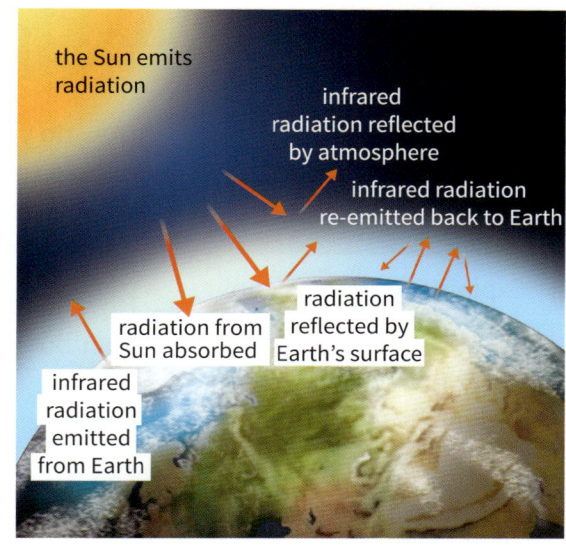

The temperature of the Earth depends on the *balance* between the radiation that is absorbed and the radiation that is emitted. If more is absorbed than is emitted, then the temperature will increase.

The atmosphere contains **greenhouse gases**.

- Greenhouse gases include carbon dioxide, methane, and water vapour.
- These gases absorb infrared and then emit it again.
- Some is re-emitted back towards Earth instead of out into space.
- This is the **greenhouse effect**.

Without an atmosphere the Earth would be very, very hot in the daytime, and very, very cold at night. However, the concentration of greenhouse gases is increasing, and this is raising the average global temperature and causing **climate change**. Climate change is altering the weather, melting the icecaps, and affecting how easy it is to grow food.

The electromagnetic spectrum

Light and infrared are just two of the types of radiation emitted by the Sun. They are two waves in the **electromagnetic spectrum**. Light and infrared are parts of the electromagnetic spectrum just like blue and green are two colours of the light spectrum.

All the waves of the electromagnetic spectrum can travel through a vacuum, and they all travel at the same speed in a vacuum. They have different uses. Some are more dangerous than others.

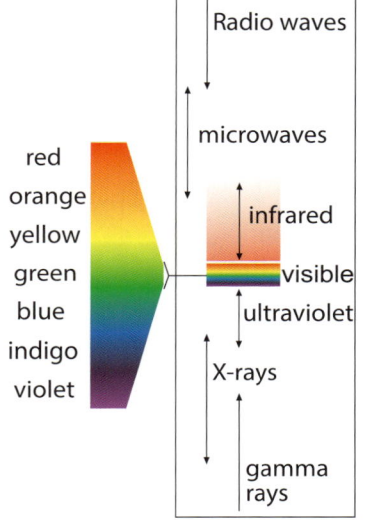

▲ *The visible spectrum on the left is part of the electromagnetic spectrum on the right.*

Key points

- All objects emit thermal radiation. They emit light if they are hot.
- Radiation, including light and infrared, is emitted by the Sun.
- Radiation does not need a medium – it can travel in a vacuum.
- The atmosphere reflects radiation like glass in a greenhouse.

Questions

1. Name some gases that produce the greenhouse effect.
2. Look at the picture of the person taken with a thermal imaging camera.
   a. Explain why this image is different from the image that we would see with our eyes.
   b. Explain why the person does *not* emit visible light.
3. Describe a situation in which the temperature of the Earth would go down.

12.6

Cooling by evaporation

Objectives

- Explain why liquids cool when they evaporate
- Describe some uses of cooling by evaporation

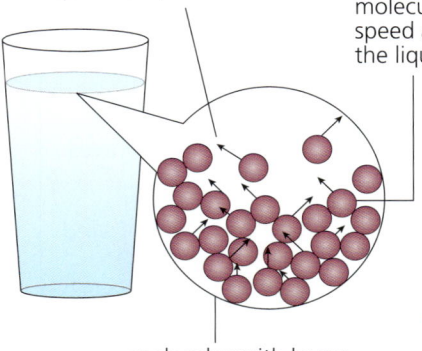

molecules at the surface with high speed escape from the liquid – evaporation

molecules with medium speed are pulled back to the liquid water

molecules with lower energy remain in the liquid water

How do you keep cool? Your body produces sweat to help to keep your body from getting too hot. You can feel the same cooling effect if you wet your hands and leave them to dry.

How does **evaporation** remove energy from your body?

▲ *Sweating removes energy from your body.*

What is evaporation?

If you leave a dish of water out, it will evaporate – dry up. We can think about the particles inside a liquid moving around in every direction. There is *not* very much space between the particles and they move over each other. Each particle is attracted to the particles around it.

When particles are travelling fast enough they can break free and move into the air. The liquid **evaporates**.

Why does evaporation produce cooling?

You can explain **evaporative cooling** using ideas about particles.

- Their **average speed** depends on the temperature of the liquid.
- At a higher temperature the average speed will be higher.
- The particles do *not* all move at exactly the same speed: some travel faster, some slower.
- The particles are colliding with each other, and energy is transferred in those collisions.
- A few of the particles at the surface will be travelling fast enough to be able to escape.
- When fast-moving particles escape, the average speed of the particles left behind is slower.
- A smaller average speed means the liquid is cooler – evaporation has cooled it.
- The cooler liquid absorbs energy from the air and more particles escape.
- Eventually all the particles escape. The liquid has evaporated. The air around the liquid has cooled.

A model for evaporative cooling

You can think of it like this. Imagine a group of 10 children.

To calculate the average height we add up all the heights and divide by 10.

1.5 m
1.1 m 1.2 m
0.8 m 0.7 m
1.2 m 1.1 m 1.4 m 0.9 m 1.5 m

▲ *A group of 10 children and their heights in metres.*

Average height
= (1.1 +1.2 + 1.5 + 0.8 + 0.7 + 1.2 + 1.1 + 1.4 + 0.9 + 1.5) / 10
= 1.14 m

If the two tallest children leave the group, the average height will be smaller.

Average height
= (1.1 +1.2 + 0.8 + 0.7 + 1.2 + 1.1 + 1.4 + 0.9) / 8
= 1.05 m

Removing tall children has reduced the average height.

Where is this cooling effect used?

Cooling animals

Your body uses cooling by evaporation. When you get very hot you sweat. Droplets of sweat, which is mainly water, form on the surface of your skin. The water evaporates as energy is transferred from your skin to the water. You feel cooler because energy is being transferred away from your body.

Dogs pant to help them cool down. Water evaporates from their tongues. Elephants use water on their skin to cool down.

▲ *Evaporation keeps animals cool.*

Evaporative coolers and refrigerators

Some people use **evaporative coolers** instead of air conditioning to cool their house. In an evaporative cooler a fan draws warm air over water. Energy is transferred to the water. It evaporates, and the air cools down. The cooled air circulates around the house.

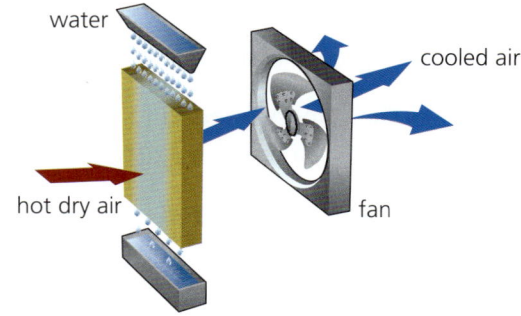
water
cooled air
hot dry air
fan

▲ *Evaporation cools air.*

> ### Science in context
>
> For thousands of years people have used water to cool their homes. Before air conditioning they used fans and water. In ancient Egypt people hung wet mats over the open doorway and used fans to cool down the air in the room.

In a **refrigerator** a special liquid called a **refrigerant** is pumped around tubes at the back. Energy from inside the refrigerator evaporates the refrigerant. This cools down the inside of the refrigerator. Air conditioning works in the same way.

Key points

- Liquids evaporate when faster particles escape.
- The average speed (the temperature) of the particles is decreased.
- Evaporative coolers transfer energy from air to a liquid to cool the room.
- Animals use liquids on their skin to cool down.

Questions

1. Describe the difference between the particles that escape when a liquid evaporates, and the particles that are left behind.
2. Explain why your hands feel cool if they get wet.
3. Explain why water evaporates faster if the air around it is warmer.
4. Modern evaporative coolers need a supply of electricity to work. Why?

1. Copy and complete these sentences.

   a. You measure _____ in degrees Celsius. [1]

   b. You measure _____ in joules. [1]

   c. A bath will have _____ energy than a cup of warm water at the same temperature. [1]

2. The amount of energy that you need to raise the temperature of something does *not* just depend on the size of the temperature rise.

   a. Name two other things the temperature rise depends on. [1]

   b. Put these in order from the smallest to the largest amount of energy required. [1]

      A The energy needed to raise the temperature of 1 kg of water by 20 °C

      B The energy needed to raise the temperature of 2 kg of water by 20 °C

      C The energy needed to raise the temperature of 1 kg of water by 10 °C

3. Explain why:

   a. Saucepans are usually made of metal. [1]

   b. Black clothes dry quicker than white clothes if they are drying in the Sun. [1]

   c. A hot air balloon rises. [1]

   d. A bird fluffs up its feathers if it is cold. [1]

4. Copy and complete the sentences below, choosing the correct bold words.

   The law of conservation of energy says that energy cannot be **created/dissipated** or **destroyed/transferred**.

   When you burn coal, you transfer energy from a **chemical/thermal** store to a **chemical/thermal** store. [4]

5. A student puts two thermometers on the desk underneath a lamp. She covers the bulb of one thermometer with black paper and the other with aluminium foil.

   a. Write down which thermometer will show the higher temperature after half an hour. [1]

   b. Explain why. [1]

6. There is a hot drink on the table.

   a. Explain how conduction changes the temperature of the tea. [1]

   b. Explain how convection changes the temperature of the tea. [1]

   c. Explain how radiation changes the temperature of the tea. [1]

   d. Explain why putting a lid on top of the cup would keep the tea hotter for longer. [1]

7. You are cooking a pizza. You place your pizza on a metal tray at room temperature and put it in the oven. When the pizza is cooked you remove the tray from the oven. Eventually it reaches room temperature again.

   Describe and explain in detail what happens to the motion of the particles in the metal tray. [6]

8. A student has left his drink outside in the Sun.

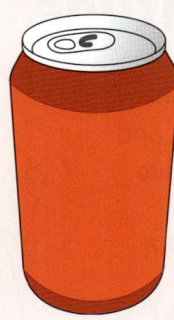

a. Explain why the can of soft drink is warmer than it would be if it was inside the house. **[1]**

The student decides to use the Sun to cool down a can of drink on a hot day. He covers the can of drink with a clay pot and pours water over the pot. He keeps pouring water over it for a few minutes. When he takes the can out it is cooler than before.

b. Explain why the can is cooler. **[2]**

c. Explain why he has to keep pouring water over the pot. **[1]**

9. Explain why:

a. Houses are painted white in hot countries. **[1]**

b. You can't find people in burning buildings with a thermal imaging camera. **[2]**

10. Some double-glazing systems trap air between two panes of glass.

panes of glass

air

a. Explain how energy is transferred through a double-glazed window from a hot room to the cold air outside. **[3]**

b. State and explain how the rate of energy transfer would change if you removed the air from the gap. **[3]**

11. Here is an experiment to demonstrate convection.

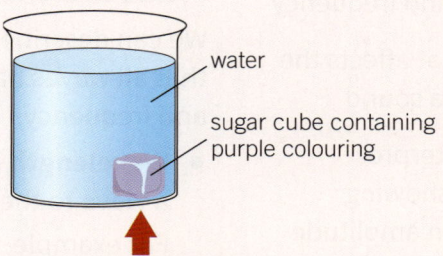

water

sugar cube containing purple colouring

Bunsen burner heats here

a. Describe what will happen to the purple colour during heating. **[3]**

b. Explain why the purple colour forms a convection current. **[3]**

c. Describe where convection currents are formed in everyday life. **[1]**

12. a. Explain why there are no convection currents in solids. **[1]**

b. If you stand near a fire on a cold night you feel warm even though the air is cold. Explain why. **[1]**

c. Write down one similarity between light and infrared radiation. **[1]**

13.1

Loudness and amplitude

Objectives

- Define wavelength, amplitude, and frequency
- Describe what affects the loudness of a sound
- Draw and interpret waveforms showing differences in amplitude

Imagine a string attached to a wall. You can make a wave in it by moving your hand up and down. Your friend takes a picture of the wave so that the wave is frozen in time. How can you describe the wave that you have made?

What are the properties of waves?

We can describe waves using three properties that all waves have: amplitude, wavelength, and frequency.

▲ Some sounds are very loud.

- **Wavelength** is the distance from one point on a wave to the *same* point on the next wave.

  For example, the distance from the *top* of one wave to the *top* of the next wave is one wavelength.

- **Amplitude** is the distance from the centre of the wave to the highest or lowest point.

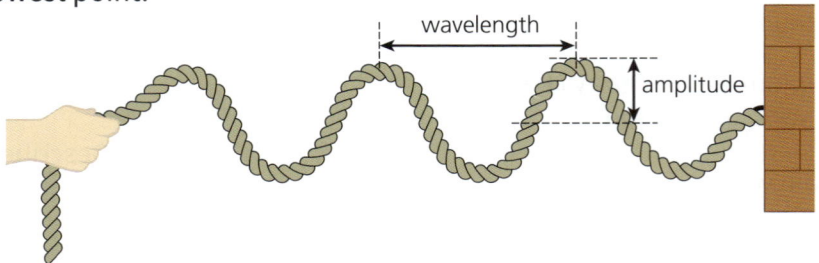

The diagram does *not* show the frequency. If you count how many waves go past in 1 second, that is the **frequency**.

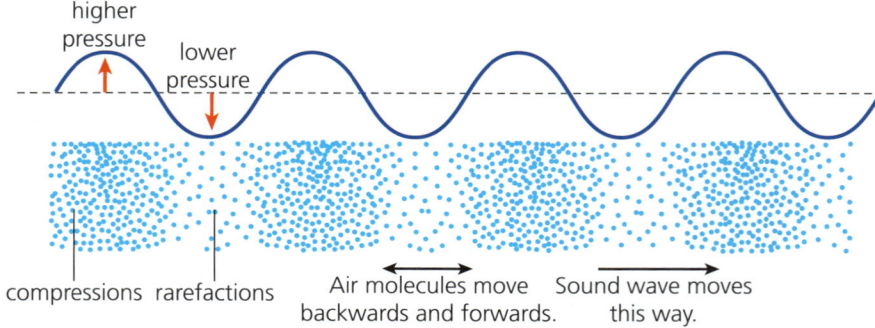

▲ A sound wave is a pressure wave.

Using an oscilloscope

To 'see' what sound waves are like we can attach a microphone to an **oscilloscope**.

- The microphone produces an electrical signal that represents the sound.
- The signal is displayed on the screen.
- The image on the screen shows how the pressure varies with time.
- Where the line on the screen goes up the particles are close together, and where it goes down they are far apart.

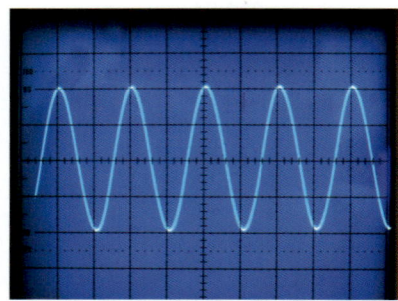

▲ The signal from a microphone.

What is loudness?

Anil sings a note into a microphone while he looks at the oscilloscope screen. Then he sings the same note, only louder. This is what he sees.

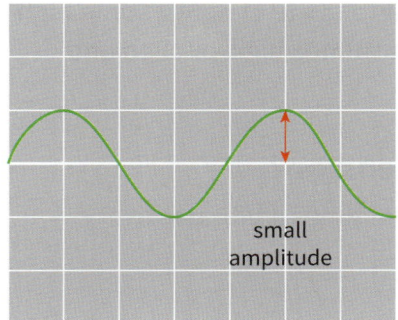

small amplitude

▲ *A soft sound.*

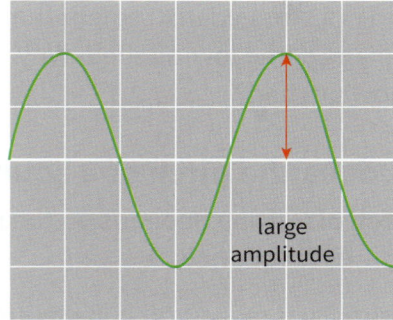

large amplitude

▲ *A loud sound.*

A soft sound has a small amplitude. A loud sound has a large amplitude.

What happens when you amplify a sound?

You might hear people say that sound 'dies away'. They think that the sound wave gets 'weaker' as it travels. The sound *spreads out* as it travels away from the source. Some energy is transferred to the medium that the sound is travelling through – energy is dissipated, so the sound becomes softer.

Singers cannot sing loud enough for everyone in a big concert hall to hear well. They use a microphone connected to an **amplifier** to make their voice louder. An amplifier increases the amplitude of the electrical signal produced by the sound wave, so that when it is broadcast through a loudspeaker it sounds louder.

▲ *This microphone is connected to an amplifier.*

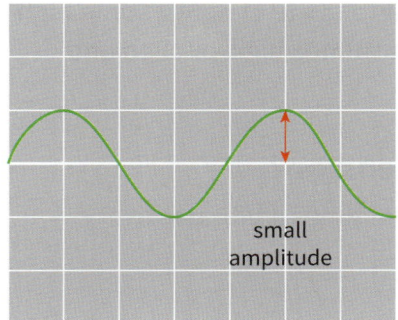

Questions

1. Copy and complete this sentence.

   A _____ sound has a larger amplitude than a _____ sound.

   You can look at a sound wave on a screen using an _____.

2. Sunil says that the amplitude of a wave is the distance from the centre to the top of a wave. Write a different definition of amplitude.

3. Which of these distances is one wavelength on a sound wave?

   A the distance from a compression to a rarefaction

   B the distance from a compression to the next compression

   C the distance from a rarefaction to the next rarefaction

4. Is the wave on the rope on the previous page a longitudinal wave or a transverse wave? Explain your answer.

5. The picture of the sound wave on the oscilloscope screen is a transverse wave. Sound waves are longitudinal waves. Describe what is being shown on the screen.

Key points

- Waves have wavelength, amplitude, and frequency.
- Wavelength is the distance from one peak to the next.
- Amplitude is the distance from the centre to the top of a wave.
- Frequency is the number of waves per second.
- An oscilloscope connected to a microphone can display a sound wave on a screen.
- The loudness of a sound depends on the amplitude.

13.2

Pitch and frequency

High and low

The oud is a stringed instrument that produces a range of notes, some high-pitched and some low-pitched. Why are some sounds high-pitched and some low pitched?

What does the pitch of a sound depend on?

The **pitch** of a string depends on the number of times it vibrates every second – its **frequency**.

- Frequency is measured in **hertz (Hz)**.
- High-pitched sounds have a high frequency and low-pitched sounds have a low frequency.
- High frequencies are measured in kilohertz (kHz): 1 kHz = 1000 Hz.

Objectives

- Describe what affects the pitch of a sound
- Draw and interpret waveforms showing differences in frequency

▲ Ouds have been played for thousands of years.

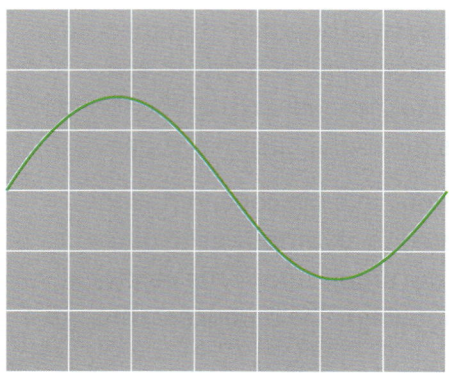

▲ A low-pitched sound has a low frequency

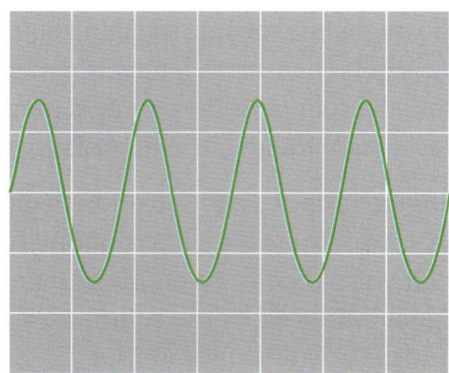

▲ A high-pitched sound has a high frequency.

- A high-frequency sound has a short wavelength.
- A low-frequency sound has a long wavelength.

Changing the frequency (and wavelength) of a sound wave does *not* have the same effect as changing the amplitude. The amplitude determines how loud the sound is. Loud sounds may be high-pitched, like a scream, or low-pitched, like an earthquake.

What can you hear?

When you listen to speech or music you are listening to a range of frequencies, from very high to very low. Young people can hear a much larger range of frequencies than older people can.

▲ A younger person has a bigger audible range.

- Humans have an **audible range** (range of hearing) from about 20 Hz to 20 000 Hz.
- As people get older they find it harder to hear very high frequencies.
- By the time you are 30 you may be able to hear only up to about 14 000 Hz.

What is ultrasound?

Ultrasound is sound with a frequency higher than 20 000 Hz, which is outside the audible range of humans. We cannot hear sounds of this frequency, or make sounds so high-pitched, but other animals can.

What can other animals hear?

Species	Audible range (Hz)
bat	2000–110 000
cat	45–8000
chicken	125–2000
cow	23–35 000
dog	67–45 000
dolphin	100–100 000
elephant	16–12 000
guinea pig	54–50 000
horse	55–33 500
mouse	1000–91 000
rat	200–76 000
whale	1000–123 000

Animals can have very different audible ranges to humans.

Some animals can make and detect ultrasound, but humans cannot. Some animals, like grasshoppers, make sounds that they cannot hear.

Questions

1. Copy and complete the sentences.

   The pitch of a sound depends on the _____ of the sound. A high-pitched sound will have waves that have a _____ wavelength.

2. Aditya plays a note on a guitar. The note has a frequency of 512 Hz. Write down the number of times a second that the string vibrates.

3. Sort these frequencies into those that are sound and those that are ultrasound.

   4500 Hz   100 000 Hz   30 000 Hz   30 Hz   1500 Hz

4. Sketch the oscilloscope trace for a:
   a. low-pitched, loud sound
   b. high-pitched, soft sound.

5. a. Name the animal in the table above with the biggest range of hearing.
   b. Name the animals that can hear ultrasound.

Key points

- Higher notes have a higher frequency.
- The range of frequencies that you can hear is the audible range.
- Different animals have different audible ranges.
- Your audible range (the range of frequencies you can hear) gets smaller as you get older.

Objectives

- Name the unit of sound intensity, or loudness

- Describe some of the risks of loud sounds and how to reduce the risks

Hearing, decibels, and risk

What is noise?

▲ We like listening to some sounds…   ▲ … but not to others.

Our ears are very sensitive to sounds. Most people can hear very quiet sounds and also very, very loud ones. We like to hear sounds like music and the voices of our family or friends. Sounds that we do not want to hear are called **noise**.

Sound levels

The **intensity** of a sound is measured with a **sound-level meter** on a scale called the **decibel** (**dB**) **scale**. A sound with a high intensity is louder.

Sound level (dB)	Activity	What this sounds like
0	threshold of hearing	cannot be heard
10	soft whisper	can just be heard
20	leaves rustling	very quiet
30		
40	talking quietly up to normal speech	quiet
50		
60		
70	noisy street	difficult to hear someone talking
80	traffic at a junction	annoying
90	truck chainsaw	can damage hearing after 8 hours per day
100	jet taking off 500 m away	can damage hearing after 2 hours per day
110	nearby thunder vuvuzela or nadaswaram	can damage hearing after 1.5 hours per day
120		
130	road drill	painful
140	jet engine 100 m away	loudest recommended level with ear protection
194	loudest sound that it is possible to make	

The decibel scale is *not* like a ruler. Every increase of 10 dB increases the intensity 10 times. A 50 dB sound is 100 times as intense as a 30 dB sound.

- The nadaswaram is an Indian musical instrument that is so loud it is usually played outside.
- The vuvuzela is a horn used at football matches in South Africa. It can emit noise of 120 dB.
- The loudest possible *continuous* sound has a loudness of 194 dB. Explosions, earthquakes, or meteor impacts can produce much louder sounds, but they do not last for long.

What can damage your hearing?

Loud sounds can damage your hearing. Sudden loud sounds can burst your eardrum. However, the risk depends not only on the number of decibels but also on the length of time you are exposed to the sound.

When scientists talk about the **risk** of an activity, they consider:

- the **probability** that something bad will happen
- the **consequence** if it did. Some activities have serious consequences, like an aeroplane crashing, but a very low probability of happening.

We often take risks when the consequences do *not* seem serious. Listening to music or going to concerts may *not* seem risky, but there is a probability that your hearing will be damaged by loud music. That probability gets bigger the longer you listen to the music.

How do you reduce the risk?

There are three main ways of reducing the risk from noise if you cannot change the sound level:

- **shielding** – putting something between the source of the sound and your ears, such as **ear defenders**
- increasing the distance – moving away from the source of the sound
- reducing the time that you spend near the source of the sound.

People can use decibel meters to help to reduce their risk.

▲ One of the world's loudest musical instruments.

▲ Ear defenders can reduce risk.

 Key points

- Sound levels are measured in decibels (dB).
- Loud sounds can damage your hearing over time.
- You can reduce the risk of damage by making the sound quieter, shielding your ears, reducing the time you are exposed, or moving away from the sound.

Questions

1. **a.** Write down what 'dB' stands for.
   b. Write down how much louder a 40 dB sound is than a 30 dB sound.
2. Jamal likes to listen to music with earphones. Write down two things he could do to reduce the risk of damage to his hearing.
3. **a.** Suggest two jobs for which people should wear ear defenders.
   b. One reason why truck drivers should *not* drive for long periods is that they may get tired and fall asleep. Suggest another reason.

13.4 Adding up and cancelling out

Objectives

- Describe what happens when waves interact
- Sketch waveforms of waves adding up and cancelling out

Some people have 'perfect pitch'. They can tell you which note you are playing on a musical instrument just by listening.

Most people cannot. The people who make sure that musical instruments are in tune use the interaction of sound waves to work out whether the correct note is being played. How do they do that?

▲ *Tuning a musical instrument.*

What happens when waves interact?

When waves interact, they can add together or cancel out. This is called **interference**.

- If the waves add up to produce a bigger wave, we say that it is **constructive interference**.
- If the waves cancel out to produce a smaller or no wave, we say that it is **destructive interference**.

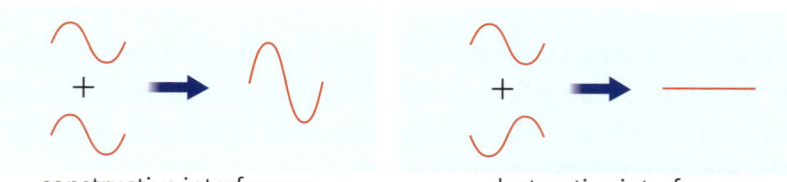

constructive interference   destructive interference

▲ *Waves add up or cancel out.*

Constructive interference	Destructive interference
The waves are in step	The waves are out of step
All the peaks are in the same place	Peaks of one wave and troughs of the other are in the same place
The total sound will be louder	The total sound will be quieter/silence

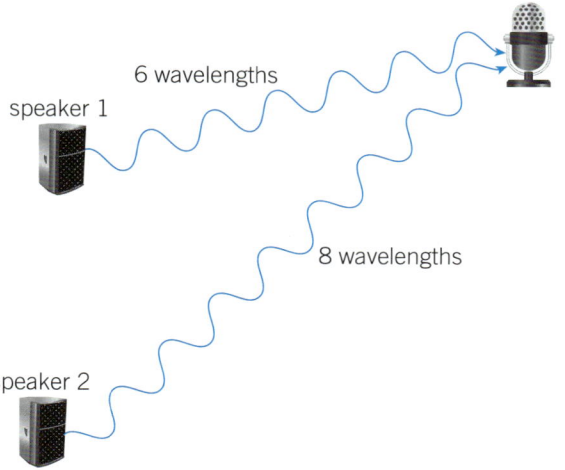

microphone

6 wavelengths

speaker 1

8 wavelengths

speaker 2

▲ *Interference between sounds from two speakers.*

For two waves to be in step and interfere constructively, the difference in position of the two waves will be zero or a whole number of wavelengths.

If the waves are out of step by half a wavelength, they will cancel out.

In destructive interference the waves need to have exactly the same amplitude to cancel out completely. You do *not* usually get silence. If the amplitude of one wave is bigger than the other, then the sound will be quieter.

How is the interference of sound useful?

People who tune pianos and guitars use the interference of sound to work out whether strings or keys are playing the correct note.

If you pluck the string of a guitar you hear a note. If you strike a tuning fork and hold it near the string you will hear a loud note if the string is in tune.

If it is *not* in tune, then the two waves will interfere. There will be times when the waves add up, and times when they cancel out. The tuner hears loud and soft sound. They adjust the string until they hear a loud note that does *not* change.

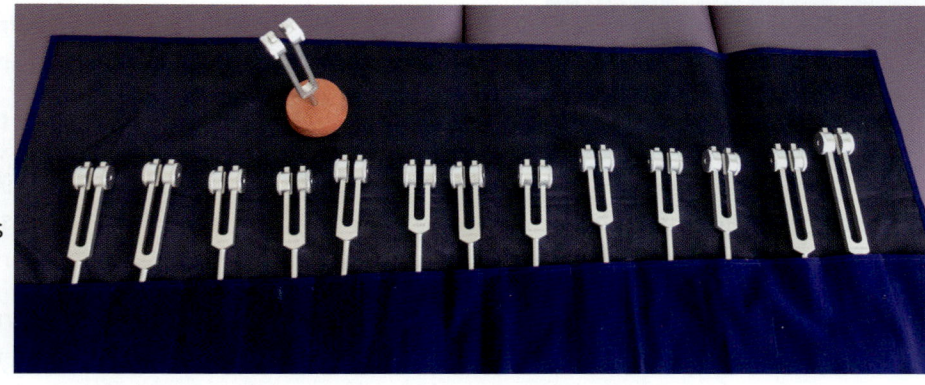

▲ *Tuning forks of different lengths make different notes.*

If you travel or work in a noisy place it can be helpful to be able to block out the noise. You can do that with noise-cancelling headphones. The headphones produce a sound that cancels out the noise.

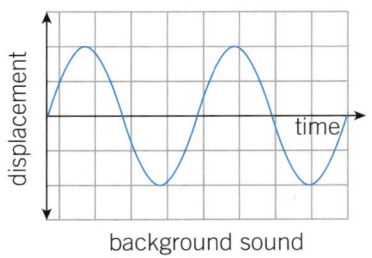

background sound

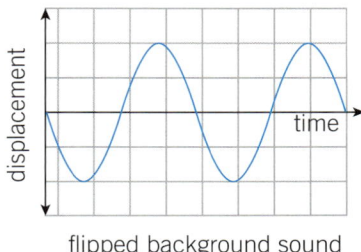

flipped background sound

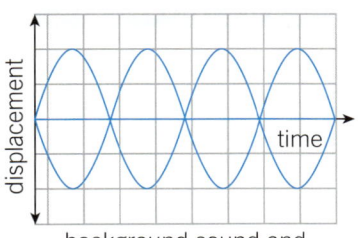

background sound and flipped sound superposed, producing cancellation

▶ *Some headphones produce sounds that cancel out noise.*

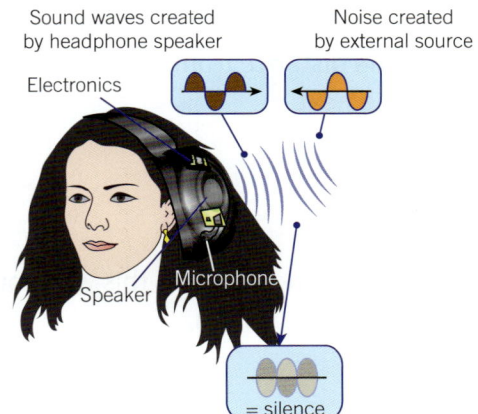

Sound waves created by headphone speaker

Noise created by external source

Electronics

Speaker   Microphone

= silence

Questions

1. Copy and complete the sentences by choosing the correct word or words in bold:
   When waves are **in step/out of step** they will add up. This is called **constructive/destructive** interference. If you listen to the sound it will sound **louder/softer**. The waves need to have exactly the same **amplitude/magnitude** and **frequency/speed** to cancel out completely.
2. Draw two waveforms that will cancel out completely when they interact.
3. Explain why some noise-cancelling headphones need a microphone as well as a loudspeaker.

📖 Key points

- Waves that interact (interfere) can add together or cancel out.
- If they add, then the total sound is louder.
- If they cancel out, then the total sound is quieter or there is silence.
- We use interference in noise-cancelling headphones and to tune musical instruments

1. For each of the quantities listed below write the letter of the definition and the unit. You need to use one letter twice.

   a. The wavelength of a wave

   b. The frequency of a wave

   c. The amplitude of a wave

   A   The number of waves per second.

   B   The distance from the middle to the top of a wave.

   C   The distance from the top of one wave to the next.

   D   hertz

   E   metres   [3]

2. A tuning fork produces this wave on an oscilloscope screen.

   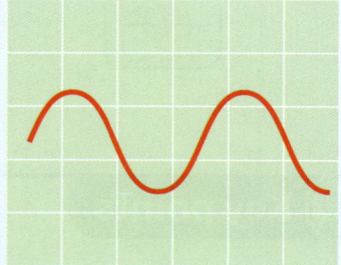

   a. Draw the wave you would see if the sound was louder.   [1]

   b. Draw the wave you would see if the sound had a higher pitch.   [1]

3. a. A note has a frequency of 400 Hz. State how many sound waves pass a point per second. [1]

   b. Another wave has a frequency that is half that of the wave in part a. Choose the number of waves per second for this wave.

      100   200   400   800   [1]

4. a. Write down the audible range of a human.   [1]

   b. Complete this sentence by choosing the correct word in bold:

      All animals have **different/the same** audible ranges.   [1]

5. Look at the arrows on the diagram below. Copy the table and tick the correct columns to show whether each arrow shows the wavelength, the amplitude, or neither.

   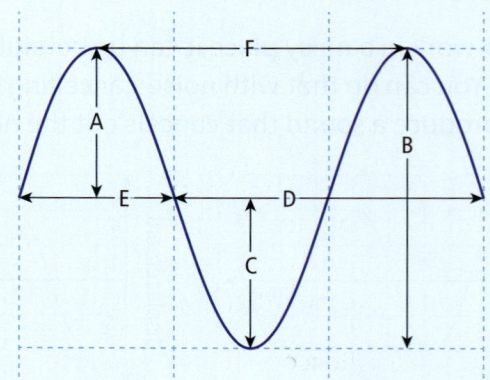

Arrow	Wavelength	Amplitude	Neither
A			
B			
C			
D			
E			
F			

   [6]

6. A boy is whistling a note that has a frequency of 1500 Hz.

   a. Explain what 'a frequency of 1500 Hz' means.   [1]

   b. Describe what would change about the sound he would hear if he whistled at 2000 Hz.   [1]

7. Rani used a sound-level meter to survey the noise at different places and times during the day. Here are her results.

Place	Noise level (dB)
by the road on the way to school	70
sitting in a classroom	50
playing with friends	60
traffic on the way home	80
reading a book	40

a. Describe what 'dB' means. [1]

b. Write down which sound is loudest. [1]

c. Rani's brother likes to listen to very loud music on his headphones. Describe two things that he could he do to reduce the risk of damaging his hearing. [2]

8. A teacher uses an oscilloscope to display some sound waves. The sounds shown are either loud or soft, and either high or low pitched. Match each number to a letter in the pictures below.

1 a low-pitched loud sound	A	
2 a high-pitched soft sound	B	
3 a low-pitched soft sound	C	
4 a high-pitched loud sound	D	

[4]

9. For each statement about ultrasound, write 'true' or 'false'.

a. Ultrasound can be used to see an unborn baby. [1]

b. Ultrasound is sound with a frequency greater than 2000 Hz. [1]

c. Ultrasound is very-high-frequency sound. [1]

d. Ultrasound cannot be heard by humans. [1]

10. a. Describe what is necessary for two waves to:

   i. add up [1]

   ii. cancel out. [1]

b. A microwave oven has a turntable to rotate the food through places where microwaves superpose. Suggest what will happen if the turntable breaks. [2]

11. A singer produces sounds that vary in pitch and loudness. Suggest and explain in detail what her vocal chords do to produce different types of sound wave. [3]

12. A student wants to investigate this question:
TWS 'How does the loudness of a sound change with the distance from a speaker?'

a. Write down the independent variable. [1]

b. Write down the dependent variable. [1]

c. Write down one control variable. [1]

d. Suggest why it might be difficult for the
TWS student to carry out this investigation. [1]

219

14.1

Current in series and parallel circuits

Objectives

- Describe the difference between a series and a parallel circuit

- Describe what happens to current in a parallel circuit

- Describe how to measure current in series and parallel circuits

- Describe the effect on the current of adding cells and lamps in series and parallel circuits

When you turn off a light in one room of a house the others will stay on. How does that **circuit** work?

What is a series circuit?

A circuit with a single loop is called a **series circuit**. The **current** will only flow if it is a complete circuit. You can turn the lamp on and off with a switch, because an open switch is a break in the circuit.

In a series circuit with several lamps:

- You cannot turn the lamps on and off separately.

- The lamps are either all on or all off.

- A switch in the circuit will turn them all on, or turn them all off.

- If one lamp breaks, or 'blows', then all the lamps go out.

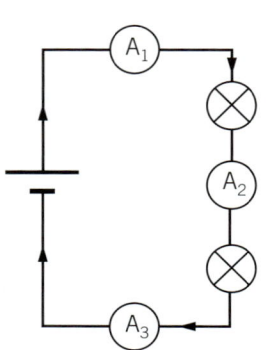

You can use an ammeter to measure the current at different points in a series circuit. The current in A_1, A_2, and A_3 is the same, so the reading on each of the ammeters is the same. Current is *not* used up in an electric circuit.

What is a parallel circuit?

Neela makes a series circuit with a battery, one lamp, and a switch. Then she adds another loop containing a lamp and a switch that uses the same battery as in the first circuit. Here are pictures of some circuits that she made.

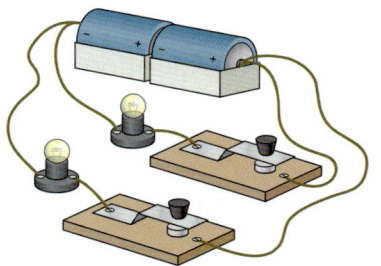

▲ *This is a parallel circuit…*

In a series circuit there is only one loop. In a **parallel circuit** there is more than one loop. The cell or battery pushes the current around each loop. Neela made a parallel circuit when she made the circuit with two loops.

- The different loops of the circuit are sometimes called branches.

- Parallel circuits are sometimes called branching circuits.

- In a series circuit, if one lamp breaks then all the lamps go off.

- In a parallel circuit, if one lamp breaks then the others still work.

- The switch in each branch of the circuit controls the lamps in that branch.

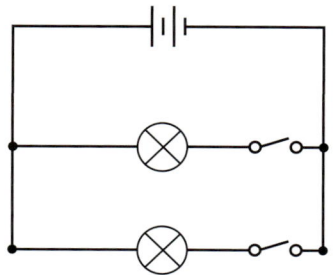

▲ *… and so is this.*

Each lamp can be turned on and off independently. This is how lights in your house are connected.

How do you measure current in parallel circuits?

Neela decided to measure the current in the branches of a parallel circuit.

She connected up a circuit with a battery, lamps, ammeters, and switches. She connected an ammeter in each branch of the circuit, and another ammeter near the battery.

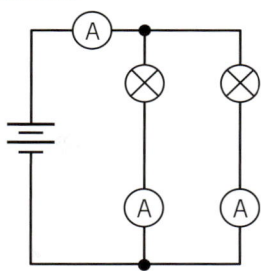

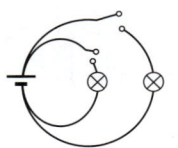 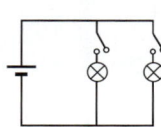

▲ *Think of parallel circuits like this.*

> I connected up my circuit and measured the current in each of the branches and next to the battery. The current in each branch was 0.4 A. The current next to the battery was 0.8 A.

Neela wondered what would happen if she added another lamp in parallel.

> This time the current near the battery was 1.2 A, but the current in each branch was still 0.4 A.

▲ *Car headlights are connected in parallel.*

Number of lamps in the circuit	Reading A$_2$	Reading A$_3$	Reading A$_4$	Reading A$_1$
1	0.4	–	–	0.4
2	0.4	0.4	–	0.8
3	0.4	0.4	0.4	1.2

As you add more branches, the current in each branch stays the same. The total current increases. You could increase the current in each branch, and the total current, by increasing the number of cells.

You can think of the current splitting when it gets to a junction of two or more branches.

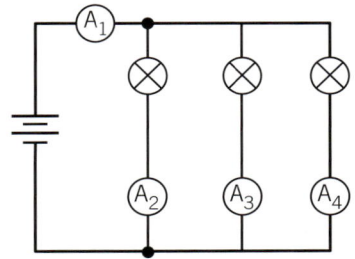

📖 Key points

- The current in a series circuit is the same everywhere.
- A parallel circuit contains more than one branch.
- Each loop (or branch) is independent.
- Components in a parallel circuit can be turned on and off independently.
- The current near the battery is equal to the sum of the currents in all the branches.
- The current splits when it gets to a junction.

❓ Questions

1. Copy and complete these sentences.
   a. An ammeter measures the _____ flowing per_____.
   b. A _____ current means that more _____ is flowing per second.
   c. If a bulb breaks in a series circuit the current will be _____.
2. Explain why you only need one switch in a series circuit.
3. A student connects up an ammeter in a circuit with a lamp and a cell. The reading is 0.5 A. She moves the ammeter to the other side of the lamp.
   a. What will the ammeter read now? Explain your answer.
   b. She then connects another lamp in parallel with the first. Describe what happens to the total current in the circuit.

Voltage in series and parallel circuits

Objectives

- Describe what is meant by voltage

- Describe how to measure voltage in series and parallel circuits

- Describe the effect on the voltage of adding cells and lamps in series and parallel circuits

New batteries in a torch make the lamp brighter.

What makes the lamps brighter?

In a circuit, the force needed to make charges move – the 'push' – is provided by the battery.

The size of the force depends on the **voltage**. The voltage also tells us about the energy that is transferred to the charge by the battery. Voltage is sometimes called **potential difference**.

- If the voltage is bigger, the push is bigger.

- More charge will pass a point in each second, so the current will be bigger.

- More energy will be transferred to the components in the circuit.

- Lamps will be brighter and buzzers will be louder.

Measuring voltage

Voltage is measured in **volts (V)** using a **voltmeter**. The circuit symbol for a voltmeter is: ─Ⓥ─

You use a voltmeter to measure the voltage *across* a component. Voltmeters are connected to each side of a component, *not* in series like ammeters.

- A voltmeter connected across the battery tells you the energy supplied *to the charge* by the battery.

- A voltmeter connected across a component tells you the energy supplied *by the charge* to the component.

The energy stored in the battery is transferred to the components by the current.

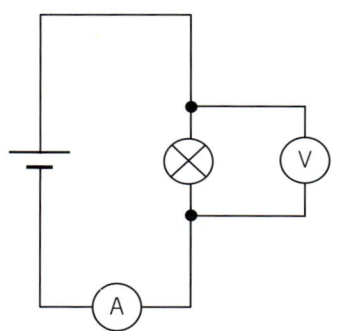

▲ *A voltmeter is connected across a component.*

What happens to voltage in a series circuit?

You can think of a battery in terms of 'lifting up' charges. When the charges go through the components they 'fall back down'.

A circuit contains two lamps and a battery. The voltage of a battery is 6 V.

- Each lamp emits light when the current flows.

- Half of the energy from the cell is transferred in each lamp, so the reading on each voltmeter is 3 V.

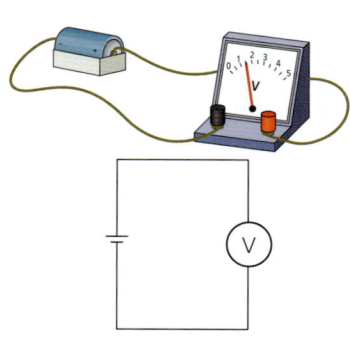

▲ *The voltage of this cell is 1.5 V.*

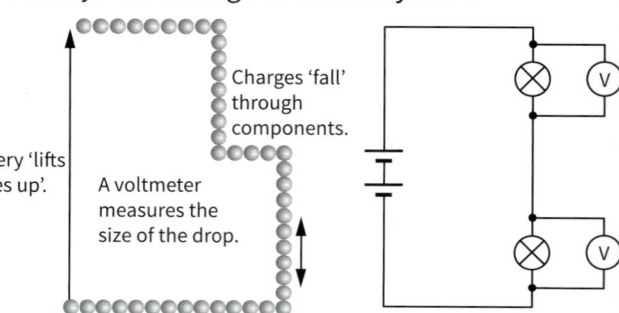

A battery 'lifts charges up'.

Charges 'fall' through components.

A voltmeter measures the size of the drop.

▲ *Measuring voltages in a series circuit.*

You can use the factory model (page 65) to help you to work out what is happening.

- A factory supplies two shops.

- Each lorry loads 6 boxes at the factory. This is like the 6 V of the battery.

- The lorry delivers 3 boxes to each shop. The lorry has 3 fewer boxes after delivering to each shop. Each lamp has a voltage of 3 V.

The voltages across the components add up to the voltage of the battery.

If the components in the circuit are *not* the same, then the voltages across them will be different.

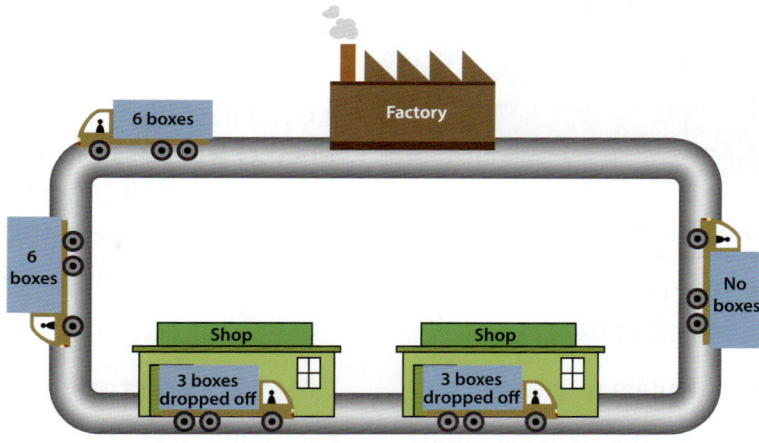

▲ *The factory model for a series circuit.*

What happens to voltage in a parallel circuit?

In a parallel circuit the voltage across each lamp is the same.

- Each branch is connected directly to the battery.

- The voltage across each component is the same as the voltage across the battery.

- Each component is independent of the other components.

We could add more lamps in parallel and the reading on the voltmeters would still be the same.

▲ *Measuring voltages in a parallel circuit.*

If the voltages are all the same, what is the effect of adding more lamps? When you add more and more lamps in parallel, the battery is drained faster.

Questions

1. **a.** Describe the difference between current and voltage.
   b. Compare how voltmeters are connected in a circuit with how ammeters are connected.
2. Calculate the number of 1.5 V cells that provide a voltage of 9 V.
3. Explain why the voltages across components in a series circuit add up to the voltage across the battery.
4. Dilip connects up a series circuit with a lamp and a buzzer. The voltage of the battery is 6 V.
   a. If the voltage across the lamp is 2 V, calculate the voltage across the buzzer.
   b. Compare the current through the lamp with the current through the buzzer.
   c. Dilip now connects the lamp and buzzer in parallel. Explain why you do not have to calculate the voltage.
5. A student says that the circuit drains the battery because the charge is used up. What would you say to him?

Key points

- Voltage tells you the energy transferred to or by a charge.
- Voltage is measured in volts (V) using a voltmeter.
- The voltages across components in a series circuit add up to the voltage of the battery.
- The voltage across each branch of a parallel circuit is the same, and is equal to the voltage of the battery.

Resistance

Objectives

- Describe how resistance affects current

- Calculate resistance

- Know the circuit symbols for fixed and variable resistors

It is *not* possible to predict whether a battery will produce a large or a small current in a circuit just from the battery. It depends on the components in the circuit.

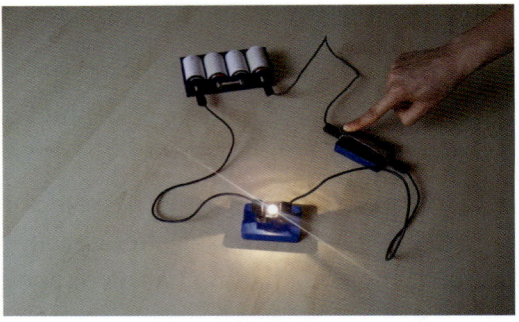

▲ *The current in a circuit depends on the components.*

What is resistance?

Each component such as a lamp provides a **resistance** to the flow of charge.

Think about the rope model of a circuit (page 64). If two people hold the rope instead of one, there is more resistance to the rope's movement.

Two lamps provide more resistance than one lamp, so less charge flows per second.

Nihal connects a circuit with one lamp. He measures a current of 0.5 A. He replaces the lamp with a buzzer.

The current is *less* than it was when the lamp was in the circuit.

The buzzer provides *more* resistance to the flow of charge. The current is less.

This is like someone squeezing the rope a bit harder.

▲ *A rope model can help you to understand resistance.*

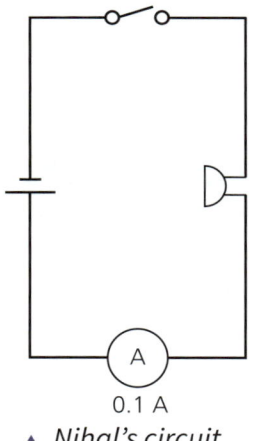

0.1 A

▲ *Nihal's circuit.*

I wonder why the current is different? There is only one component in the circuit.

How do you calculate resistance?

In the rope model, the speed of the rope depends on:

- how hard the 'battery' person is pulling the rope (the voltage)

- how hard the 'component' person is squeezing the rope (the resistance).

The current in a circuit is like the speed of the rope. It depends on the voltage and the resistance. Resistance is measured in **ohms**, which has the symbol Ω.

$$\text{current (A)} = \frac{\text{voltage (V)}}{\text{resistance (Ω)}}$$

This means that you can calculate the resistance of a component using the current and the voltage.

$$\text{resistance (Ω)} = \frac{\text{voltage (V)}}{\text{current (A)}}$$

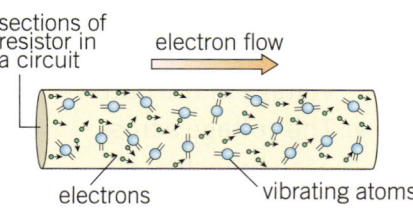

sections of resistor in a circuit   electron flow

electrons   vibrating atoms

▲ *Flowing charge has to get past obstacles that offer resistance: vibrating atoms slow electrons down.*

Nihal decides to calculate the resistance of each component. He sees that the voltage of the battery is 6 V.

$$\text{resistance of the lamp} = \frac{\text{voltage (V)}}{\text{current (A)}}$$

$$= \frac{6\ V}{0.5\ A}$$

$$= 12\ \Omega$$

$$\text{resistance of buzzer} = \frac{\text{voltage (V)}}{\text{current (A)}}$$

$$= \frac{6\ V}{0.1\ A}$$

$$= 60\ \Omega$$

The buzzer has a bigger resistance so the current is smaller.

What are the circuit symbols for resistors?

The symbols for two different types of resistor are shown below.

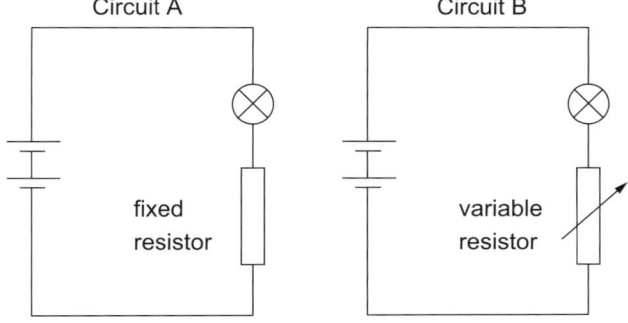

In circuit A the resistance of the resistor does *not* change.

In circuit B the resistance of the resistor changes. As the resistance increases the lamp will get dimmer. This is how a dimmer switch works.

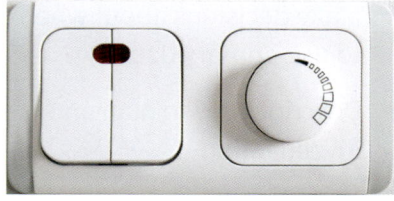

▲ *You can adjust the light level with the variable resistor on the right.*

Science in context

Our understanding of voltage, current, and resistance along with other technologies has enabled us to develop sophisticated electronic systems such as phones, computers, and the Internet.

Questions

1. A student measures the current in a circuit with a variable resistor. Describe what she must do to the variable resistor to increase the current in the circuit.
2. Calculate the resistance of a resistor that carries a current of 0.25 A when there is a voltage of 12 V across it.
3. A parallel circuit has a motor and a buzzer connected across the same battery. The current through the battery is 1.5 A, and through the buzzer is 0.4 A.
   a. Deduce the current through the motor.
   b. Suggest which of the two components has the bigger resistance.
   c. The battery has a voltage of 6 V.
      Justify your answer to part **b** with two calculations.
4. Explain why adding an additional branch to a parallel circuit *reduces* the overall resistance of the circuit.

Key points

- If the resistance increases the current decreases (if the voltage is kept the same)
- resistance (Ω)

  $$= \frac{\text{voltage (V)}}{\text{current (A)}}$$

- Both fixed and variable resistors can be used in circuits.

Objectives

- Make decisions about when to use primary data
- Describe how to plan an investigation

▲ *Chetana*

> I wonder how the resistance of the wire depends on the thickness of the wire.

▲ *Dipali*

> I wonder how the resistance of the wire depends on the length of the wire.

Planning investigations: resistance of a wire

Chetana, Dipali, and Lakshima have been learning about current and voltage. Their teacher explained that the current flowing in a component depends on the voltage of the battery and the resistance of the component. If the voltage stays the same, then increasing the resistance will make the current smaller.

What affects the resistance of a wire?

Asking questions

Chetana has noticed that the filaments of the lamps that they use in their experiments are made of very thin wire.

Dipali looks at a filament. She notices that it is curled up and thinks that it must be very long.

Lakshima wonders what the filament is made of.

The students list the **variables** in their investigation:

- the length of the wire
- the thickness of the wire
- the material that the wire is made of
- the voltage of the battery.

▲ *Lakshima*

> I wonder how the resistance of the wire depends on the material that the wire is made of.

When you are choosing ideas to test, you need to make sure they can be tested by collecting data. All these questions involve variables that the students can change. They can control the other variables so that the data show the link between their chosen variable and the resistance of the wire.

A student might wonder if some bulbs look brighter than others to different people. This is an interesting question, but *not* one that you can easily collect data to test.

The students decide to investigate the length of the wire. They will:

- measure the current that flows through different lengths of wire
- use the current and voltage to calculate resistance
- keep the thickness of the wire, the material that the wire is made of, and the voltage of the battery the same.

Planning investigative work: preliminary work

The students need to work out which type of wire to use, which thickness of wire, and which voltage to use. They do some preliminary work to help them decide how to do the experiment safely, and how to get a good range of results.

They use **secondary sources** to find the most suitable wire. They try out different voltages for different lengths of wire. They notice that the wire gets very hot for high voltages, so they choose a lower voltage of 1.5 V.

Voltage (V)	Current for 10 cm of wire (A)	Current for 1 m of wire (A)
1.5		
3.0		
6.0		

▲ *Students recorded their preliminary work in this table.*

Planning investigative work: choosing equipment

The students use a digital ammeter to measure the current in the wire. The ammeter will measure the current to the nearest 0.1 A.

They decide to change the length using clips to connect to different positions on the wire. They will use a ruler to measure the length of wire between the clips. They can measure the length to the nearest 1.0 mm.

Planning investigative work: assessing hazards and controlling risk

The students know that the wire can get very hot. They put the wire on a heatproof mat and do not leave the battery connected for very long.

Obtaining and considering evidence

The students collect evidence by measuring the current for different lengths of wire. They repeat each experiment three times to make their results more reliable. Before calculating any averages they discard any **anomalous results**.

They plot a graph to try to find a pattern in their results. This will help them make a conclusion about the link between the current and the length of the wire.

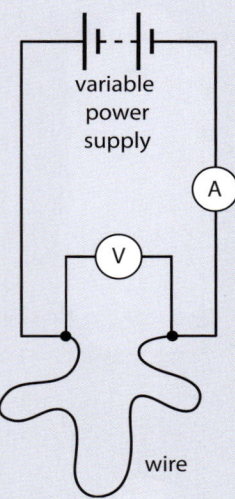

variable power supply

wire

▲ *This circuit can be used to measure the current through the wire.*

 Questions

1. List two measuring instruments the students will need.
2. Look at Chetana's question about the thickness of the wire. She decides to carry out an investigation to answer this question.
   a. Identify the variables she should change, measure, and control.
   b. Describe a difficulty she might have in this investigation.
3. Describe the safety issues in the investigation into length of wire.
4. Lakshima completes an investigation into her question about the material of the wire.
   a. What would be the difference between Lakshima's graph and Chetana's graph? Explain your answer.
   b. Draw a table that Lakshima could use to collect her data.

📖 Key points

Planning an investigation may involve:
- identifying variables
- doing preliminary work
- choosing equipment
- assessing hazards and controlling risk.

Objectives

- Describe the difference between energy and power
- Calculate power

▲ *Top: CFL lamp; middle: LED lamp; bottom: filament lamp.*

Energy and power

The Kovalam lighthouse in Kerala, India, has a very bright light at the top of a tall cylindrical tower. The light flashes 4 times every minute and can be seen nearly 40 km away. Ships use the light to navigate to the harbour at Kovalam.

▲ *The light from a lighthouse has to be seen from a long distance.*

What is power?

Some lamps are brighter than others. The lamp in the lighthouse is much, much brighter than the lamps used in torches or in houses or offices.

An electric current transfers energy from the **mains supply** to the lamp.

A bright lamp transfers more energy per second. It emits more light per second than a dim lamp.

- The rate at which energy is transferred is called the **power**.
- Power is measured in **watts** (W) or **kilowatts** (kW).
- 1 kW = 1000 W.

How do you calculate power?

You can calculate power using an equation:

$$\text{Power (W)} = \frac{\text{energy transferred (J)}}{\text{time taken (s)}}$$

The lighthouse lamp transfers 24 000 J of energy each minute.

The lamp in a torch transfers energy more slowly than the lighthouse lamp so the power is much less. The torch lamp transfers energy at a rate of 120 J each minute.

To calculate the power you have to convert the time to seconds.

For the lighthouse:

$$\text{Power} = \frac{24\,000\ \text{J}}{60\ \text{s}} = 400\ \text{W}$$

For the torch:

$$\text{Power} = \frac{120\ \text{J}}{60\ \text{s}} = 2\ \text{W}$$

A lighthouse lamp is 200 times more powerful than a torch lamp.

How do light bulbs differ?

For over 100 years light bulbs have been widely used.

- **Filament** (or incandescent) lamps contain a piece of wire that is very thin and glows when it gets hot.
- Compact fluorescent lamps (CFLs) are energy-saving lamps first sold in 1981.
- **Light-emitting diode (LED)** lamps are very **efficient**.

The table below shows how much power, or energy per second, each type of lamp needs to produce different light levels.

Light intensity (lumens)	Power of incandescent light bulb (W)	Power of CFL (energy-saving lamp) (W)	Power of LED lamp (W)
800	60	15	10
1800	100	25	18
2800	150	45	26

- LEDs produce a lot of light for their size.
- LEDs can be made small enough to sew into your clothes.
- Doctors use the intense light in cancer treatments.
- LEDs were used in the London and Rio de Janeiro Olympics to make spectacular patterns inside the Olympic stadium.

How do you pay for electricity?

When people pay their electricity bill they pay for fuel to be burned to produce a current to transfer energy. The amount of energy transferred depends on the power of the appliances and how long they are used for.

We can rearrange the equation for power like this:

Energy (J) = power (W) × time (s)

People transfer lots and lots of joules of energy each month. So, instead, the energy is measured in a different unit, the kilowatt-hour (kWh).

Energy (kWh) = power (kW) × time (h)

For example, if you use a 10 W (0.01 kW) LED for 10 hours in one month,

energy transferred = 0.01 kW × 10 h = 0.1 kWh

To produce the same amount of light using a filament light bulb you would need to use a 60 W light bulb. You would pay for 0.6 kWh instead. Using new light bulbs can save a lot of money. LEDs are a lot more expensive to purchase than filament or CFL light bulbs, but they do last a long time.

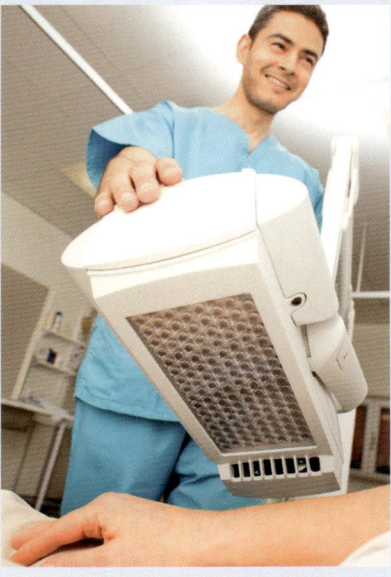
▲ Cancer treatment with LEDs.

▲ LEDs are used to make impressive light displays.

Questions

1. Calculate the number of watts in 2 kW.
2. A CFL transfers 1200 J of energy each minute. Calculate its power.
3. Look at the table and the calculations above for the energy cost of using an LED and a filament bulb for 10 hours each month.
   a. Calculate how much energy in kWh you would need for a CFL lamp to produce the same amount of light.
   b. The electricity company charges 10 rupees per kWh. Calculate how much it would cost to use each lamp for 10 hours each month.
   c. Calculate the money you would save each month if you used an LED instead of a filament bulb.

Key points

- Power is the energy transferred per second.
- Power is measured in watts (W) or kilowatts (kW).
- Energy (J) = power (W) × time (s)
- LEDs have a wide range of uses.

1. a. Draw a circuit diagram for the circuit below. **[2]**

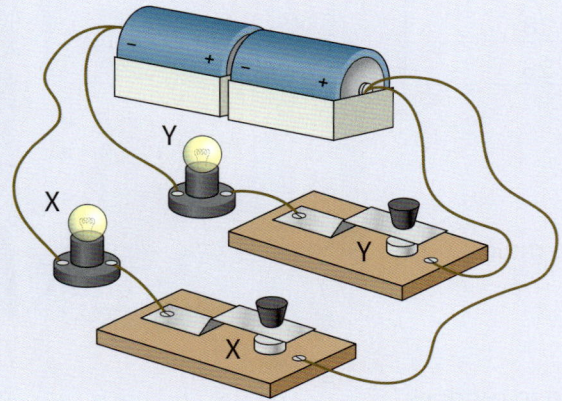

b. State whether this is a series or a parallel circuit. **[1]**

c. Copy and complete the table to show what would happen when you press the switches. **[3]**

Switches closed	Bulbs lit
X	
Y	
X and Y	

d. Describe how you could measure the current flowing in each bulb. **[1]**

2. Look at the sentences below. Name the circuit component or type of circuit that each sentence describes.

a. This is the energy source for an electric circuit. **[1]**

b. In this circuit all the components are in a single loop. **[1]**

c. This enables you to turn a component on or off. **[1]**

d. In this circuit all the components can be turned on or off independently. **[1]**

3. A student uses four bulbs and a battery pack to make a circuit. He puts a box over the circuit so the connections cannot be seen.

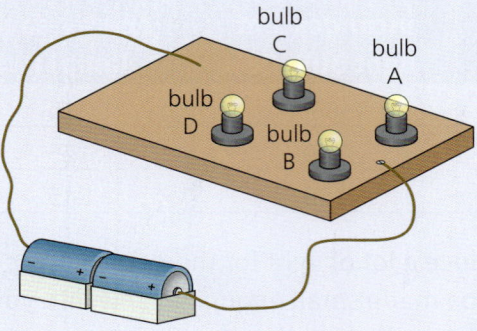

Another student unscrews each bulb in turn and watches the effect on the other three bulbs.

If you take this bulb out...	... this is what happens
A	Bulbs B, C, and D go out.
B	Bulb C goes out, bulbs A and D stay on.
C	Bulb B goes out, bulbs A and D stay on.
D	Bulbs A, B, and C stay on.

a. Draw a diagram to show how the bulbs must be connected. **[3]**

b. Name the bulb or bulbs that will be the brightest. Explain your answer. **[2]**

c. Name the bulb or bulbs that will be the dimmest. Explain your answer. **[2]**

4. Here is a circuit diagram. All the lamps are the same.

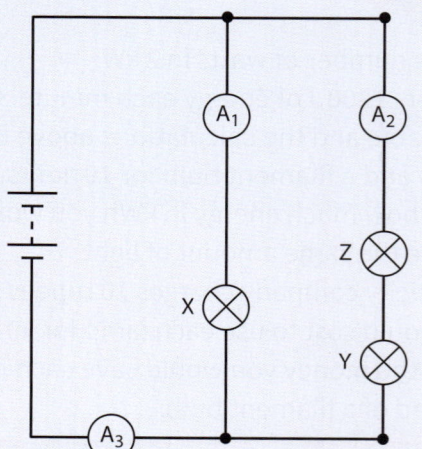

a. Are the lamps connected in series, parallel, or both? Explain your answer. **[2]**

b. Which ammeter will show the lowest reading? Explain your answer. **[2]**

c. Describe the link between the readings on the three ammeters. **[1]**

d. Which lamp will be brightest? Explain your answer. **[2]**

e. Describe what would happen to the other lamps if:

    i. lamp X broke **[1]**

    ii. lamp Y broke **[1]**

    iii. lamp Z broke. **[1]**

5. A circuit contains a 6 V cell and two lamps connected in series.

a. Draw a circuit diagram showing how you would connect a voltmeter to measure the voltage across each bulb. **[1]**

b. Both bulbs are identical. Write down the reading on each voltmeter. **[1]**

c. Explain your answer to part **b**. **[1]**

6. A student is planning how to collect some data to answer this question:

TWS

'Which circuit component has the biggest resistance?'

He writes down his ideas, but they are *not* in the right order. Put the statements in the right order using the letters. **[1]**

A	Think about how I can make sure that the experiment is safe.
B	Connect up the circuit and measure the current through each component.
C	Select equipment so that I can measure the current.
D	Do some preliminary work to find the best voltage to use.
E	Write the results in a table.

TWS

7. A student wires up a circuit to measure the current through a buzzer and a lamp. He connects a voltmeter to measure the voltage across the buzzer.

He draws the circuit in his notebook.

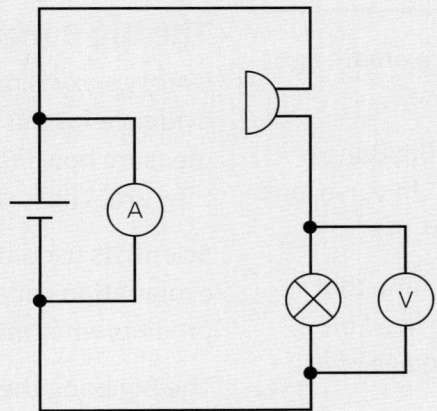

a. List the things that are wrong with this circuit diagram. **[2]**

b. Draw the correct circuit diagram. **[2]**

8. A student connects one bulb, an ammeter, and a cell in series. He connects a voltmeter across the bulb. The current through the bulb is 0.4 A. The voltage across the bulb is 3 V.

a. Explain what is meant by current. **[1]**

b. Explain what is meant by voltage. **[1]**

c. Calculate the resistance of the lamp. **[2]**

d. The lamp transfers 72 J in one minute. Calculate the power. **[1]**

9. A circuit contains two lamps and a cell. They are connected in series.

a. Describe and explain what happens to the brightness of the lamps if another lamp is added in series. **[2]**

b. Describe how the brightness would change if the lamps were connected in parallel, *not* in series. Explain your answer. **[2]**

c. Describe how the brightness would change if another cell is added to both the series and the parallel circuits. **[1]**

The origin of the Universe

An **astronomer** looks at galaxies in the night sky. She works out that they are all moving away from the Earth. You might think that the Earth is special, or that it is the centre of the Universe, but that is *not* the case.

Objectives

- Give the approximate age of the Universe

- Describe the Big Bang theory of the Universe, and evidence for it

- Compare the time that humans have lived on Earth with the age of the Earth

The Big Bang

Every galaxy is moving away from every other galaxy. Astronomers found evidence for this movement in the 1920s, when they used telescopes to measure how fast galaxies are moving. They also measured how far away they are. The further away they were, the faster they were moving.

Scientists used the evidence from their observations to develop an explanation called the **Big Bang**. It is an idea that explains observations and predicts what might happen in the future.

The Big Bang theory says:

- The Universe began about 14 billion years ago.
- The whole Universe **expanded** from something smaller than an **atom** and hotter than anything we can imagine.
- In a fraction of a second the Universe grew to the size of a galaxy, and it has continued to expand ever since.
- As it expanded, it cooled, and energy changed into particles, making atoms and molecules of hydrogen and helium.
- Eventually there were stars, planets, moons, and galaxies, and the Universe we see today.

▲ *An astronomer using a telescope in an observatory.*

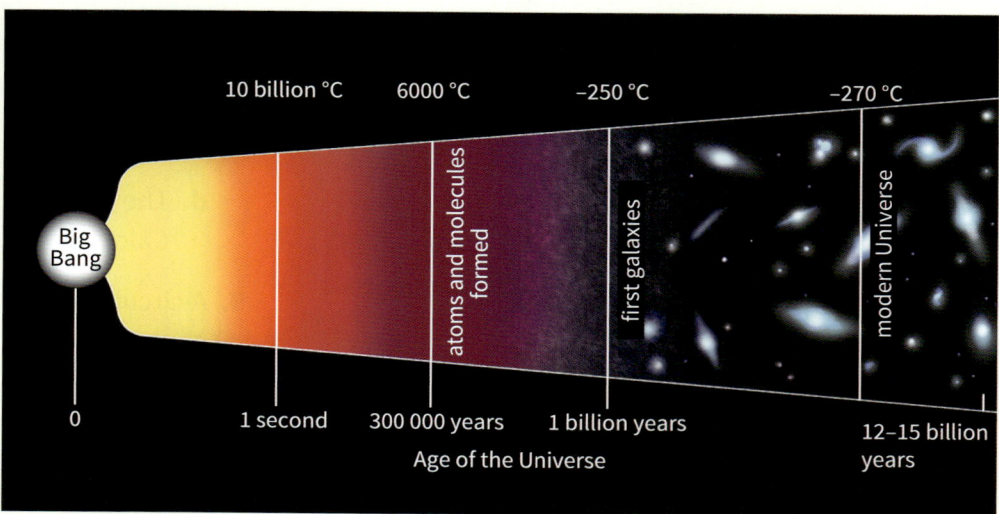

▲ *The Big Bang theory explains how the modern universe was formed over billions of years.*

The galaxies are moving away from us because the space in between all galaxies is expanding. There is nothing special about our galaxy, the Milky Way. If we lived in the **Andromeda** galaxy it would still appear that all the other galaxies, including the Milky Way, were moving away from us.

A timeline for the Universe

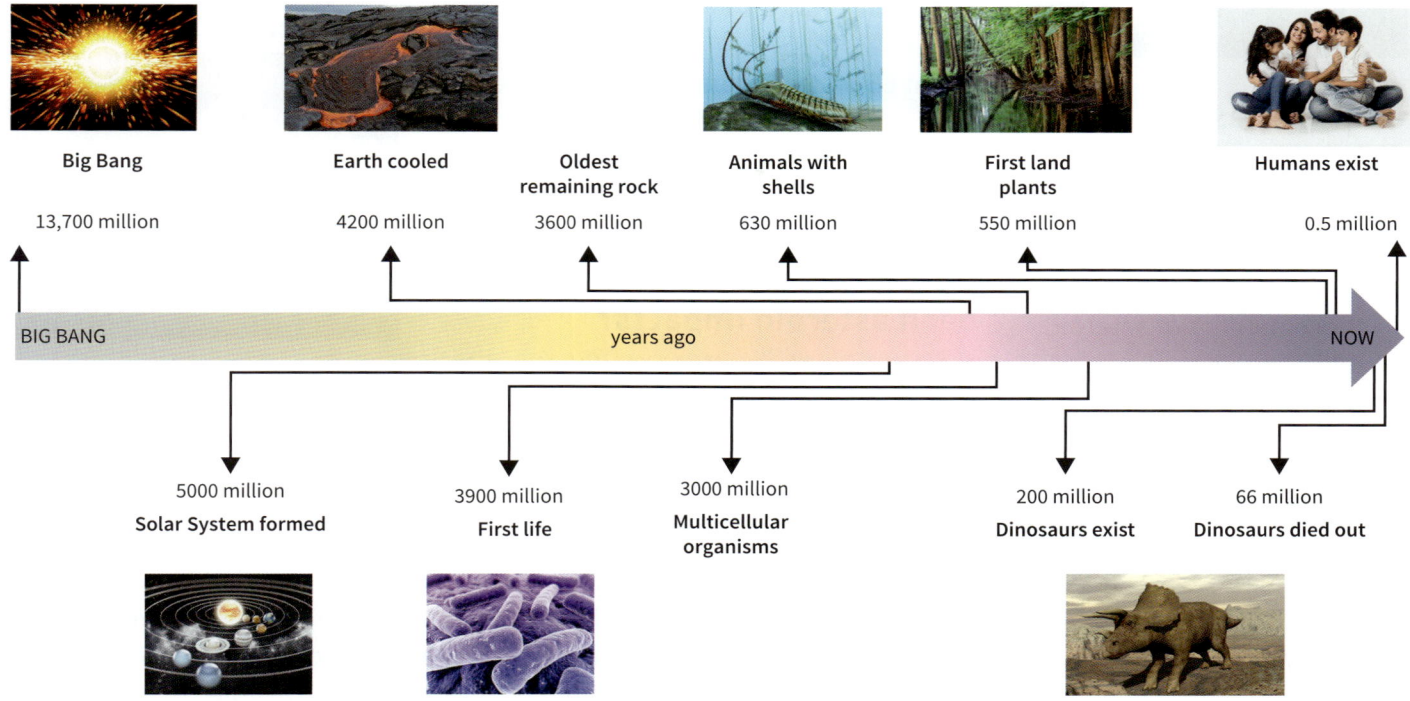

Big Bang	Earth cooled	Oldest remaining rock	Animals with shells	First land plants	Humans exist
13,700 million	4200 million	3600 million	630 million	550 million	0.5 million

BIG BANG — years ago — NOW

5000 million	3900 million	3000 million	200 million	66 million
Solar System formed	First life	Multicellular organisms	Dinosaurs exist	Dinosaurs died out

Sometimes it is easier to picture the timescale with an **analogy**.

- If the Universe started 24 hours ago, then the Earth would have formed 9 hours ago.
- The first animals would have appeared about an hour ago.
- Human beings would have existed for less than the time that it takes to blink your eyes.

The end of the Universe

No one knows what will happen to the Universe.

- It may expand forever.
- It may expand, and the expansion could get faster.
- Gravity may pull it back in again.
- It may expand to a certain size and stay that way.

To know what is going to happen we would need to be able to measure the mass of all the objects in the Universe.

This is *not* possible with the technology that we have at the moment.

Questions

1. **a.** Write down the age of the Universe.
   b. Write down the age of the Solar System.
2. Why is it difficult to predict what will happen to the Universe?
3. Could humans have seen a dinosaur? Explain your answer.

Key points

- Scientists think that the Universe began with a Big Bang about 14 billion years ago.
- Some evidence for the Big Bang is that all the galaxies are moving away from each other.
- Earth has existed for a fraction of the time that the Universe has existed.
- Humans have existed for a tiny fraction of the time that the Earth has existed.

Collisions, asteroids, and mass extinctions

Objectives

- Describe evidence for asteroid collisions
- Describe some consequences of asteroid collisions

▲ *There are millions of craters on the Moon.*

▲ *The Lonar crater in India is now a lake.*

About 66 million years ago a huge asteroid hit the Earth. Scientists think that this impact contributed to the extinction of the dinosaurs. Has this happened before?

▲ *An asteroid impact happened at the time the dinosaurs died out.*

How do we know there have been asteroid impacts?

Asteroids can be pulled out of their orbit in the asteroid belt by Jupiter's gravity. Their new direction may take them on a collision course with the Earth. You learned about asteroids and **meteorites** on page 183.

There is evidence that very many asteroids have collided with objects in the Solar System in the past. If you look at the Moon you see it covered in **craters**. The Moon has been hit by lots of pieces of rock in the past. The conditions on the Moon mean that the craters are still visible.

What is the evidence for asteroid impacts on Earth?

▶ *Meteor Crater in Arizona, USA, was formed when a piece of iron 40 m in diameter hit the Earth.*

There are some visible craters on the Earth. The crater produced by the asteroid that may have caused the extinction of the dinosaurs is deep off the coast of Mexico, and was not discovered until 1978.

What happens when an asteroid hits the Earth?

An asteroid with a diameter of 50 m can make a crater over a kilometre wide.

On impact, the ground and the asteroid can be **vaporised**. This sends material into the air in the form of dust, ash, and gas. If the asteroid hits the ocean it can cause a tsunami.

The effect of the impact depends what the asteroid is made of, and what it hits.

A large, metallic asteroid hitting soft rock will make a large crater.

▲ *Some craters, like this one in Canada, are only visible from space.*

What effect does an asteroid impact have?

If the asteroid is very large there will be enough dust to block out light from the Sun. The material that ends up in the Earth's atmosphere can change climate, and cause mass extinctions.

The climate depends on the balance between the energy that reaches the Earth from the Sun, and the energy that the Earth radiates into space.

If radiation from the Sun is blocked, then the temperature of the Earth would fall very quickly. This is called an **impact winter**. If the Sun was blocked out for a long time, then plants that rely on photosynthesis and animals that eat the plants would die. An impact winter that continued for over a year would make it difficult for humans to survive.

An event that causes a large percentage (75%–90%) of all species to disappear in a relatively short time is called a **mass extinction event.** Scientists think that these events happen on average approximately every 25 million years.

The most famous was the event that probably killed off most of the dinosaurs when an asteroid about 9 km in diameter hit the Earth.

However, most mass extinction events are thought to result from climate change due to volcanic emissions or changes to the Earth's oceans.

How often do significant asteroid impacts happen?

More than 100 tons of dust hits the Earth every day, but it gets burned up in the atmosphere.

An object the size of...	... hits the Earth every...	... and produces...
a car (5 m)	year	a fireball
a football field (100 m)	5000 years	a large crater, significant damage, tsunami
half (400 m) the tallest building in the world	100 000 years	climate change, impact winter, possible mass extinction

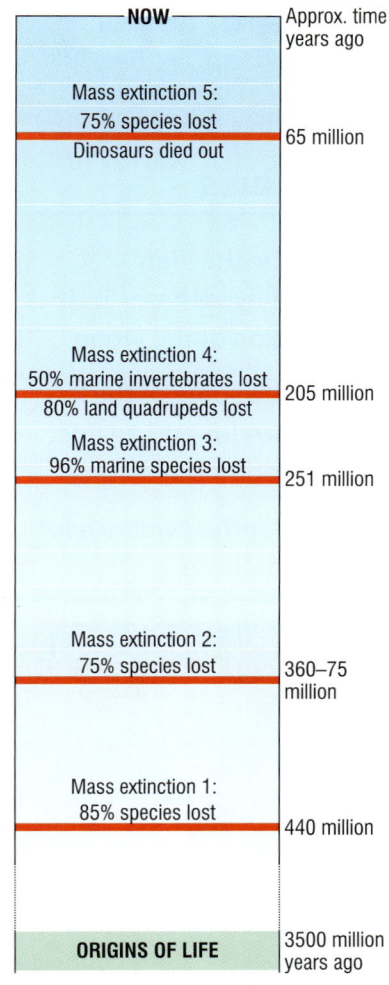

▲ There have been five mass extinction events in Earth's history.

Key points

- Craters and meteorites show that asteroids have hit Earth in the past.
- Small asteroid impacts can cause small craters and fireballs.
- Large asteroid impacts can make craters, throw material into the atmosphere, and cause climate change and mass extinction.

Questions

1. Describe the two main effects of large asteroid impacts.
2. Explain why craters are easily seen on the Moon but *not* on Earth.
3. Suggest whether there are any objects in the Solar System that do *not* have craters.
4. Suggest why it is difficult to know the exact cause of a mass extinction event.

15.3 Collisions and the Moon

When you look up in the night sky you can usually see the Moon. It was *not* there when the Earth formed. How did it get there?

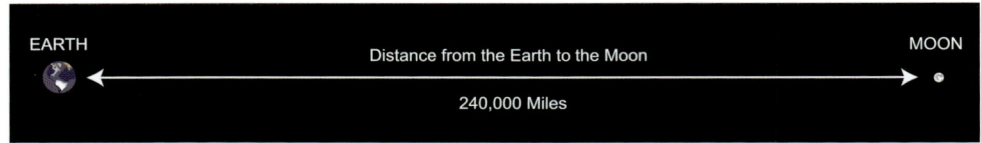

▲ *The Earth and Moon to scale.*

Objectives

- Describe the giant-impact hypothesis for the formation of the Moon
- Describe some of the evidence for and against the hypothesis
- Describe the evidence for other theories

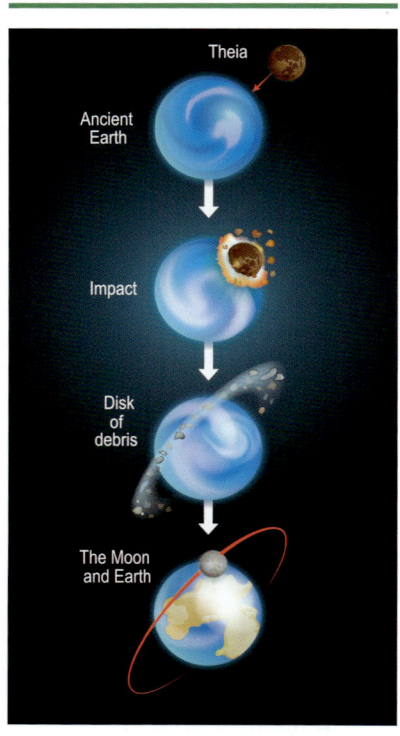

▲ *Theia collides with the Earth.*

What is the 'giant-impact' hypothesis?

Most astronomers think that the Moon was formed when a giant object smashed into the Earth.

The Earth had formed by about 4.54 billion years ago from the gas and dust left from the formation of the Sun. In addition, various objects had hit the Earth later and been absorbed into it. Other objects, including planets, were made from the leftover gas and dust too.

- Gravity pulled together dust and gas.
- The dust contained iron that had been made in other stars.
- The iron formed the central core of the objects that formed.
- Debris rained down on the surface of the young Earth and probably kept the surface molten for a long time.

The **giant-impact hypothesis** says that the Moon formed at some time before 4 billion years ago.

- An object the size of Mars called Theia collided with the Earth.
- The energy of the collision heated the Earth and Theia.
- The molten iron cores of the two objects merged to form the core that the Earth has today.
- The lighter rock was thrown out into orbit and formed the Moon.
- After the Moon formed, the surfaces of both the Earth and the Moon cooled to produce a crust.

Astronomers can make predictions based on this hypothesis. Here are some of the predictions and some of the evidence for and against them.

Prediction	Evidence
The Earth should have an iron core.	Evidence suggests it does.
Some of the material of the Moon should be the same as the Earth, and some should be different because it came from Theia.	The material brought back from the Moon by the Apollo astronauts suggests that there is very little difference between the composition of the Earth and the Moon.
The Moon should orbit the Earth at a distance we can predict from models.	It does orbit at that distance.
The Moon's orbit should align with the Earth.	The orbits of the Earth and Moon do align.

Scientists have no direct evidence that the core of the Earth is made of iron. They think that it is made of iron because:

- it would account for the density of the Earth (if the core is made of rock then the density would be too low)
- the Earth has a magnetic field around, so it is likely that some metal element is moving in the core.

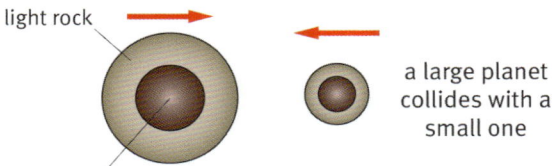

light rock

heavy iron

a large planet collides with a small one

all the iron ends up in the large planet - Earth

Other theories for the formation of the Moon

Co-formation theory

In one version of the **co-formation theory**, the Moon was formed alongside the Earth by gravity pulling together dust and gas. In another version, two objects the size of Mars collided, and the Earth and Moon formed from them.

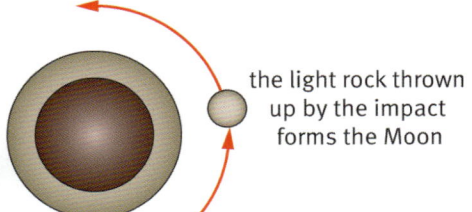

the light rock thrown up by the impact forms the Moon

Evidence for this theory	Evidence against this theory
• The Moon has a similar composition to the Earth. • The Moon would be in orbit at its present location.	• The Moon is less dense than the Earth. • The Moon does *not* seem to have a heavy core.

▲ *Did the molten iron cores merge?*

Capture theory

In the **capture theory** the Moon formed elsewhere in the Solar System. Earth's gravity captured the Moon as it was passing by.

Evidence for this theory	Evidence against this theory
• The Moon is less dense than the Earth.	• Objects that are captured are *not* usually spherical. • The orbits of objects captured this way do *not* align.

▲ *Some astronomers think that the two Martian moons (Phobos and Deimos) are captured asteroids.*

🌐 Science in context

There have been other theories about how the Moon formed. We are still not completely sure that the giant impact hypothesis is correct. Theories change over time. The co-formation and capture theories have been around for a long time.

📖 Key points

- The giant-impact hypothesis says that an object called Theia collided with a young Earth.
- The iron cores of the objects merged and the lighter rock formed the Moon.
- A lot of evidence supports the hypothesis, but not all.
- There are other theories, with less evidence.

⊙ Questions

1. Describe one piece of evidence that supports the giant-impact hypothesis, and one that does *not* support it.
2. Suggest why most scientists don't believe the other theories that have been proposed.
3. Suggest why astronomers might think that Mars's two moons are captured asteroids.

The life cycle of stars

When we look into the night sky we are looking back in time. The light from some of the objects has taken millions or billions of years to reach us. Some of the stars we are seeing are very young, and others have reached the end of their life cycle.

▲ *The Eagle Nebula is a stellar nursery.*

What is a nebula?

A **nebula** is a cloud of dust and gas. In some nebulae gravity pulls the gas together to make a star. These are called **stellar nurseries.**

When does a star shine?

Our Sun has a life cycle, as an animal does. Like all stars, it is born and it will die.

- The Sun was born from a cloud of gas and dust.
- Gravity pulled the gas together to form a **main-sequence star**, like our star is now.

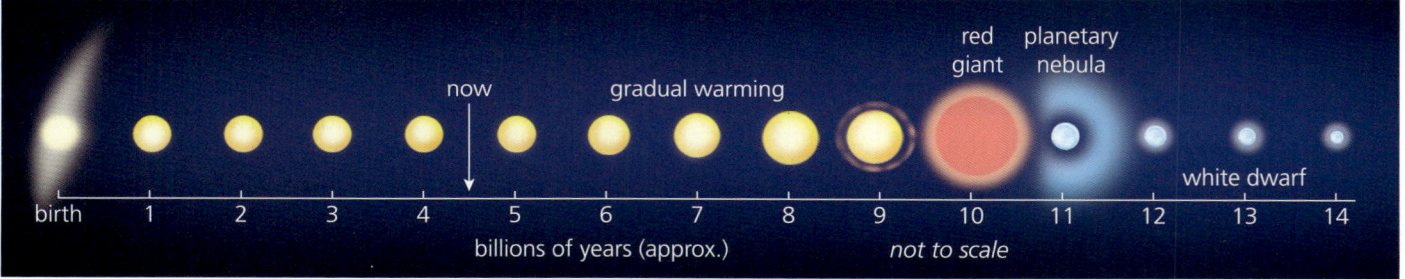

▲ *The life cycle of our Sun.*

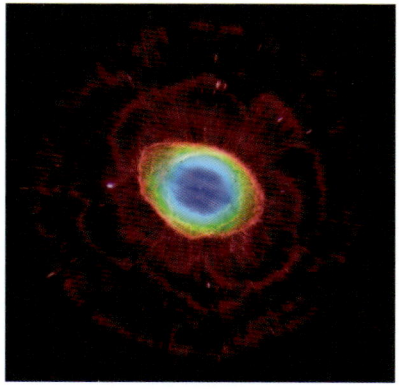

▲ *The Ring Nebula is a planetary nebula, not a stellar nursery.*

Nuclear fusion reactions produce the energy that makes the Sun shine. In nuclear fusion, hydrogen atoms fuse to make helium, and energy is released. Helium then fuses to make other elements. Elements up to iron in the Periodic Table are made in stars like our Sun. The temperatures and pressures needed for nuclear fusion are found in the centre of stars.

The Sun has enough fuel to shine like this for another 5 billion years. Eventually all of the hydrogen will be used up and it will go through the final stages of its life.

- It will grow to become a **red giant** and will swallow up Mercury, Venus, and possibly Earth as well.
- The outer layers will be thrown out into space to form clouds of gas called a **planetary nebula**. (A planetary nebula has nothing to do with planets. Astronomers thought they looked like planets.)
- The centre will shrink and become a hot **white dwarf**, and then cool down to become an invisible **black dwarf** star.

How is a stellar nursery formed?

Stars that are much bigger than our Sun are called **massive stars**. A massive star will turn into a **red supergiant**, explode to form a **supernova**, and then form a **neutron star** or a **black hole**.

Massive stars eventually become supernovae. These are some of the most energetic explosions in the Universe. Elements that are heavier than iron in the Periodic Table are made in supernova explosions.

The remnants of the star are flung out into the universe to form nebulae. These nebulae can produce new stars and planets, like our Sun and our planet Earth.

You are made of stardust.

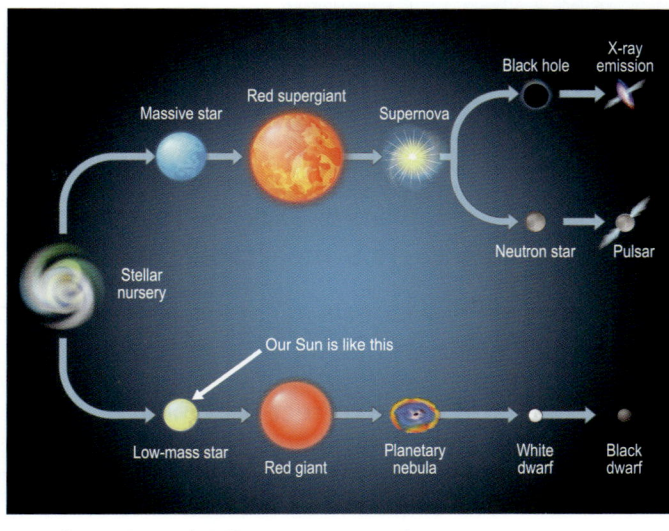

▲ *Life cycles of different types of stars.*

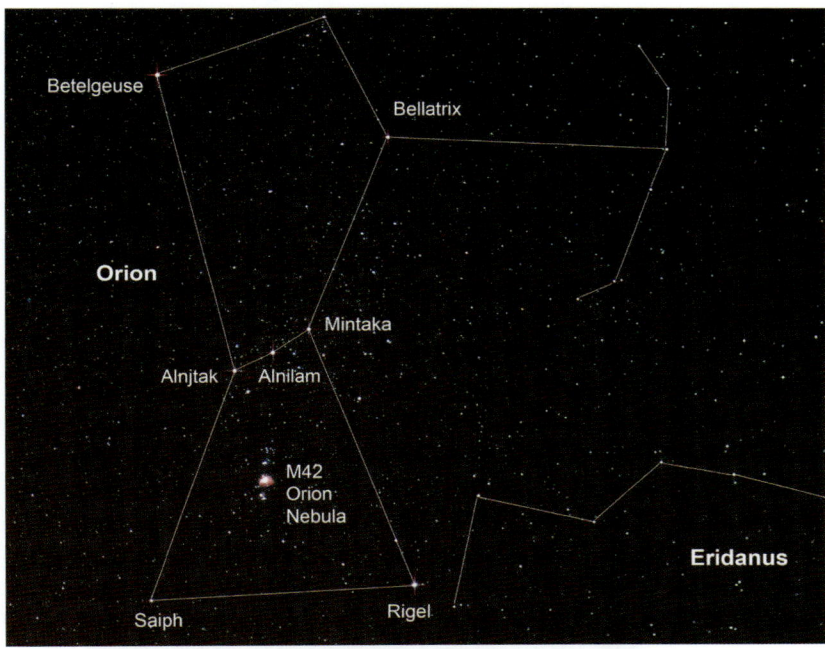

▲ *The Orion Nebula can be seen with the naked eye.*

▲ *The Orion Nebula looks beautiful through a large telescope.*

Key points

- Stars have two different possible life cycles depending on their mass.
- Low-mass stars like our Sun have a life cycle that ends up as a white dwarf, then a black dwarf.
- Massive stars end up as black holes or neutron stars.
- Both types of stars produce nebulae, which can go on to form new stars.
- Heavy elements are only made in supernova explosions.

Questions

1. Match the words to their definitions:

A	Nebula	1	The type of star our Sun is now
B	Nuclear fusion	2	Cloud of dust and gas
C	Main sequence	3	The process that produces energy in stars
D	Red giant	4	What our Sun will become in about 5 billion years

2. Describe one similarity and one difference between the life cycle of a low-mass and a massive star.
3. Suggest why gold is more precious than diamond.

1. a. Describe one piece of evidence that there have been asteroid impacts on Earth. **[1]**

   b. Explain why scientists might disagree about the evidence. **[1]**

2. Give the letter of the correct statement. **[1]**

   A The size of an impact crater does not depend on the type of ground an asteroid hits.

   B The size of an impact crater does not depend on the type of material of the asteroid.

   C A bigger crater will be formed if the asteroid hits softer rock.

   D The Earth has only been hit by a few asteroids.

3. Which of the following is a piece of evidence for the giant-impact hypothesis? **[1]**

   A The Moon is orbiting the Earth at the expected distance.

   B People have landed on the Moon.

   C There is no atmosphere on the Moon.

   D The Moon has less gravity than the Earth.

4. Put these statements in order to describe the giant-impact hypothesis. **[5]**

   A The lighter rock was thrown out into orbit.

   B After the Moon formed the surfaces of Moon and Earth cooled to produce crusts.

   C The molten iron cores of the two objects merged to form the core that the Earth has today.

   D The energy of the collision heated the Earth and Theia.

   E The lighter rock came together to form the Moon.

   F An object the size of Mars called Theia collided with the Earth.

5. You can use primary and secondary data to
TWS develop explanations in science.

   a. Describe the difference between primary and secondary data. **[1]**

   b. An astronomer uses some data from a science book to develop an explanation about the location of the Moon's orbit. Are the data primary or secondary data? **[1]**

   c. A student uses a table of numbers that they found in a magazine to find out about the number of moons around all the planets. Are the data primary or secondary data? **[1]**

   d. A teacher uses a telescope to make some observations of the craters on the Moon. Are the data primary or secondary data? **[1]**

   e. Which type of data are likely to be more
TWS reliable? Explain your answer. **[2]**

6. Put these in order of age from youngest to oldest. **[3]**

   A the Sun   C the Earth

   B the Universe   D the Moon

7. An asteroid hit the remote Tunguska region of Russia in 1908, producing a fireball that flattened 80 million trees. No one was hurt.

   a. Explain why the complete asteroid was *not* found on the ground. **[1]**

   b. Define 'mass-extinction event'. **[1]**

   c. Describe how an asteroid produces a mass extinction event. **[3]**

   d. Explain why mass extinctions are rare. **[1]**

8. A student wants to investigate impact craters.

They get a deep tray and fill it with flour.

When they drop a marble onto the tray it makes a crater.

marble

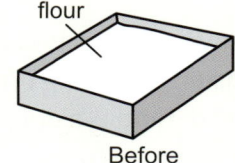
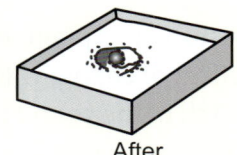

flour

Before   After

a. Complete this sentence for a hypothesis about forming craters that they could test.

Craters are formed because the _____ of the falling marble is transferred to the flour, which makes the flour move. **[1]**

 b. Complete this sentence for a question that they could investigate.

I wonder how changing the _____ that I drop the marble from affects the _____ of the crater. **[2]**

c. Name the independent variable. **[1]**

d. Name the dependent variable. **[1]**

e. Name one control variable. **[1]**

f. Describe and explain the type of graph that they could plot. **[2]**

9. Here are some statements that would provide evidence for different theories for how the Moon was formed.

A The Earth should have an iron core.

B Some of the material of the Moon should be the same as the Earth, and some should be different because it came from Theia.

C The Moon should orbit the Earth at a distance we can predict from models.

D The Moon's orbit should align with the Earth.

a. Give the letter or letters of the statements that are true. **[1]**

b. Give the letter of a statement that supports the giant-impact theory over other theories. **[1]**

c. Describe an alternative theory for the formation of the Moon. **[2]**

d. Explain why the theory you have described has *not* overtaken the giant-impact theory. **[1]**

10. A student is trying to explain the timeline for the formation of the Universe.
They use the idea of condensing the length of time that the Universe has been in existence to one year.

a. Copy the table and add the following events at the correct time. **[2]**

dinosaurs became extinct
Milky Way formed   Earth formed

Date	Event
January 1	Big Bang
May	
Early September	Sun formed
Middle September	
December 7th	Dinosaurs lived
December 30th	
December 31st 9.25pm	Humans walk upright

b. Describe what happened to the temperature of the Universe over time. **[1]**

c. The Universe has been expanding for about 14 billion years. Suggest the maximum distance, in light years, that the most powerful telescope possible could see into the Universe. Explain your answer. **[2]**

1. A girl is playing with some oil and some water. She pours the oil and water into a glass and the oil floats on the water.

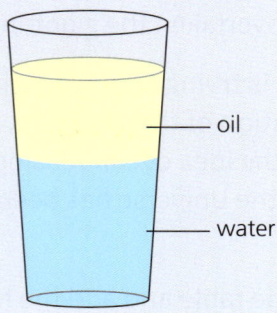

oil

water

a. Write down which liquid is more dense, the oil or the water. Explain your answer. **[2]**

She cuts up a cube-shaped piece of banana and puts it into the glass.

It floats *between* the layer of oil and water.

b. Write down what can you say about the density of the banana compared to the density of the oil and the density of the water. **[2]**

She decides to measure the density of the banana.

c. Write down what equipment she will need to use. **[2]**

TWS **d.** Explain how she could use that equipment to measure the density. **[3]**

2. During the day in the desert it can get very hot. Some people even run marathons in the desert. There is an 'ultra-marathon' that takes place in the Sahara desert.

a. The temperature in the Sahara can reach 45 °C. Describe the difference between temperature and energy. **[2]**

b. Describe how energy from the Sun reaches a runner in the desert. **[1]**

c. A runner will perspire (sweat) when he gets hot. Explain how that helps to keep him cool. **[2]**

d. At the end of the race he goes inside a cool room. It feels cold. In which direction is energy being transferred: from the room to him, or from him to the room? **[1]**

e. Is the energy being transferred by conduction, convection, or radiation, or more than one of these methods? Explain your answer. **[2]**

3. A student is planning an investigation into the **TWS** effect of changing the tension in a string on the pitch of the sound produced. Here is a diagram of the experiment that he is planning to do.

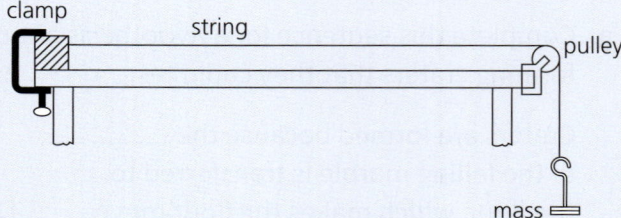

clamp   string   pulley

mass

He plans to change the masses hanging on the string, pluck it to make a sound, and use a microphone to look at the wave on the screen of the oscilloscope. He knows that the screen shows the number of waves in 10 milliseconds. His teacher explains that he can count how many waves there are on the screen and divide by 0.01 to get the frequency in Hz.

a. Give the name of the independent variable. **[1]**

b. Give the name of the dependent variable. **[1]**

c. Give the name of the variable or variables he will need to control. **[1]**

d. Draw the table he needs to use to record his results. **[2]**

e. Give the name of the type of graph he should plot from his results. Explain your answer. **[2]**

f. He does the experiment and finds that as he adds more masses the frequency gets bigger. **TWS** Name a musical instrument where this would be important. **[1]**

4. A referee needs to blow a whistle in a football game to stop play. This is a picture of the whistle sound wave on the screen of an oscilloscope:

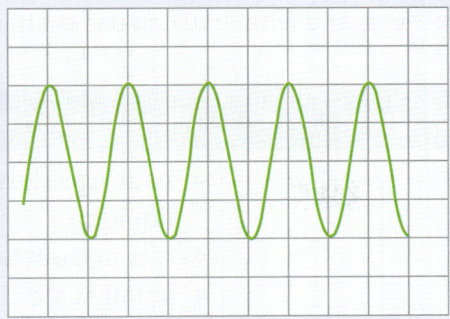

He blows his whistle again, but this time with a lower pitch and quieter.

a. Draw a wave to show what this sound would look like. **[2]**

b. Explain why you have drawn the wave in the way that you have. **[1]**

c. On a sketch of the wave draw the wave that would superpose destructively with the wave on the screen. **[1]**

d. Suggest where it is useful to have waves cancelling out. **[1]**

5. A student has connected a parallel circuit.

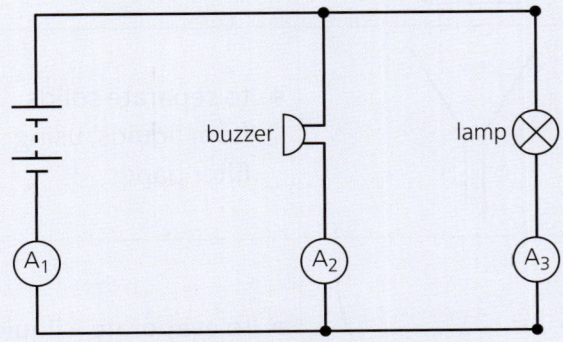

a. The reading on ammeter 1 is 0.5 A, and the reading on ammeter 2 is 0.2 A. Give the reading on ammeter 3. **[1]**

b. Describe what would happen to the ammeter readings if you swapped the buzzer and the lamp. **[2]**

c. Describe what would happen if you used a battery with a bigger voltage. Explain your answer. **[2]**

The student replaces the buzzer with a different lamp. The reading on ammeter 2 is now 0.1 A.

d. Deduce the reading on the other two ammeters. **[2]**

e. Describe where to put a voltmeter to measure the voltage across the lamp. **[1]**

f. Describe where to put a voltmeter to measure the voltage across the battery. **[1]**

g. Would the two voltmeter readings be the same or different? Explain your answer. **[2]**

h. What would happen to the reading on ammeter 3 if you added another bulb in that branch? Explain your answer. **[2]**

6. The Earth has a magnetic field.

a. Describe one theory for why the Earth has a magnetic field. **[1]**

b. Explain why a student might think that the 'bar magnet' producing the Earth's magnetic field is the wrong way round. **[2]**

7. An astronomer observes a planet around a distant star. It is not possible to see whether there are craters on the planet.

a. Suggest why the astronomer might think there would be craters. **[1]**

b. Suggest what might have caused the craters, and where they come from. **[2]**

c. A crater on Earth was formed at about the time of the last mass-extinction event. Name one of the types of animals that died out in that event. **[1]**

d. An object is thought to have collided with the Earth to produce the Moon. Name this theory. **[1]**

Choosing apparatus

There are many different types of scientific apparatus. The table below shows what they look like, how to draw them, and what you can use them for.

Apparatus name	What it looks like	Diagram	What you can use it for
test tube			• heating solids and liquids • mixing substances • small-scale chemical reactions
boiling tube			• a boiling tube is a big test tube; you can use it for doing the same things as a test tube
beaker			• heating liquids and solutions • mixing substances
conical flask			• heating liquids and solutions • mixing substances
filter funnel			• to separate solids from liquids, using filter paper
evaporating dish			• to evaporate a liquid from a solution
condenser			• to cool a substance in the gas state, so that it condenses to the liquid state

stand, clamp, and boss			• to hold apparatus safely in place
Bunsen burner			• to heat the contents of beakers or test tubes • to heat solids
tripod			• to support apparatus above a Bunsen burner
gauze			• to spread out thermal energy from a Bunsen burner • to support apparatus, such as beakers, over a Bunsen burner
pipette			• to transfer liquids or solutions from one container to another
syringe			• to transfer liquids and solutions • to measure volumes of liquids or solutions
spatula			• to transfer solids from one container to another
tongs and test tube holders			• to hold hot apparatus, or to hold a test tube in a hot flame

Working accurately and safely

You need to make accurate measurements in science practicals. You will need to choose the correct measuring instrument, and use it properly.

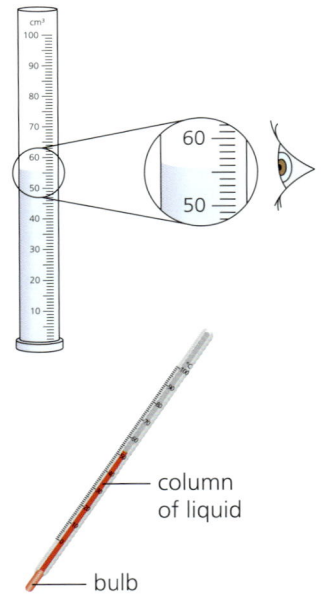

▲ *The different parts of a thermometer.*

Measuring cylinder

Measuring cylinders measure volumes of liquids or solutions. A measuring cylinder is better for this job than a beaker because it measures smaller differences in volume.

To measure volume:

1. Place the measuring cylinder on a flat surface.

2. Bend down so that your eyes are level with the surface of liquid.

3. Use the scale to read the volume. You need to look at the bottom of the curved surface of the liquid. The curved surface is called the **meniscus**.

Measuring cylinders measure volume in cubic centimetres, cm^3, or millilitres, ml. One cm^3 is the same as one ml.

Thermometer

The diagram to the left shows an alcohol thermometer. The liquid expands when the bulb is in a hot liquid and moves up the column. The liquid contracts when the bulb is in a cold liquid.

To measure temperature:

1. Look at the scale on the thermometer. Work out the temperature difference represented by each small division.

2. Place the bulb of the thermometer in the liquid.

3. Bend down so that your eyes are level with the liquid in the thermometer.

4. Use the scale to read the temperature.

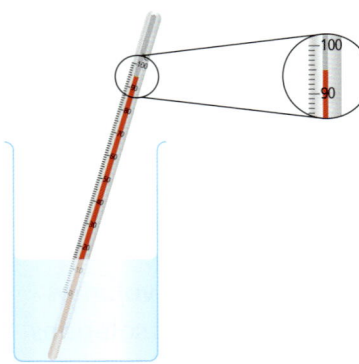

▲ *The temperature of the liquid is 95 °C.*

Most thermometers measure temperature in degrees Celsius, °C.

Balance

A **balance** is used to measure mass. Sometimes you need to find the mass of something that you can only measure in a container, like liquid in a beaker. To use a balance to find the mass of liquid in a beaker:

1. Place the empty beaker on the pan. Read its mass.

2. Pour the liquid into the beaker. Read the new mass.

3. Calculate the mass of the liquid like this:

(mass of liquid) = (mass of beaker + liquid) – (mass of beaker)

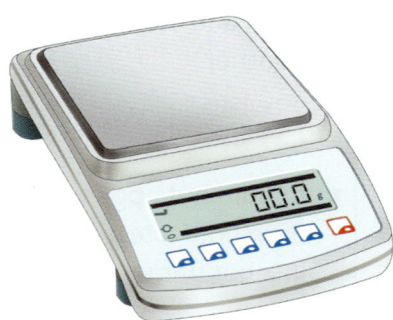

▲ *The balance measures mass.*

Balances normally measure mass in grams, g, or kilograms, kg.

Working safely

Hazard symbols

Hazards are the possible dangers linked to using substances or doing experiments. Hazardous substances display **hazard symbols**. The table shows some hazard symbols. It also shows how to reduce risks from each hazard.

Hazard symbol	What it means	Reduce risks from this hazard by...
	Corrosive – the substance attacks and destroys living tissue, such as skin and eyes.	• wearing eye protection • avoiding contact with the skin
	Irritant – the substance is not corrosive, but will make the skin go red or form blisters.	• wearing eye protection • avoiding contact with the skin
	Toxic – can cause death, for example, if it is swallowed or breathed in.	• wearing eye protection • wearing gloves • wearing a mask, or using the substance in a fume cupboard
	Flammable – catches fire easily.	• wearing eye protection • keeping away from flames and sparks
	Explosive – the substance may explode if it comes into contact with a flame or heat.	• wearing eye protection • keeping away from flames and sparks
	Dangerous to the environment – the substance may pollute the environment.	• taking care with disposal

Other hazards

The table does not list all the hazards of doing practical work in science. You need to follow the guidance below to work safely. Always follow your teacher's safety advice, too.

- Take care not to touch hot apparatus, even if it does not look hot.
- Take care not to break glass apparatus – leave it in a safe place on the table, where it cannot roll off.
- Support apparatus safely. For example, you might need to weigh down a clamp stand if you are hanging heavy loads from the clamp.
- If you are using an electrical circuit, switch it off before making any change to the circuit.
- Remember that wires may get hot, even with a low voltage.
- Never connect wires across the terminals of a battery.
- Do not look directly at the Sun, or at a laser beam.
- Wear eye protection – whatever you are doing in the laboratory!

Using ammeters and voltmeters

Measuring current and voltage

In your experiments with electrical circuits you will measure current and voltage using meters. There are two types of meter: analogue and digital.

- Analogue meters have a needle that moves and shows the current on a scale.
- Digital meters have a display of numbers. A digital meter is usually a multimeter, which means that you can use it to measure current or voltage.

Using an analogue ammeter

An ammeter measures the current. You might want to find the current flowing through a component such as a lamp. To do this you place the ammeter *in series* with the lamp.

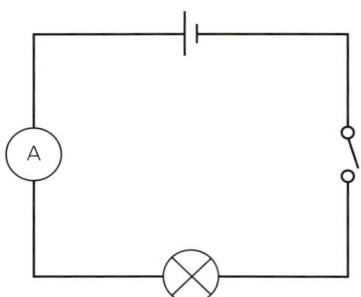

▲ An analogue ammeter.

▲ An ammeter connected to a circuit with a cell or battery, a lamp, and a switch.

For most components in a circuit it does not matter which way round you connect them. This is not true for an ammeter. It needs to be connected so that, if you follow the wires back to the battery or cell, the wire in the *black* terminal on the ammeter is connected to the *negative* terminal of the battery. The wire in the *red* terminal should be connected to the *positive* terminal of the battery.

This is how you connect the ammeter into a circuit with a cell or battery, a lamp, and a switch.

1. Disconnect the cell or open the switch.

2. Follow the wire from the negative terminal of the cell until you get to the lamp. Disconnect the lead from the lamp.

3. Plug that lead into the black terminal on your ammeter.

4. Use another lead to connect the lamp to the red terminal of the ammeter.

An analogue meter may have two different scales. Start by connecting it up using the red terminal labelled with the higher value of current. If the ammeter does not show a current in your circuit using that scale, then move the lead to the terminal with the lower value of current.

Using an analogue voltmeter

A voltmeter measures the voltage. You might want to find the voltage across a component such as a lamp. To do this you place the voltmeter *in parallel* with the lamp.

Like an ammeter, a voltmeter needs to be connected the right way round. This is how you connect the voltmeter into a circuit with a cell or battery, a lamp, and a switch.

1. Disconnect the cell or open the switch.

2. Follow the wire from the negative terminal of the cell until you get to the lamp. Do not disconnect the lamp but plug another lead into the terminal on that side of the lamp.

3. Connect the other end of this lead to the black terminal on your voltmeter.

4. Use another lead to connect the other side of the lamp to the red terminal of the voltmeter.

▲ *An analogue voltmeter.*

Connecting meters

You connect an ammeter in series, and a voltmeter in parallel. It is important not to connect an ammeter in parallel with a component, or directly across a battery or cell. The resistance of the ammeter is very small so a very large current would flow that could damage the ammeter and drain the cell.

The resistance of a voltmeter is very, very high. If you connect it in series rather than in parallel no current will flow in the circuit.

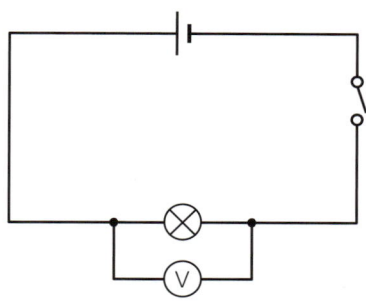

▲ *A voltmeter connected to a circuit with a cell or battery, a lamp, and a switch.*

Using a multimeter

You can use a multimeter to measure current *or* voltage.

The multimeter has a dial that you turn to select current (A or mA) or voltage (V). You select whether you want to measure a large current (10 A) or a small current. Some examples of how to connect a multimeter are shown below.

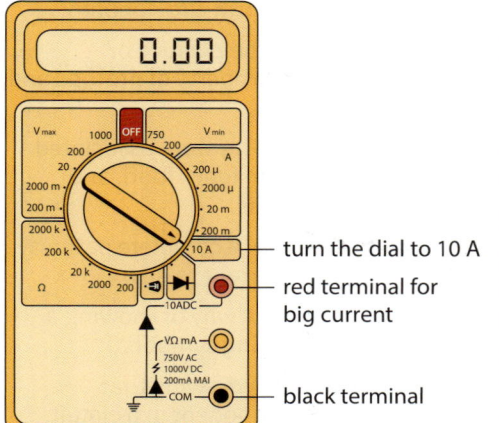

▲ *Using a multimeter to measure a big current.*

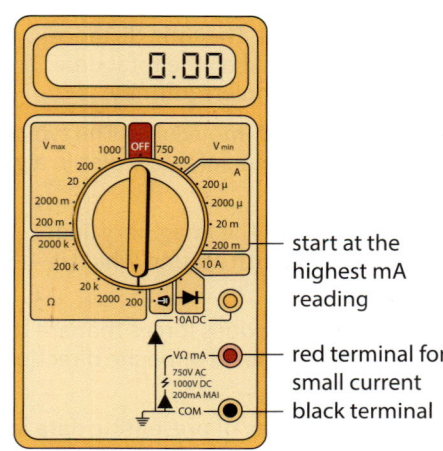

▲ *Using a multimeter to measure a small current.*

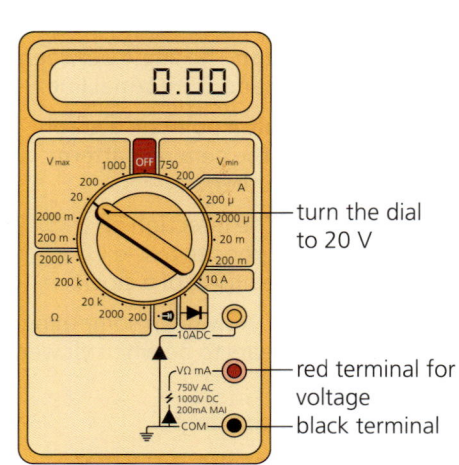

▲ *Using a multimeter to measure voltage.*

Glossary

Absorbed What happens energy of electromagnetic radiation (e.g. light) or sound is transferred to a thermal store on passing through or into a medium.

Accelerating Speeding up, getting faster.

Acceleration The rate of change of increasing speed (the amount by which speed increases in one second).

Accuracy (of a measurement) How correct a measurement is – how close it is to its true value.

Air resistance The force on an object that is moving through the air, causing it to slow down (also known as drag).

Alloy A material made of a mixture of metals, or of carbon with a metal.

Ammeter A device for measuring electric current in a circuit.

Ampère (amp) The unit of measurement of electric current, symbol A.

Amplifier A device for making a sound louder.

Amplitude The distance from the middle to the top or bottom of a wave.

Analogy A way of explaining something by saying that it is like something else.

Andromeda The nearest galaxy to the Milky Way.

Angle of incidence The angle between the incident ray and the normal line.

Angle of reflection The angle between the reflected ray and the normal line.

Angle of refraction The angle between the refracted ray and the normal line.

Anomalous result A point on a graph that does not fit the general pattern (also called an outlier).

Anticlockwise The direction of rotation that is opposite to the movement of the hands of a clock.

Apparent depth How deep something underwater appears to be when viewed from above.

Area The size of a surface.

Archimedes' principle A law that states that the upthrust on an object in a fluid is equal to the weight of fluid displaced.

Armature The coil of wire in an electromagnetic device such as a generator.

Artificial satellites Spacecraft made by people to orbit the Earth for various purposes.

Asteroid An irregularly shaped lump of rock in orbit around the Sun.

Asteroid belt A large number of asteroids between Mars and Jupiter.

Astronomer A scientist who studies space.

Atmosphere The layer of air above Earth's surface.

Atmospheric pressure The force on an area of the Earth's surface due to the weight of air above it, or the pressure in the atmosphere.

Atom The smallest particle of an element that can exist.

Attract Pull together, for example opposite poles of a magnet or positive and negative charges attract each other.

Audible Able to be heard.

Audible range The range of frequencies that can be heard.

Auditory canal The passage from the outer ear to the eardrum.

Auditory nerve Signals are sent from your ear to the brain along this nerve.

Aurora A display of lights in the sky due to the interaction of the solar wind with the magnetic field of a planet.

Average Found by adding a set of values together and dividing by the number of values (also called the mean).

Average speed The total distance travelled divided by the total time taken for a complete journey.

Average speed (molecules) The typical value of the speed of molecules in a gas or liquid.

Axis, of Earth The imaginary line through the Earth around which it spins.

Balanced Describes forces that are the same size but act in opposite directions on an object.

Bar chart A way of presenting data in which one of the variables is categoric (words).

Battery Two or more electrical cells joined together.

Big Bang The expansion of space which we believe started the Universe.

Biodegradable Able to be broken down by bacteria or other living organisms.

Biodiesel A biofuel made from plant oils.

Bioethanol A biofuel made from carbohydrates such as sugar.

Biofuel A fuel produced from renewable resources.

Biogas Gas produced from waste products, usually methane, used to generate electricity.

Biomass Material from plants used for fuel, e.g. wood.

Bioplastics Plastics made from starch.

Black dwarf A remnant of a star like our Sun that no longer gives out light.

Black hole A remnant of a star much bigger than our Sun from which nothing can escape, not even light.

Capture theory A theory that says that the Moon is an object that was formed outside the Solar System and captured by the Earth's gravity.

Categoric Describes a variable whose values are words not numbers.

Cell (electrical) A device that uses a chemical reaction to produce a voltage/potential difference.

Centre of mass (or centre of gravity) The point in an object where the mass appears to be concentrated or weight acts.

Centripetal force The force directed towards the centre that causes a body to move in a uniform circular path.

Charge This can be positive or negative. It is a property of protons and electrons.

Chemical store The store associated with the energy in fuels, food, and electrical batteries.

Circuit (electric) A complete pathway for an electric current to flow.

Circuit diagram A way of showing a circuit clearly, using symbols.

Circuit symbol A drawing that represents a component in a circuit.

Climate change Changes to long-term weather patterns as a result of global warming.

Clockwise The direction of rotation that is the same as the movement of the hands of a clock.

Coal A fossil fuel formed from dead plants that have been buried underground over millions of years.

Coal-fired power station A place where the fossil fuel coal is burned to generate electricity.

Cobalt A metallic element that is magnetic.

Cochlea A snail-shaped tube in the inner ear where the sensory cells that detect sound are.

Co-formation theory A theory that says that the Moon formed in the same way and at the same time as the Earth.

Colour blindness Someone with colour blindness cannot tell certain colours apart, because some cone cells in the retina of the eye do not work properly.

Comets Bodies in space made of dust particles frozen in ice, which orbit the Sun.

Communicate To share and exchange information.

Compact disc A metal disc that can store high-quality digital recordings.

Compass A device containing a small magnet that is used for finding directions.

Component An item used in an electric circuit, such as a lamp.

Compressed Squashed into a smaller space, or deformed by making smaller.

Compression (in a sound wave) The part of a sound wave where the air particles are close together.

Conclusion A statement about what the results of an investigation tell you.

Conduction (of energy) The way in which energy is transferred through solids (and, to a much lesser extent, liquids and gases).

Conductor A material such as a metal or graphite that conducts charge or energy well.

Cone A specialised cell in the retina that is sensitive to bright light and colour.

Consequence (risk) What can happen as a result of something you do.

Conservation (of energy) A law that says that energy is never created or destroyed but is is always transferred from one store to another.

Constant Not changing.

Constructive interference The effect of adding waves that are in step to produce a larger amplitude wave.

Contact force A force that acts when an object is in contact with a surface, air, or water.

Continuous Describes a variable that can have any value across a range, such as time, temperature, length.

Convection The transfer of energy by the movement of a gas or liquid.

Convection current The way in which energy is transferred through liquids and gases by the movement of their particles.

Core A rod of a magnetic material placed inside a solenoid to make the magnetic field of an electromagnet stronger.

Cornea The transparent layer at the front of the eye.

Correlation A link between two things; it does not necessarily mean that one thing causes the other.

Craters Holes in the ground caused when meteors or asteroids hit the Earth.

Creative thinking Thinking in a new way.

Crescent moon The shape that we see when a thin section of the Moon is lit, a few days after a new moon.

Critical angle The smallest angle of incidence at which total internal reflection occurs.

Current The flow of electric charge (electrons) around a complete circuit.

Data Measurements taken from an investigation or experiment.

Day The period of time when one section of the Earth (or other planet) is facing the Sun.

Decelerating Slowing down, getting slower.

Deceleration The amount by which speed decreases in one second.

Decibel (dB) A commonly used unit of sound intensity or loudness.

Deform To change shape.

Degrees Celsius (°C) A temperature scale with 0 °C fixed at the melting point of ice and 100 °C fixed at the boiling point of water.

Demagnetise To destroy a magnet by heating it up, hitting it, or putting it in an alternating current.

Density The mass of a substance in a certain volume.

Dependent variable The variable that changes when you change the independent variable.

Destructive interference The effect of adding waves that are out of step to produce a smaller amplitude wave.

Detector Something that absorbs electromagnetic radiation or sound to produce a signal.

Diffuse Describes reflection from a rough surface.

Diffusion The movement of particles in gases and liquids from where there are a lot of the particles to where there are fewer.

Directly proportional A relationship in which one quantity increases in the same way as another.

Discrete Describes a variable that can only have whole-number values.

Dispersion The splitting up of a ray of light of mixed wavelengths by refraction into its components.

Dissipated When transferred energy is not useful/is wasted.

Distance multiplier A type of lever that uses a larger force to produce a smaller force at a larger distance.

Distance–time graph A graph showing how the distance travelled varies with time.

Domain A small region inside a magnetic material that behaves like a tiny magnet.

Drag A force on an object moving through air or water, causing it to slow down.

Dwarf planet A lump of rock in orbit around the Sun that is nearly spherical but has other objects around it.

Dynamo A device that produces a potential difference when it spins (a small generator).

Ear The organ of the human body that detects sound.

Ear defenders A device used to protect the ears from noise.

Eardrum A membrane that transmits sound vibrations from the outer ear to the middle ear.

Earth A rocky inner planet, the third planet from the Sun.

Earth (charge) To connect a metal wire from an object to the ground to take any charge away.

Earthing The process of connecting objects to the ground.

Echo A reflection of a sound wave by an object.

Echolocation The process of finding out the position of something using echoes.

Eclipse The Sun or Moon is blocked from view on Earth (see also lunar eclipse or solar eclipse).

Efficient Describes something that does not waste much energy.

Effort The amount of force that you use to push down when using a lever.

Elastic Describes a type of material that can be stretched and will return to its original length when the pulling force is removed.

Elastic limit The point beyond which a spring will never return to its original length when the pulling force is removed.

Elastic potential energy (EPE) Energy stored in an elastic object that is stretched or squashed.

Elastic store The energy store associated with objects that are deformed

Electric car A car powered by electric batteries.

Electric circuit A complete pathway for an electric current to flow.

Electric current A flow of electric charge (electrons) around a complete circuit.

Electric field A region around a charged object where other charged objects feel a force.

Electrical signal (ear) Information is transferred from the ear to the brain as an electrical signal (nerve signal).

Electromagnet A temporary magnet produced using an electric current.

Electromagnetic spectrum A range of radiation with electrical and magnetic properties that can travel through a vacuum (for example, the Sun's radiation). It also refers to the range of wavelengths of electromagnetic radiation produced by the Sun and other sources.

Electron A tiny sub-atomic particle with a negative charge that flows through a wire to create an electric current.

Electrostatic attraction The force between two particles of objects that have opposite charge

Electrostatic force The force between two charged objects.

Electrostatic phenomena Things that happen because objects have become charged.

Electrostatic repulsion The force between two particles of objects that have the same charge

Element A substance that is made of one type of atom that cannot be split into other substances.

Emit To send something out (such as heat, light, vapour).

Endoscope A medical instrument for seeing inside the human body.

Energy A quantity that can be stored or transferred; a way of calculating which processes are possible.

Energy conservation Energy is never made or lost but can be transferred, .

Energy store An object of collection of objects about which a quantity of energy can be calculated

Energy transfer Energy can be transferred by forces, electricity, heating, or waves.

Energy transfer diagram A diagram that shows how energy is transferred or changed in a process or device.

Equator An imaginary line round the middle of the Earth at an equal distance from both the North and South Poles.

Equilibrium Balanced (as in a lever or see-saw).

Evaporate To turn from a liquid to a vapour (gas).

Evaporation The change of state from liquid to gas that can happen at any temperature.

Evaporative coolers Equipment used in hot countries that uses the cooling effect of evaporation to cool houses.

Evidence Observations and measurements that support or disprove a scientific theory.

Exoplanet A planet in orbit around a star other than our Sun.

Expand To increase in size, get bigger.

Explanation A statement that gives a reason for something using scientific knowledge.

Extension The distance by which an object gets longer when you stretch it.

Eye The organ of sight, which focuses and detects light.

Fair test A controlled investigation in which only one variable at a time is changed, while all other conditions are kept the same.

Field study An experiment or observations made in the natural habitat of an organism.

Filament The very thin, coiled piece of wire that glows inside a light bulb.

Filter A piece of material that allows some radiation (colours) through but absorbs the rest.

First quarter/third quarter The shape that we see when about half the Moon is lit, about a week before or after a new moon.

Floating An object floats when the upthrust from the water is equal to the downwards force of the object's weight.

Force A push or a pull that acts on an object to affect its movement or shape.

Force multiplier A lever or hydraulic machine that can lift or move heavy weights using a force smaller than the weight.

Forcemeter A device used to measure forces; also called a newtonmeter or spring balance

Fossil fuel Fuel made from the decayed remains of animals and plants that died millions of years ago. Fossil fuels include coal, oil, and natural gas.

Frequency The number of complete

waves or vibrations produced in one second (measured in hertz).

Friction A force that resists movement because of contact between surfaces.

Fuel A material that contains a store of energy and can be burned, e.g. gas, oil, coal, petrol (gas or gasoline).

Fulcrum The point about which a lever or see-saw turns, also called the pivot.

Full moon The shape that we see when the whole disc is lit, when the Moon is opposite the Sun.

Fundamental (sound) The lowest frequency of sound produced by an instrument or object.

Galaxy A number of stars and the solar systems around them grouped together.

Gas (natural) A fossil fuel that collects above oil deposits underground.

Gas pressure (air pressure) The force exerted by air particles when they collide with 1 square metre (1 m^2) of a surface.

Gemstone A stone made from minerals that can be cut and used in jewellery.

Generator A device that produces a voltage.

Geocentric model A model of the Universe with the Earth at the centre.

Geothermal An energy source that uses water heated to steam underground to produce electricity.

Giant impact hypothesis A hypothesis that suggests that the Moon formed from an object produced by a collision between a very young Earth and a Mars-sized planet.

Gibbous moon The shape that we see when most of the Moon is lit, a few days after a full moon.

Global positioning system (GPS) A system that pinpoints the position of something using signals from a satellite.

Gradient The slope or steepness of a graph.

Gravitational field A region in which there is a force on a mass due to its attraction to another mass.

Gravitational field strength The force on a mass of 1 kg, measured in N/kg.

Gravitational potential energy (GPE) Energy stored in an object because of its height above the ground.

Gravity (gravitational force) The force of attraction between two objects because of their mass.

Gravity store The energy store associated with moving objects away from the Earth.

Greenhouse effect The effect of heating the Earth because the energy from the Sun is reflected back to the Earth from a layer of greenhouse gases in the atmosphere.

Greenhouse gases Gases that reflect energy back to the surface of the Earth, warming it and maintaining a temperature suitable for life, such as carbon dioxide, water vapour, and methane.

Hazard symbol A warning symbol on a substance that shows what harm it might cause if not handled properly.

Heating The energy transfer process that involves changing the temperature or state of a material.

Heat pump A device that transfers heat from the ground to a building on the surface.

Heliocentric (model) A model of the Universe with the Sun at the centre.

Hertz (Hz) The unit of frequency.

Hooke's law A law that says that the extension is proportion to the force applied up to the elastic limit.

Hydroelectricity Electricity generated using water falling downhill to turn generators.

Hypothesis A scientific theory or proposed explanation made on the basis of evidence which can be further tested.

Image The point from which rays of light entering the eye appear to have originated.

Impact winter The period of cold weather that scientist think would follow an impact by a very large asteroid or meteor.

Incident ray The ray coming from a source of light.

Incompressible Describes something that cannot be compressed (squashed).

Independent variable The variable that you change, that causes changes in the dependent variable.

Induced (voltage) A voltage produced when a conductor is in a changing magnetic field.

Inertia The tendency of an object to resist a change in speed caused by a force.

Infinite Without end.

Infrared (radiation) A type of electromagnetic radiation that transfers energy from a hotter to a colder place, which can be known as heat.

Inner ear The part of the ear made up of the cochlea and semi-circular canals.

Inner planets Mercury, Venus, Earth, and Mars.

instantaneous speed Speed at a particular moment.

Insulator A material that does not conduct energy or electricity very well.

Intensity (sound) How loud a sound is, measured in decibels.

Interference (of waves) The effect of combining two or more waves.

International Space Station (ISS) A research station in orbit around the Earth.

Interstellar space The space between stars or solar systems.

Inversely proportional A relationship in which one quantity decreases as the other increases.

Inverted Upside down.

Investigation An activity such as an experiment or set of experiments designed to produce data to answer a scientific question or test a theory.

Ions Atoms that have gained or lost electrons.

Iris The coloured part of your eye that controls the size of the pupil.

Iron A metallic element that is the main substance in steel. Iron is a magnetic material.

Joule The unit of energy, symbol J.

Jupiter A large outer planet made of gas, the fifth from the Sun.

Kilogram The unit of mass, symbol kg.

Kilojoule (kJ) 1000 joules.

Kilometres per hour The unit of speed, km/h.

Kilowatt 1000 watts.

Kinetic energy Energy associated with moving objects.

Kinetic store The energy store associated with moving objects.

Kuiper belt The region outside the Solar System where astronomers think that some comets come from.

Last quarter The shape that we see when about half the Moon is lit, about a week after a full moon.

Laterally inverted The type of reversal that occurs with an image formed by a plane mirror.

Law of conservation of energy The law that says that energy cannot be created or destroyed but can be transferred.

Law of reflection The law that says that the angle of incidence is equal to the angle of reflection.

Lens A device made of shaped glass which focuses light rays from objects to form an image.

Lever A simple machine consisting of a rigid bar supported at a point along its length.

Life cycle (of a star) The process that describes how a star is formed and what will happen to it.

Light A form of electromagnetic radiation that comes from sources like the Sun, and transfers energy.

Light source An object that emits visible light, also called a luminous object.

Light year The distance light travels in one year.

Light-emitting diode (LED) A low-energy lamp.

Lightning conductor A piece of metal connected to tall buildings to conduct lightning to the ground.

Line graph A way of presenting results when there are two numerical variables.

Line of best fit A smooth line on a graph that travels through or very close to as many of the points plotted as possible.

Liquid pressure The pressure produced by collisions of particles in a liquid.

Load An external force that acts over a region of length, surface, or area.

Lodestone A naturally occurring magnetic rock.

Longitudinal Describes a wave in which the vibrations are in the same direction as the direction in which the wave moves.

Loudspeaker A device that changes an electrical signal into a sound wave.

Lubrication Reducing friction between surfaces when they rub together.

Luminous Describes something that gives out light.

Lunar eclipse Occurs when the light from the Sun is blocked by the Earth because it is in a direct line between the Earth and the Sun, and the Moon is in shadow.

Magnet An object that attracts magnetic materials and repels other magnets.

Magnetic field An area around a magnet where there is a force on a magnetic material or another magnet.

Magnetic field lines Imaginary lines that show the direction of the force on a magnetic material in the magnetic field.

Magnetic force The force between the poles of two magnets, or between a magnet and a magnetic material such as iron.

Magnetic material A material that is attracted to a magnet, such as iron, steel, nickel, or cobalt.

Magnetic resonance imaging (MRI) scanner A machine that uses strong magnetic fields to produce images of the inside of the human body.

Magnetised Made into a magnet.

Magnetism The property of attracting or repelling magnets or magnetic materials.

Main sequence star The longest stage of a star's life cycle; the current stage of our Sun.

Mains supply Electricity generated in power stations and available through power sockets in buildings.

Mars A rocky planet, the fourth from the Sun.

Mass The amount of matter in an object. The mass affects the acceleration for a particular force.

Mass extinction event An event that causes the extinction of a large number of species over a relatively short (geological) timescale.

Massive (star) Having lots of mass (massive does not mean large).

Matter The scientific word for materials or objects which can be solid, liquid or gas

Measuring cylinder A cylinder used to measure the volume of a liquid.

Medium (sound/light) The material that affects light or sound by slowing it down or transferring the wave.

Meniscus The curved upper surface of a liquid.

Mercury The rocky inner planet nearest the Sun.

Meteor A piece of rock or dust that makes a streak of light in the night sky.

Meteorite A stony or metallic object that has fallen to Earth from outer space without burning up.

Metres per second The unit of speed, m/s.

Microphone A device for converting sound into an electrical signal.

Middle ear The eardrum and ossicles (small bones) that transfer vibrations from the outer ear to the inner ear.

Milky Way The galaxy containing our Sun and Solar System.

Milliamp One thousandth of an amp.

Minerals Chemicals in rocks.

Model A way of representing something that you cannot see or experience directly. A model may be a physical model built on a different scale to the original system, or it may take the form of equations.

Moment A measure of the ability of a force to rotate an object about a pivot.

Moon A rocky body orbiting Earth; it is Earth's only natural satellite.

Moons The natural satellites of planets.

Natural satellite A moon in orbit around a planet.

Neap tide The lower tide that occurs when the Moon is not aligned at all with the Sun.

Nebula A region of dust and gas where stars are born.

Negative charge Describes the charge on an electron, or the charge on an object that has had electrons transferred to it.

Negatively charged Describes an object that has had electrons transferred to it.

Neptune A large outer planet made of gas, eighth from the Sun.

Neutral Describes an object that has no charge; its positive and negative charges cancel out.

Neutral point (magnetic field) A point where there is no force on a magnet or magnetic material because two or more magnetic fields cancel out.

Neutralise To cancel out, when you add an equal amount of positive charge to negative charge.

Neutron A tiny sub-atomic particle with no charge that is found in the nucleus of an atom. The relative mass of a neutron is 1.

Neutron star A very small, massive star.

New moon The shape that we see when the Moon is in between the Earth and the Sun.

Newton the unit of force including weight, symbol N.

Newton's first law of motion A law that says that a resultant force is needed to change the motion of an object.

Newtonmeter A meter that is used to measure the magnitude of a force; also called a forcemeter or spring balance

Newtonmetre the unit of moment, symbol Nm.

Nickel A metal that is magnetic.

Night The period on one section of the Earth or other planet when it is facing away from the Sun.

Noise Any undesired or unwanted sound.

Non-contact force A magnetic, electrostatic, or gravitational force that acts without being in contact with something.

Non-luminous Describes objects that produce no light; objects that are seen by reflected light.

Non-renewable Describes energy sources that will run out eventually (such as fossil fuels).

Normal An imaginary line at right angles to a surface where a light ray strikes it.

Normal (force) The force from a solid object that pushes at 90° to the surface.

North pole The pole of a magnet that points north. A north pole repels another north pole.

Northern hemisphere The half of the Earth between the Equator and the North Pole.

Nuclear fission The process of splitting atoms that releases energy in a nuclear power station

Nuclear fusion The process of joining hydrogen together in the Sun and other stars that releases energy.

Nuclear store The energy store associated with the process of nuclear fusion.

Nucleus The central part of an atom, made up of protons and neutrons.

Object Something that can be seen or touched.

Observations The results of looking carefully at something and noticing properties or changes.

Oil (or crude oil) A thick black liquid formed underground from the remains of prehistoric plants and animals that died millions of years ago. It is used to make fuels such as petrol/gasoline and diesel, and many plastics.

Oort cloud A cloud of comets and dust outside the Solar System.

Opaque Describes objects that absorb, scatter, or reflect light and do not allow any light to pass through.

Optic nerve A sensory nerve that runs from the eye to the brain.

Optical fibre A very fine tube of plastic or glass that uses total internal reflection to transmit light.

Orbit The path taken by one body in space around another (such as the Earth around the Sun).

Oscilloscope A device that enables you to see electrical signals that change, such as those made in a microphone.

Ossicles The small bones of the middle ear (hammer, anvil, and stirrup) that transfer vibrations from the eardrum to the oval window.

Outer ear The pinna and auditory canal.

Outer planets Jupiter, Saturn, Uranus, and Neptune.

Oval window The membrane that connects the ossicles to the cochlea in the ear.

Oxide A compound made when an element combines with oxygen.

Parallel circuit An electric circuit in which there are two or more paths for an electric current.

Partial eclipse Occurs when part of the light from the Sun is blocked by the Moon or the Earth.

Particles The tiny pieces of matter that everything is made from.

Pascal The unit of pressure, symbol Pa, equal to 1 N/m^2.

Pendulum Any rigid body that swings about a fixed point

Penumbra The area of blurred or fuzzy shadow around the edges of the umbra.

Period The time taken to complete one cycle of motion.

Periscope A tube with mirrors or prisms that enables you to see over objects.

Permanent magnet A piece of metal that stays magnetic.

Permanently extended The irreversible extension of a spring when loaded beyond its elastic limit.

Petrol (gas or gasoline) A hydrocarbon fuel (containing hydrogen and carbon) that comes from crude oil.

Phases of the Moon Parts of the Moon that we see as it orbits the Earth.

Photosynthesis The process by which plants make their own food from carbon dioxide and water, using light.

Pie chart A way of presenting data in which only one variable is a number.

Pinna The outside part of the ear that we can see.

Pitch A property of sound determined by its frequency.

Pivot A support on which a lever turns or oscillates.

Plane mirror A mirror with a flat reflective surface .

Planet Any large body that orbits a star in a solar system.

Planetary nebula A cloud of dust and gas from which planets are formed.

Plastic A type of material that can be stretched and does not return to its original length.

Pluto Used to be regarded as the ninth and last planet from the Sun; now called a dwarf planet together with others of the same size that are beyond Pluto's orbit.

Pole, of Earth The north and south points of the Earth connected by its axis of tilt.

Poles, of magnet The opposite and most strongly attractive parts of a magnet.

Positive charge Describes the charge on a proton, or the charge on an object that has had electrons transferred away from it.

Potential difference Also known as voltage, a measure of the energy that each charge transfers.

Potential energy Stored energy as a result of a object's change of position or change of shape.

Power The rate of transfer of energy, measured in watts.

Power station A place where fuel is burned to produce electricity.

Precision The number of decimal places given for a measurement.

Prediction A statement saying what you think will happen.

Preliminary work The work that you do before or during the planning stage of an investigation, to work out how to do it.

Pressure The force applied by an object or fluid divided by the area of surface over which it acts.

Pressure gauge An instrument for measuring pressure in a liquid or gas.

Primary colours For light these are red, blue, and green.

Primary data Data collected directly by scientists during a particular investigation.

Primary source Sources of data that you have collected yourself.

Primary sources (of energy) Energy sources from the environment or underground, such as coal, uranium, or the wind.

Principle of moments The law that says that the sum of the clockwise moments is equal to the sum of the anticlockwise moments about a fixed point.

Prism A triangular-shaped piece of glass used to produce a spectrum of light.

Probability (risk) The chance that something will happen.

Proportional A relationship in which two variables increase at the same rate, for example when one is doubled the other doubles too.

Proton A tiny sub-atomic particle with a positive charge that is found in the nucleus of an atom. The relative mass of a proton is 1.

Proxima Centauri The nearest star to our Sun.

Pupil The hole in the front of your eye where the light goes in.

Question In science, a problem that is stated in a form that you can investigate.

Rainbow An optical phenomenon that appears as the colours of the spectrum when falling water droplets are illuminated by sunlight.

Rarefaction The part of a sound wave where the air particles are most spread out.

Ray diagram A model of what happens to light, shown by drawing selecting rays.

Reaction The force from a solid object that pushes at 90° to the surface.

Reaction time In humans, the time the brain takes to process information and act in response to it.

Real Describes an image that you can put on a screen, or the image formed in your eyes.

Real depth The depth underwater that an object actually is.

Receiver (sonar) A device that absorbs sound waves.

Red giant Part of the life cycle of a star like our Sun when it becomes much bigger and cooler.

Red supergiant The next stage in the life cycle of our Sun.

Reed switch A switch that uses a magnet to work.

Refinery A place where crude oil is refined and separated into fuels.

Reflects When a surface causes light or sound to it bounce off it, it reflects it.

Reflected ray The ray that is reflected from a surface.

Reflection The change in direction of a light ray or sound wave after it hits a surface and bounces off.

Refraction The change in direction of a light ray as a result of its change in speed.

Refractive index A measure of how much light slows down when it goes from one medium to another.

Refrigerant A liquid used in a refrigerator.

Refrigerator a machine for keeping things cold using evaporation.

Relay An electrical device that allows current flowing through it in one circuit to switch on and off a larger current in a second circuit.

Reliable Describes an investigation in which very similar data would be collected if it was repeated under the same conditions.

Renewable Describes energy resources that are constantly being replaced and are not used up, such as falling water or wind power.

Repel To push away.

Reservoir A large amount of water behind a dam; it is used in hydroelectric power.

Resistance How difficult it is for current to flow through a component in a circuit.

Resultant force The single force equivalent to two or more forces acting on an object.

Retina The layer of light-sensitive cells at the back of the eye.

Retrograde motion The apparent 'backwards' motion of planets due to the motion of the Earth and the planet around the Sun.

Reverberation The persistence of a sound for a longer period than normal.

Risk The chance of injury from a hazard. A combination of the probability that something will happen and the consequence if it did.

Risk assessment A statement of what could cause injury or damage, and the action to be taken to reduce the probability that it will happen.

Rod A specialised cell in the retina that is sensitive to bright light.

Satellite Any body that orbits another (such as the Moon or a weather satellite around Earth).

Saturn A large outer planet made of gas, the sixth from the Sun.

Scatter graph A graph that shows all the values in a set of measurements.

Seasons Changes in the climate during the year as the Earth moves around its orbit.

Secondary colours Colours that can be obtained by mixing two primary colours.

Secondary data Data collected by other scientists and published.

Secondary source Sources of data collected by others that you use.

Secondary sources (of energy) Sources of energy that are produced from primary sources, such as electricity produced from coal, or petrol/gasoline produced from crude oil.

Semicircular canals The part of the ear that helps you to balance.

Series circuit An electrical circuit in which the components are joined in a single loop.

Shadow An area of darkness on a surface produced when an opaque object blocks out light.

Shielding Putting something in between a source and a receiver, for example, sound is shielded by ear defenders.

Signal (electrical) A current or voltage that changes over time.

Significant figures The number of digits in a decimal number.

Solar cells Devices that use light to produce a voltage

Solar eclipse Occurs when the Moon blocks out the light from the Sun because it is in a direct line between the Earth and the Sun, and part of the Earth is in shadow.

Solar energy Energy from the Sun which can be used directly to heat water or to make electricity.

Solar panels Devices that use light and infrared radiation to heat water.

Solar System The Sun (our star) and the planets and other bodies in orbit around it. There are other solar systems in the Universe as well as our own.

Solar wind The continuous flow of charged particles from the Sun.

Solenoid A core of wire used to make an electromagnet.

Sonar A system that uses ultrasound to detect underwater objects or to determine the depth of the water.

Sound Vibrations of molecules produced by vibrating objects and detected by your ear that transfers energy.

Sound-level meter A device for measuring the intensity (loudness) of a sound.

Sound wave A series of compressions and rarefactions that moves through a medium.

Source (of light/sound) Something that emits (gives out) light or sound.

South pole The pole of a magnet that points south. A south pole attracts a north pole.

Southern hemisphere The half of the Earth between the Equator and the South Pole.

Spark A flash of light that you see when the air conducts electricity.

Spectrum A band of colours produced when light is spread out by a prism.

Speed The distance travelled in a given time, usually measured in metres per second, m/s.

Speed of light The distance light travels in one second (300 million m/s).

Speed–time graph A graph that shows how the speed of an object varies with time.

Spring A metal wire wound into spirals that can store elastic potential energy.

Spring balance A device for measuring forces; sometimes called a forcemeter or a newtonmeter.

Spring tide The higher tide that occurs when the Earth, Moon and Sun are aligned.

Stable Describes an object in equilibrium that cannot easily be toppled.

Star A body in space that gives out its own light. The Sun is a star.

Static (charge) Charge on an insulator that does not move.

Steady speed A speed that doesn't change.

Steel An alloy of iron with carbon and other elements. Steel is a magnetic material.

Stellar nurseries Regions of interstellar space containing dust and gas where stars form.

Streamlining Designing the shape of an object so as to reduce resistance to motion from the air or a liquid.

Stretch The extension when an elastic material such as a spring is pulled outwards or downwards.

Sun The star at the centre of our Solar System.

Sunlight Light from the Sun.

Sunspots Dark spots on the surface of the Sun.

Supernova An exploding star.

Supersonic Describes a speed that is faster than the speed of sound.

Symbol A sign that represents something (see also circuit symbols, and hazard symbols).

Tangent A straight line that touches a curve or circle.

Telescope A device made with lenses that allows distant objects to be seen clearly.

Temperature A measure of how hot something is, which is related to the average speed of the particles.

Tension A stretching force.

Terminal (of a cell) The positive or negative end of a cell or battery.

Terminal velocity The highest velocity an object reaches when moving through a gas or a liquid; it happens when the drag force equals the forward or gravitational force.

Thermal A rising current of heated air.

Thermal store The energy associated with objects or systems when there is a change in temperature.

Thermal equilibrium The state of two objects at the same temperature.

Thermal image An image made using thermal or infrared radiation.

Thermal imaging camera A device that forms an image using thermal or infrared radiation so that different temperatures appear as different colours.

Thermometer A device used to measure temperature.

Thought experiment The process of thinking through what might happen in an experiment without actually doing it.

Thrust The force from an engine or rocket.

Tidal energy/power Energy from the movement of water in tides which can be used to generate electricity.

Total eclipse Occurs when all of the light from the Sun is blocked out by the Earth or the Moon.

Total internal reflection The complete reflection of light at a boundary between two media.

Transducer A device that changes an electrical signal into light or sound, or changes light or sound into an electrical signal.

Transfer (of energy) Shifting energy from one place to another.

Translucent Describes objects that transmit light but diffuse (scatter) the light as it passes through.

Transmitted Light or other radiation passed through an object.

Transmitter A device that gives out a signal, such as sound in a sonar transmitter.

Transparent describes objects that transmit light; you can see through transparent objects.

Transverse Describes a wave in which the vibrations are at right angles to the direction in which the wave moves.

Turbine A component that spins due to steam, water or wind, and which is used to turn a generator.

Turning effect A force causing an object to turn.

Turning force The moment of a force.

Ultrasound Sound at a frequency greater than 20 000 Hz, beyond the range of human hearing.

Umbra The area of total shadow behind an opaque object where no light has reached.

Unbalanced Describes forces on an object that are unequal.

Universe Everything that exists.

Upright Describes an image that is the right way up.

Upthrust The force on an object in a liquid or gas that pushes it up.

Uranium A metal used in nuclear power stations.

Uranus A large outer planet made of gas, the seventh from the Sun.

Useful energy The energy that you want from a process.

Vacuum A space that has no particles, and so no matter.

Vaporized (meteor) What can happen to a meteor or asteroid on collision, when solid material is turned into dust.

Variable A quantity that can change, such as time, temperature, length, or mass. In an investigation you should change only one variable at a time to see what effect it has.

Venus A rocky inner planet, the second from the Sun.

Vibrate To move continuously and rapidly to and fro.

Vibration Motion to and fro of the parts of a liquid or solid.

Virtual Describes an image that cannot be focused onto a screen.

Volt The unit of measurement of voltage, symbol V.

Voltage A measure of the strength of a cell or battery used to send a current around a circuit; it is measured in volts.

Voltmeter A device for measuring voltage.

Volume (of space) The amount of space that something takes up, which is related to its mass by its density.

Waning Describes the Moon that we see when the amount of the lit side is decreasing.

Wasted energy Energy transferred to non-useful forms; often energy transferred to the surroundings.

Water (energy from) Using water to generate electricity from tides, water behind dams and waves.

Water resistance The force on an object moving through water that causes it to slow down (also known as drag).

Watts The unit of power, symbol W.

Wave A variation that transfers energy or information.

Wave energy/power Using energy from waves to generate electricity.

Wavelength The distance between two identical points on the wave, such as two adjacent peaks or two adjacent troughs.

Waxing Describes the Moon that we see when the amount of the lit side is increasing.

Weight The force of the Earth on an object due to its mass.

White dwarf A small, very dense star; part of the life cycle of our Sun.

Wind energy/power Energy from wind that can be used to generate electricity.

Wind farm A collection of wind turbines.

Wind turbine A turbine and generator that uses the energy of the wind to generate electricity.

Year The length of time it takes for a planet to orbit the Sun.

Index

Index

OXFORD
UNIVERSITY PRESS

Great Clarendon Street, Oxford, OX2 6DP, United Kingdom

Oxford University Press is a department of the University of Oxford. It furthers the University's objective of excellence in research, scholarship, and education by publishing worldwide. Oxford is a registered trade mark of Oxford University Press in the UK and in certain other countries

British Library Cataloguing in Publication Data

Data available

978-1-38-201901-9

Digital edition: 978-1-38-201907-1

10 9 8

Paper used in the production of this book is a natural, recyclable product made from wood grown in sustainable forests. The manufacturing process conforms to the environmental regulations of the country of origin.

Printed in India by Manipal Technologies Limited

Author's acknowledgments

I would like to thank my editors for their support and feedback, and my wonderful friends Michele, Rob, Lesa, and Bill for all their support, long walks and tea. I would also like to thank Oleksiy, my dance instructor, for providing continuing encouragement for my writing and for my foxtrot.

Acknowledgements

The publisher and authors would like to thank the following for permission to use photographs and other copyright material:

Cover: Jacobs Stock Photography Ltd/Getty Images. Photos: p18: Arcady/Shutterstock; p20(t): Vera Larina/Shutterstock; p20(m): Brian A Jackson/Shutterstock; p20(bl): Image Professionals GmbH/Alamy Stock Photo; p20(br): Monkey Business Images/Shutterstock; p21(t): ZouZou/Shutterstock; p21(b): Valentyn Volkov/Shutterstock; p24(tl): ESB Professional/Shutterstock; p24(tr): bodom/Shutterstock; p24(ml): Vaclav Mach/Shutterstock; p24(mr): Will Thomass/Shutterstock; p24(b): Roman Sigaev/Shutterstock; p25(tl): Realstock/Shutterstock; p25(tr): Dean Drobot/Shutterstock; p25(ml): Gumenyuk Dmitriy/Shutterstock; p25(mr): StockImageFactory.com/Shutterstock; p25(b): Triff/Shutterstock; p26(t): PhotoAlto sas/Alamy Stock Photo; p26(b): bunnyphoto/Shutterstock; p27: ezhenaphoto/Shutterstock; p28(t): Lim JinKang/Shutterstock; p28(m): Ketkar/Shutterstock; p28(b): Kim Weenink/Shutterstock; p29: Tom Hirtreiter/Shutterstock; p30(l): Petrenko Andriy/Shutterstock; p30(r): Eric Broder Van Dyke/Shutterstock; p32(t): Viacheslav Lopatin/Shutterstock; p32(b): l i g h t p o e t/Shutterstock; p36(t): Tyler Panian/Shutterstock; p36(bl): GIPhotostock/Science Photo Library; p36(br): New Africa/Shutterstock; p37: Melissa Madia/Shutterstock; p38(t): dedek/Shutterstock; p38(b): Semjonow Juri/Shutterstock; p39: NASA Pictures/Alamy Stock Photo; p40(t): paul prescott/Shutterstock; p40(b): travelview/Shutterstock; p41(t): MarcelClemens/Shutterstock; p41(bl): mari2d/Shutterstock; p41(br): Huebi/Shutterstock; p42(t): Joggie Botma/Shutterstock; p42(bl): ktsdesign/Shutterstock; p42(br): Action Sports Photography/Shutterstock; p43(t): artfotoxyz/Shutterstock; p44: Willyam Bradberry/Shutterstock; p46(t): John Fox/Getty Images; p46(b): WILLIAM WEST/Staff/Getty Images; p47: lexaarts/Shutterstock; p50(l): Pressmaster/Shutterstock; p50(r): Jake Lyell/Alamy Stock Photo; p51: Rene Frederick/Getty Images; p52(t): Hitdelight/Shutterstock; p52(b): Gang Liu/Shutterstock; p53: Anton Gvozdikov/123RF; p54: Cavan Images/Getty Images; p55: Mikael Damkier/Shutterstock; p58: sciencephotos/Alamy Stock Photo; p60: Monkey Business Images/Shutterstock; p62(t): Andrei Nekrassov/Shutterstock; p62(b): Madarakis/Shutterstock; p65(b): yevgeniy11/Shutterstock; p66(b): Iakov Filimonov/Shutterstock; p67(t): Olegusk/Shutterstock; p67(b): MikeDotta/Shutterstock; p70(t): michelecaminati/Shutterstock; p70(b): NASA; p71: Celli07/Shutterstock; p73: Pekka Parviainen/Science Photo Library; p74(t): Natalia Kirsanova/Shutterstock; p74(m): Diego Barucco/Shutterstock; p74(bl): Tristan3D/Shutterstock; p74(br): Onkamon/Shutterstock; p75(t): Giovanni Benintende/Shutterstock; p75(b): 3Dsculptor/Shutterstock; p76: Peter Hermes Furian/Shutterstock; p78(t): NASA; p78(m): Warachai Krengwirat/Shutterstock; p78(b): Igor Kovalchuk/Shutterstock; p79(t): Ianmyrza/Shutterstock; p79(b): Primož Cigler/Shutterstock; p80(t): Puwadol Jaturawutthichai/Shutterstock; p80(b): saiko3p/Shutterstock; p81(t): Mike P Shepherd/Alamy Stock Photo; p81(b): World History Archive/Alamy Stock Photo; p84: Hulton Archive/Stringer/Getty Images; p85: Babak Tafreshi/Science Photo Library; p90: Boleslaw Kubica/Shutterstock; p92(tl): Clive Rose/Staff/Getty Images; p92(tr): Per-Anders Pettersson/Getty Images; p92(m): yogesh_more/Shutterstock; p92(b): Zvonimir Orec/Shutterstock; p94(t): Quinn Rooney/Getty Images; p94(b): YanLev/Shutterstock; p95: Vladimir Wrangel/Shutterstock; p96: Dmitry Yashkin/Shutterstock; p98(b): INDRANIL MUKHERJEE/Staff; p98(t): Jamie McDonald - FIFA/Getty Images; p100: David Acosta Allely/Shutterstock; p103: Gary Blakeley/Shutterstock; p104(tr): wk1003mike/Shutterstock; p104(mr): Jeannette Meier Kamer/Shutterstock; p104(ml): Barnaby Chambers/Shutterstock; p104(b): Christian Mueller/Shutterstock; p105(t): Kevin Norris/Shutterstock; p105(b): Vit Kovalcik/Shutterstock; p106: Holos/Getty Images; p108(l): Vitalii Nesterchuk/Shutterstock; p108(r): Greg Epperson/Shutterstock; p109: Gerard Koudenburg/Shutterstock; p110: OLIINYK INNA/Shutterstock; p112: Waj/Shutterstock; p113(tl): Shcherbakov Ilya/Shutterstock; p113(tr): Erkki Makkone/Shutterstock; p113(r): SINITAR/Shutterstock; p114: Suzanne McGowan/Alamy Stock Photo; p116: Matt Humphrey/Alamy Stock Photo; p117: Imaginechina Limited/Alamy Stock Photo; p118(l): Graeme Shannon/Shutterstock; p118(r): GIRODJL/Shutterstock; p119: Nattawit Khomsanit/Shutterstock; p120(tl): Brent Parker Jones/Oxford University Press; p120(tr): Brent Parker Jones/Oxford University Press; p120(b): I WALL/Shutterstock; p121(r): Monkey Business Images/Shutterstock; p121(l): Ljupco Smokovski/Shutterstock; p121(m): luchschenF/Shutterstock; p122: Istvan Csak/Shutterstock; p123: Pool IFREMER/MAIOFISS/Getty Images; p124: Ted Kinsman/Science Photo Library; p125(tl): Charles D. Winters/Science Photo Library; p125(tm): Charles D. Winters/Science Photo Library ;p125(r): szefei/Shutterstock; p125(b): ChiccoDodiFC/Shutterstock; p126(t): Wirestock Creators/Shutterstock; p126(b): Andrew Mcclenaghan/Science Photo Library; p127: Andrew Lambert Photography/Science Photo Library; p130(r): photomaster/Shutterstock; p130(l): Ed Connor/Shutterstock; p131(l): imagenavi/Getty Images; p131(m): Yammy8973/Shutterstock; p131(r): Vladitto/Shutterstock; p132: ER Degginger/Science Photo Library; p133(l): The Mariner 4291/Shutterstock; p133(r): thomas koch/Shutterstock; p134(t): GUDKOV ANDREY/Shutterstock; p134(m): Jaggat Rashidi/Shutterstock; p134(bl): zimmytws/Shutterstock; p134(br): Robert Pernell/Shutterstock; p135: Karavanov_Lev/Shutterstock; p136: Pat_Hastings/Shutterstock; p137: Milos Stojiljkovic/Shutterstock; p139: John Greim/Science Photo Library; p141(t): manjik/Shutterstock; p141(b): Arhangell/Shutterstock; p142(l): Mopic/Shutterstock; p142(r): bogdan ionescu/Shutterstock; p144(t): OmMishra/Shutterstock; p144(m): ponsulak/Shutterstock; p145: Gino Santa Maria/Shutterstock; p146(t): Raksha Shelare/Shutterstock; p146(bl): Zurijeta/Shutterstock; p146(br): Atstock Productions/Shutterstock; p148(t): Kriangkrai Thitimakorn/Shutterstock; p148(b): Dmitry Naumov/Shutterstock; p149(t): Andrew Lambert Photography/Science Photo Library; p149(b): nienora/Shutterstock; p150: Edgar Martirosyan/Shutterstock; p151(t): Phanie/Alamy Stock Photo; p151(b): Marko Rupena/Shutterstock; p154(t): Science Stock Photography/Science Photo Library; p154(m): Gabe Palmer/Alamy Stock Photo; p154(b): RealVector/Shutterstock; p156(t): LuYago/Shutterstock; p156(bl): Alchemy/Alamy Stock Photo; p156(bm): Alchemy/Alamy Stock Photo; p156(br): Alchemy/Alamy Stock Photo; p157: sciencephotos/Alamy Stock Photo; p158: Andrea Leone/Shutterstock; p159(t): Marmaduke St. John/Alamy Stock Photo; p159(b): Q'ju Creative/Shutterstock; p160(t): Tyler Olson/Shutterstock; p160(m): Nikolay Nemchinov/Shutterstock; p160(b): Simon Turner/Alamy Stock Photo; p161(l): Mehau Kulyk/Science Photo Library; p161(r): Geoff Tompkinson/Science Photo Library; p162: GIPhotostock/Science Photo Library; p166(t): 24K-Production/Shutterstock; p166(m): Sinisa Botas/Shutterstock; p166(b): berna namoglu/Shutterstock; p169(tl): Roger Brown Photography/Shutterstock; p169(tr): Awe Inspiring Images/Shutterstock; p169(b): Bloomberg/Getty Images; p170(t): Per-Anders Pettersson/Getty Images; p170(b): Kupriyanova Tanya/Shutterstock; p170(b): Jacek Fulawka/Shutterstock; p170(l): Andrey Armyagov/Shutterstock; p172(r): CHAIYA/Shutterstock; p173(t): Designua/Shutterstock; p173(b): SIMON MAINA/Stringer/Getty Images; p174: khmad Dody Firmansyah/Shutterstock; p175: Yonhap News Agency/EPA/Shutterstock; p176(t): Jenny Lord/Shutterstock; p176(m): Steve Heap/Shutterstock; p176(bl): papillondream/Shutterstock; p176(br): Roman Mikhailiuk/Shutterstock; p177(l): Henna Hayrynen/Shutterstock; p176(ml): Pawarun Chitchirachan/Shutterstock; p176(mr): photokup/Shutterstock; p176(r): Muk Photo/Shutterstock; p180(t): NASA; p180(m): Wolfgang Kloehr/Shutterstock; p180(b): oceanfishing/Shutterstock; p181: mironov/Shutterstock; p182(t): Dotted Yeti/Shutterstock; p182(m): Paul Fleet/123RF; p182(b): Mopic/Shutterstock; p183(l): Migel/Shutterstock; p183(r): I. Pilon/Shutterstock; p184(l): Olga Popova/Shutterstock; p184(r): ALPA PROD/Shutterstock; p185(t): koya979/Shutterstock; p185(b): polarman/Shutterstock; p187(tl): Tristan3D/Shutterstock; p187(tr): sdecoret/Shutterstock; p187(bl): Michael Rosskothen/Shutterstock; p187(br): Dotted Yeti/Shutterstock; p190(t): Nicram Sabod/Shutterstock; p190(mr): lenisecalleja.photography/Shutterstock; p190(ml): Ethan Daniels/Shutterstock; p190(bl): Turtle Rock Scientific/Science Source/Science Photo Library; p190(br): Laurie Neilsen; p191: GuilhermeMesquita/Shutterstock; p192(t): Lawrence Lawry/Science Photo Library; p192(ml): Artisticco/Shutterstock; p192(m): Shark_749/Shutterstock; p192(mr): MXW Stock/Shutterstock; p193: Cultura Creative RF/Alamy Stock Photo; p195: Dani Simmonds/Shutterstock; p196(t): T-Service/Science Photo Library; p196(b): Skryl Sergey/Shutterstock; p197: Africa Studio/Shutterstock; p198: Ivan Smuk/Shutterstock; p199: Tony Mcconnell/Science Photo Library; p200(t): ggw/Shutterstock; p200(m): Nrel/Us Department Of Energy/Science Photo Library; p200(b): Andy.M/Shutterstock; p201(b): TED ALJIBE/Staff/Getty Images; p202(b): Tony Mcconnell/Science Photo Library; p203: Gail Johnson/Shutterstock; p204(t): Smit/Shutterstock; p204(b): Image Point Fr/Shutterstock; p206: Blend Images/Shutterstock; p207: Riaan van den Berg/Shutterstock; p209: stocksolutions/Shutterstock; p210: Albert Lozano/Shutterstock; p211: mirjana simeunovich/Shutterstock; p212: nexus 7/Shutterstock; p213(tl): GeoStock/Getty Images; p213(bl): Christopher Meder/Shutterstock; p213(tr): Sally Wallis/Shutterstock; p213(br): EcoPrint/Shutterstock; p214(l): john angerson/Alamy Stock Photo; p214(r): S-F/Shutterstock; p215(t): reddees/Shutterstock; p215(b): Fuse/Getty Images; p216: Diego Cervo/Shutterstock; p217(t): BatuhanPehlivan/Shutterstock; p217(b): dennizn/Shutterstock; p221: Ivan Kurmyshov/Shutterstock; p224: fen deneyim/Shutterstock; p225: Green Jo/Shutterstock; p228(t): Dmitry Rukhlenko/Shutterstock; p228(mt): stoupa/Shutterstock; p228(mb): Claudio Divizia/Shutterstock; p228(b): Somchai Som/Shutterstock; p229(t): Science Photo Library/Alamy Stock Photo; p229(b): Michael Regan/Staff/Getty Images; p232: SPUTNIK/Alamy Stock Photo; p233(t1): IgorZh/Shutterstock; p233(t2): Robert Crow/Shutterstock; p233(t3): AuntSpray/Shutterstock; p233(t4): Alex Stemmer/Shutterstock; p233(t5): StockImageFactory.com/Shutterstock; p233(b1): cigdem/Shutterstock; p233(b2): SciePro/Shutterstock; p233(b3): Elenarts/Shutterstock; p234(t): Esteban De Armas/Shutterstock; p234(mt): NASA; p234(mbr): Alex Savin/Shutterstock; p234(mbl): Purva Joshi/Shutterstock; p234(b): Elena11/Shutterstock; p236(t): Designua/Shutterstock; p236(b): Designua/Shutterstock; p237: Elena11/Shutterstock; p238(t): Mohamed Elkhamisy/Shutterstock; p238(b): NASA images/Shutterstock; p239(bl): bluecrayola/Shutterstock; p239(t): Designua/Shutterstock; p239(br): Antares_StarExplorer/Shutterstock; p240: trialhuni/Shutterstock; p248: Andrei Nekrassov/Shutterstock; p249: Andrei Nekrassov/Shutterstock.

Artwork by Integra Software Services, Q2A Media Services Pvt. Ltd, Aptara, Phoenix Photosetting, Tech Graphics, IFA Design, Plymouth, UK, Hardlines, Wearset Ltd, HL Studios, GreenGate Publishing, Peter Bull Art Studio, Tech-Set Ltd, Kamae Design, Mark Walker, Erwin Haya, Jeff Bowles, Roger Courthold, Mike Ogden, Jeff Edwards, Russell Walker, Clive Goodyer, Jamie Sneddon, Dave Russell, James Stayte, Edward Fullick, Oxford University Press China, and Oxford University Press Fajar, Oxford University Press ANZ, and Oxford University Press UK.

Every effort has been made to contact copyright holders of material reproduced in this book. Any omissions will be rectified in subsequent printings if notice is given to the publisher.

This Student Book refers to the Cambridge Lower Secondary Science (0893) published by Cambridge Assessment International Education.

This work has been developed independently from and is not endorsed by or otherwise connected with Cambridge Assessment International Education.

The manufacturer's authorised representative in the EU for product safety is Oxford University Press España S.A. of el Parque Empresarial San Fernando de Henares, Avenida de Castilla, 2 – 28830 Madrid (www.oup.es/en).